INSECT LIVES

INSECT LIVES

Stories of Mystery and Romance from a Hidden World

Edited by
Erich Hoyt and Ted Schultz

Harvard University Press
Cambridge, Massachusetts

For my father, Robert Emmet Hoyt,
and my mother, Betty Shutrump Hoyt,
with love and admiration.

For my father, Robert A. Schultz,
and my mother, Reiko June Schultz,
with thanks for putting up with all the insects,
amphibians, reptiles, and rodents.

Printed in the United States of America

First Harvard University Press paperback edition, 2002.
Published by arrangement with John Wiley & Sons, Inc.

Acknowledgments continue on p. 349.

Library of Congress Cataloging-in-Publication Data

Insect lives : stories of mystery and romance from a hidden world /
[compiled by] Erich Hoyt and Ted Schultz.
p. cm.
ISBN 0-674-00952-5 (pbk.)
1. Insects. I. Hoyt, Erich. II. Schultz, Ted.
QL463.I7224 1999
595.7—dc21 99-24788

CONTENTS

Editors' Note

The selections herein are, for the most part, presented exactly as written and published. In a few cases, we have made some concessions on behalf of readability, removing a few references in the scientific papers or adding a word or two of explanation in brackets. However, to retain a sense of authenticity, we have kept all the original spellings, grammar, and punctuation, even if peculiar to one era or another, entirely intact.

Introduction

Alien creatures have overrun planet Earth. They wear their skeletons on the outside, bite sideways, smell with antennae, taste with their feet, and breathe through holes in the sides of their bodies. Their eyes are placid, unmoving orbs; when we humans look into them, we experience neither recognition nor empathy. They are the insects.

From a human point of view, insects are aliens, denizens of another world, shadow opposites with whom we share planet Earth. But, by nearly any objective measure, it is really we who are the aliens and they who are the Earth natives. Insects have been here for 400 million years; we have been here for 1 million. They occupy nearly every nook and cranny of the terrestrial environment and number anywhere from 5 million to 30 million species: nobody really knows. We and all our large and lumbering terrestrial vertebrate kin together constitute no more than 24,000 species.

Insects are arguably as complicated as humans and yet as different from humans as any complex animal that we know. If we trace the chain of species leading to humans backward in time, and if we do the same for insects, the two separate lineages do not converge until the early Cambrian, more than 500 million years ago. The common ancestor is unknown, but it is thought to have been something like a simple flatworm. From this humble beginning two mighty stocks went their separate ways, evolving down different tracks, solving life's problems in radically different ways. They still do.

The list of differences between insects and humans—indeed, between insects and all vertebrates—is worth considering, for these differences represent alternate body plans that achieve the same goals. Insect jaws open and shut horizontally; their heart is a tube running the length of their bodies. With their skeletons on the outside, insects have the appearance, at least in the eyes of those who appreciate them, of intricate, miniature sculptures, sometimes more baroque than our wildest human imaginings. In contrast, our soft and squishy flesh hangs from internal skeletons. Insects smell (that is, they sample airborne molecules) with antennae, structures for which we do not even possess an analogue. With two antennae, spaced apart, insects have the olfactory equivalent of our three-dimensional sight and hearing (both of which they also have); some species can home in on faint odors from miles away, sensing and responding to single molecules of substance. In contrast, we snort volumes of atmosphere up hairy nostrils and miss most of it with our crude olfactory sense. Most of the time we don't even stop to smell the flowers. Insect vision is modeled along completely different lines from that of humans. Their eyes are spheres composed of thousands of separate facets, each with its own set of lenses and light and color detectors. Truly a redundant system! And insects hear through ears in their legs, or their chests, or their sides, depending on the species.

In short, they're different. But that difference makes the similarities between humans and insects all the more uncanny. It is especially among the social insects—for example, termites, ants, bees, and wasps—that we find the most astonishing parallels with humans. They live in large, organized societies, practice agriculture, keep pets and cattle, sacrifice themselves for the good of the colony, stage tournaments and all-out wars, enslave one another, take drugs, and communicate with symbolic language. These surprising parallels with humans have made the social insects among the most admired and studied of all insects from ancient times to the present.

In this book, we offer a sweeping tour of the human fascination with insects—the diabolical as well as the divine—from the Bible and Aristotle to Darwin and the great nineteenth-century naturalists sending accounts from the rain forest. Some of the finest nature and science writers have been entomologists—scientists who study insects—but much of their writing has been underappreciated. We have selected contemporary works from scientists who write well such as E. O. Wilson, May Berenbaum, and Tom Eisner. Human fascination with insects has hardly been limited to scientists, and *Insect Lives* also includes excerpts from a horror film screenplay, the product notes for Mexican jumping beans (actually moth larvae), the odd Hunkin and *Far Side* cartoon, and dozens of rare and beautiful insect illustrations.

Once we started to look for good writing on insects, we found it everywhere and in the most unusual places. The final selections used in the book were chosen and edited from a rather long shortlist. We must interject a brief advisory here: We have not attempted to be representative of all insect orders. With so many orders and such uneven reportage among the diversity of species, that would clearly be an impossible task. In fact, we must confess to a bias toward ants and social insects. Still, we have endeavored to keep our particular passions appropriately reined in. There is enough great ant and bee writing to make a large anthology in itself; we were forced to exclude treasured passages. At the other extreme, we have occasionally strayed from insects to other arthropods—that is, to eight- or ten-legged creatures—but again, not in a truly representative way. Therefore, what you

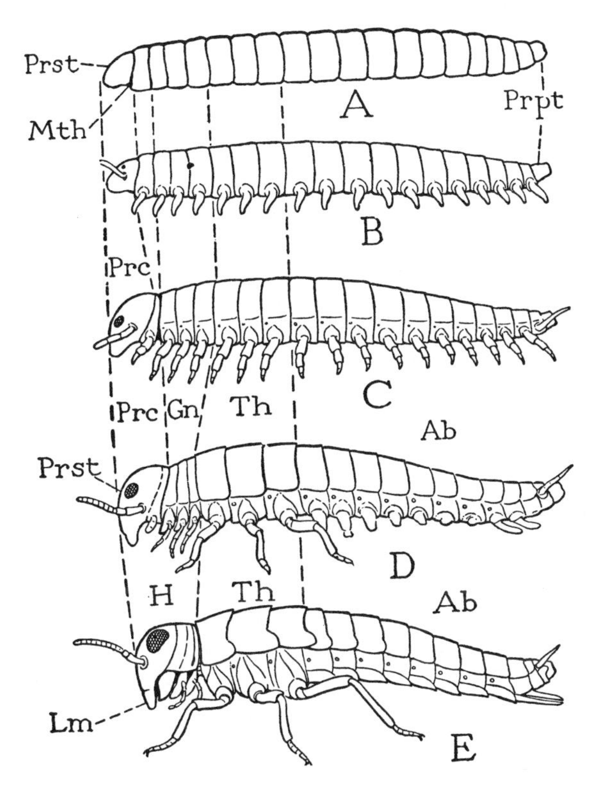

From worm to insect. A famous 1935 drawing depicting the evolution of insects from a hypothetical earthworm-like ancestor, by the great entomologist Robert Evans Snodgrass. Notice how groups of body segments that are separated in the primitive animal fuse to form the head (H), thorax (Th), and abdomen (Ab). Labels refer to anatomical parts: Prst = prostomium, the primitive head; Mth = mouth; Prpt = peripoct; Prc = protocephalon; Gn = gnathal segments; Lm = labrum (*Principles of Insect Morphology* by Robert Evans Snodgrass, The McGraw-Hill Book Co., New York, 1935).

have here is a somewhat eclectic assortment, arranged along various themes to help bring insects entertainingly to light, and each introduced with a minimum of fanfare and annotation to frame the author, subject, and chosen selection.

Our ultimate goals? To give enlightening pause to the steppers, swatters, and screamers who live in fear or dread of six legs—that would be reason enough. But we also hope that this book will illuminate insect lives in such a way that it transports and frees the curious general reader from the constraints of being human—for at least a mayfly's brief lifetime or two—in suspended appreciation of that other, hidden world beneath our feet and beyond our rolled-up newspapers.

1

Wonders of Creation: Insects Praised

Since the beginning of human history, we have carried on an up-and-down, hot-and-cold relationship with this planet's much more ancient inhabitants, the insects. On the positive side, we have long appreciated the obvious beauty of some insects, for example, butterflies and ladybugs. We have erected gardens devoted to the evening enjoyment of firefly watching, we have caged crickets just to listen to their songs, and we have ornamented our bodies with the metallic elytra, or wing covers, of buprestid beetles. We have worshipped the archetypal image of the scarab beetle and used it and other insects—damselflies, wasps, and katydids among them—as motifs in jewelry, sculpture, and even architecture. We have, since ancient times, entered into alliances with ants to keep pests out of our citrus groves, we have tamed bees for their honey and wax, we have gathered the secretions of scale insects to make shellac, and we have pampered caterpillars for their luxurious silk. In most human cultures, we have even enjoyed insects from a culinary perspective.

In this chapter, we celebrate the biology, aesthetics, and practical, utilitarian value of insects with writings from authors who appreciate insects as objects of extraordinary complexity, elegance, and beauty, and who have tried in one way or another to tell us more about them.

So Great the Excitement

Alfred Russel Wallace

Along with Charles Darwin, Alfred Russel Wallace (1823–1913) was one of the architects of the theory of evolution by natural selection. Wallace spent many years collecting plants and animals in South America and in the islands of the Malay Archipelago, supporting himself by selling specimens to avid natural history buffs back in Britain. (How times have changed!) Wallace's two book-length accounts of his travels are engrossing natural and human histories of a more pristine world. More than anything else, Wallace's enthusiasm for the natural world, and especially for birds, beetles, and butterflies, radiates from every page.

On our way back in the heat of the day I had the good-fortune to capture three specimens of a fine *Ornithoptera,* the largest, the most perfect, and the most beautiful of butterflies. I trembled with excitement as I took the first out of my net and found it to be in perfect condition. The ground color of this superb insect was a rich shining bronzy black, the lower wings delicately grained with white, and bordered by a row of large spots of the most brilliant satiny yellow. The body was marked with shaded spots of white, yellow, and fiery orange, while the head and thorax were intense black. On the under side the lower wings were satiny white, with the marginal spots half black and half yellow. I gazed upon my prize with extreme interest, as I at first thought it was quite a new species. It proved, however, to be a variety of *Ornithoptera remus,* one of the rarest and most remarkable species of this highly esteemed group. I also obtained several other new and pretty butterflies. When we arrived at our lodging-house, being particularly anxious about my insect treasures, I suspended the box from a bamboo on which I could detect no sign of ants, and then began skinning some of my birds. During my work I often glanced at my precious box to see that no intruders had arrived, till after a longer spell of work than usual I looked again, and saw to my horror that a column of small red ants were descending the string and entering the box. They were already busy at work at the bodies of my treasures, and another half-hour would have seen my whole day's collection destroyed. As it was, I had to take every insect out, clean them thoroughly as well as the box, and then seek for a place of safety for them. As the only effectual one, I begged a plate and a basin from my host, filled the former with water, and standing the latter in it, placed my box on the top, and then felt secure for the night; a few inches of clean water or oil being the only barrier these terrible pests are not able to pass. . . .

I have rarely enjoyed myself more than during my residence here. As I sat taking my coffee at six in the morning, rare birds would often be seen on some tree close by, when I would hastily sally out in my slippers, and perhaps secure a prize I had been seeking after for weeks. The great hornbills of Celebes *(Buceros cassidix)* would often come with loud-flapping wings and perch upon a lofty tree just in front of me; and the black baboon-monkeys *(Cynopithecus nigrescens)* often stared down in astonishment at such an intrusion into their domains; while at night herds of wild pigs roamed about the house, devouring refuse, and obliging us to put away every thing eatable or breakable from our little cooking-house. A few minutes' search on the fallen trees around my house at sunrise and sunset would often produce me more beetles than I would meet with in a day's collecting, and odd moments could be made valuable, which when living in villages or at a distance from the forest are inevitably wasted. Where the sugar-palms were dripping with sap, flies congregated in immense numbers, and it was by spending half an hour at these when I had the time to spare that I obtained the finest and most remarkable collection of this group of insects that I have ever made.

Then what delightful hours I passed wandering up and down the dry river-courses, full of water-holes and rocks and fallen trees, and overshadowed by magnificent vegetation! I soon got to know every hole and rock and stump, and came up to each with cautious step and bated breath to see what treasures it would produce. At one place I would find a little crowd of the rare butterfly *(Tachyris zarinda),* which would rise up at my approach, and display their vivid orange and cinnabar-red wings, while among them would flutter a few of the fine blue-banded Papilios. Where leafy branches hung over the gully, I might expect to find a grand *Ornithoptera* at rest, and an easy prey. At certain rotten trunks I was sure to get the curious little tiger-beetle *(Therates flavilabris).* In the denser thickets I would capture the small metallic blue butterflies *(Amblypodia)* sitting on the leaves, as well as some rare and beautiful leaf-beetles of the families Hispidae and Chrysomelidae.

I found that the rotten jack-fruits were very attractive to many beetles, and used to split them partly open and lay them about in the forest near my house to rot. A morning's search at these often produced me a score of species—Staphylinidae, Nitidulidae, Onthophagi, and minute Carabidae being the most abundant. Now and then the "sagueir" makers brought me a fine rosechafer *(Sternoplus schaumii)* which they found licking up the sweet sap. Almost the only new birds I met with for some time were a handsome ground-thrush *(Pitta celebensis)*, and a beautiful violet-crowned dove *(Ptilonopus celebensis)*, both very similar to birds I had recently obtained at Aru, but of distinct species.

About the latter part of September a heavy shower of rain fell, admonishing us that we might soon expect wet weather, much to the advantage of the baked-up country. I therefore determined to pay a visit to the falls of the Máros River, situated at the point where it issues from the mountains—a spot often visited by travellers, and considered very beautiful. Mr. M. lent me a horse, and I obtained a guide from a neighboring village; and taking one of my men with me, we started at six in the morning, and after a ride of two hours over the flat rice fields skirting the mountains which rose in grand precipices on our left, we reached the river about half-way between Máros and the falls, and thence had a good bridle-road to our destination, which we reached in another hour. The hills had closed in round us as we advanced; and when we reached a ruinous shed which had been erected for the accommodation of visitors, we found ourselves in a flat-bottomed valley about a quarter of a mile wide, bounded by precipitous and often overhanging limestone rocks. So far the ground had been cultivated, but it now became covered with bushes and large scattered trees.

As soon as my scanty baggage had arrived and was duly deposited in the shed, I started off alone for the fall, which was about a quarter of a mile further on. The river is here about twenty yards wide, and issues from a chasm between two vertical walls of limestone over a rounded mass of basaltic rock about forty feet high, forming two curves separated by a slight ledge. The water spreads beautifully over this surface in a thin sheet of foam, which curls and eddies in a succession of concentric cones till it falls into a fine deep pool below. Close to the very edge of the fall a narrow and very rugged path leads to the river above, and thence continues close under the precipice along the water's edge, or sometimes in the water, for a few hundred yards, after which the rocks recede a little, and leave a wooded bank on one side, along which the path is continued, till in about half a mile a second and smaller fall is reached. Here the river seems to issue from a cavern, the rocks having fallen from above so as to block up the channel and bar further progress. The fall itself can only be reached by a path which ascends behind a huge slice of rock which has partly fallen away from the mountain, leaving a space two or three feet wide, but disclosing a dark chasm descending into the bowels of the mountain, and which, having visited several such, I had no great curiosity to explore.

Crossing the stream a little below the upper fall, the path ascends a steep slope for about five hundred feet, and passing through a gap enters a narrow valley, shut in by walls of rock absolutely perpendicular and of great height. Half a mile further this valley turns abruptly to the right, and becomes a mere rift in the mountain. This extends another half mile, the walls gradually approaching till they are only two feet apart, and the bottom rising steeply to a pass which leads probably into another valley but which I had no time to explore. Returning to where this rift had begun, the main path turns up to the left in a sort of gulley, and reaches a summit over which a fine natural arch of rock passes at a height of about fifty feet. Thence was a steep descent through thick jungle with glimpses of precipices and distant rocky mountains, probably leading into the main river valley again. This was a most tempting region to explore, but there were several reasons why I could go no further. I had no guide, and no permission to enter the Bugis territories, and as the rains might at any time set in, I might be prevented from returning by the flooding of the river. I therefore devoted myself during the short time of my visit to obtaining what knowledge I could of the natural productions of the place.

The narrow chasms produced several fine insects quite new to me, and one new bird, the curious *Phlaegenas tristigmata,* a large ground-pigeon with yellow breast and crown and purple neck. This rugged path is the highway from Máros to the Bugis country beyond the mountains. During the rainy season it is quite impassable, the river filling its bed and rushing between perpendicular cliffs many hundred feet high. Even at the time of my visit it was most precipitous and fatiguing, yet women and children came over it daily, and men carrying heavy loads of palm-sugar of very little value. It was along the path between the lower and the upper falls, and about the margin of the upper pool, that I found most insects. The large semi-transparent butterfly *(Idea tondana)* flew lazily along by dozens, and it was here that I at length obtained an insect which I had hoped but hardly expected to meet with—the magnificent *Papilio androcles,* one of the largest and rarest known swallow-tailed butterflies. During my four days' stay at the falls I was so fortunate as to obtain six good specimens. As this beautiful creature flies, the long white tails flicker like streamers, and when settled on the beach it carries them raised upward, as if to preserve them from injury. It is scarce even here, as I did not see more than a dozen specimens in all, and had to follow many of them up and down the river's bank repeatedly before I succeeded in their capture. When the sun shone hottest about noon, the moist beach of the pool below the upper fall presented a beautiful sight, being dotted with groups of gay butterflies—orange, yellow, white, blue, and green—which on being disturbed rose into the air by hundreds, forming clouds of variegated colors. . . .

The part of the village in which I resided was a grove of cocoa-nut-trees, and at night, when the dead leaves were sometimes collected together and burnt, the effect was most magnificent—the tall stems, the fine crowns of foliage, and the immense fruit-clusters, being brilliantly illuminated against a dark sky, and appearing like a fairy palace supported on a hundred columns, and groined over with leafy arches. The cocoa-nut-tree, when well grown, is certainly the prince of palms both for beauty and utility.

During my very first walk into the forest at Batchian, I had seen sitting on a leaf out of reach, an immense butterfly of a dark color marked with white and yellow spots. I could not capture it as it flew away high up into the forest, but I at once saw that it was a female of a new species of *Ornithoptera*, or, "bird-winged butterfly," the pride of the Eastern tropics. I was very anxious to get it and to find the male, which in this genus is always of extreme beauty. During the two succeeding months I only saw it once again, and shortly afterward I saw the male flying high in the air at the mining village. I had begun to despair of ever getting a specimen, as it seemed so rare and wild; till one day, about the beginning of January, I found a beautiful shrub with large white leafy bracts and yellow flowers, a species of Mussaenda, and saw one of these noble insects hovering over it, but it was too quick for me, and flew away. The next day I went again to the same shrub and succeeded in catching a female, and the day after a fine male. I found it to be as I had expected, a perfectly new and most magnificent species, and one of the most gorgeously colored butterflies in the world. Fine specimens of the male are more than seven inches across the wings, which are velvety black and fiery orange, the latter color replacing the green of the allied species. The beauty and brilliancy of this insect are indescribable, and none but a naturalist can understand the intense excitement I experienced when I at length captured it. On taking it out of my net and opening the glorious wings, my heart began to beat violently, the blood rushed to my head, and I felt much more like fainting than I have done when in apprehension of immediate death. I had a headache the rest of the day, so great was the excitement.

Wallace, Alfred R. 1869. *The Malay Archipelago: The Land of the Orang-utan, and the Bird of Paradise.* London: Macmillan & Company.

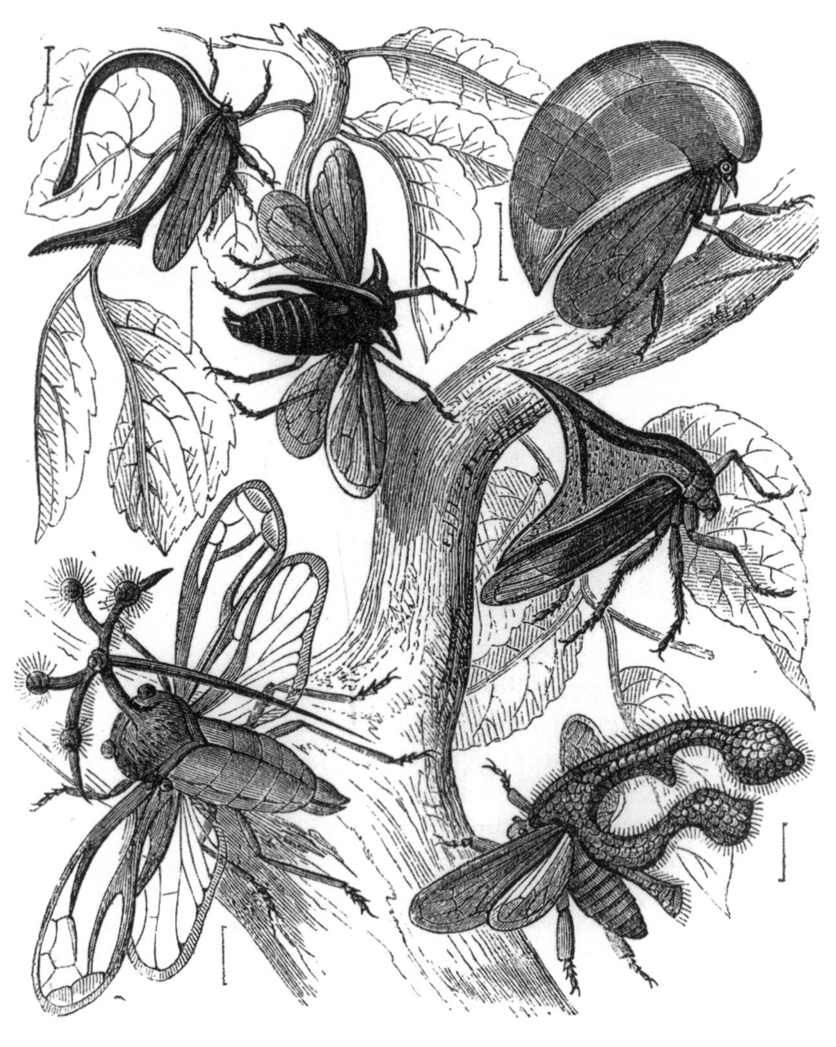

Strange and wonderful treehoppers, by Mm. E. Blanchard (*The Insect World* by Louis Figuier, Chapman and Hall, London, 1869).

To a Butterfly

William Wordsworth

The English poet, William Wordsworth (1770–1850), in "To a Butterfly," evokes the delight of having one's own garden where butterflies come to alight and spend time and provide metaphors for the full yet ephemeral nature of summer. In recent years, butterfly "gardening" has become popular. The idea is that by planting certain combinations of flowering plants, butterflies in great number and diversity can be attracted to visit.

I've watched you now a full half-hour,
Self-poised upon that yellow flower;
And, little Butterfly! indeed
I know not if you sleep or feed.
How motionless!—not frozen seas
More motionless! and then
What joy awaits you, when the breeze
Hath found you out among the trees,
And calls you forth again!

This plot of orchard-ground is ours;
My trees they are, my Sister's flowers;
Here rest your wings when they are weary;
Here lodge as in a sanctuary!
Come often to us, fear no wrong;
Sit near us on the bough!
We'll talk of sunshine and of song,
And summer days, when we were young;
Sweet childish days, that were as long
As twenty days are now

Wordsworth, William. "To a Butterfly."

The Sacred Beetle

Jean-Henri Fabre

The feature documentary Microcosmos*—the popular 1996 film that created a new, appreciative audience of insect watchers—was the work of love of two French naturalists-filmmakers, Claude Nuridsany and Marie Perranou. One of the longest and most lyrical sequences in the film shows a beetle rolling its precious dung ball across uneven terrain—a Sisyphean task rendered larger than life on the big screen. In the credits, the filmmakers acknowledged their debt to the nineteenth-century French naturalist Jean-Henri Fabre (1823–1915), who wrote hundreds of tales of the insect world, based on his keen observations. Fabre's contemporary Charles Darwin called him "an incomparable observer" and made references to his work in* On the Origin of Species.

The son of an illiterate mother and a father who was a failed innkeeper, Fabre managed to work his way through school, eventually becoming professor at the lycée in Avignon, France. He taught there on a modest salary for nearly twenty years before being discharged for his unpopular ideas on teaching; among other things, he admitted girls to his science classes. Following his teaching career, he took up writing full time, supporting his wife and children with superb stories based on his adventures and work with insects. Self-educated in entomology, Fabre wrote about what he knew and loved: the insects found in the fields surrounding his home in the Rhône region of the south of France.

The following selection is the original story of the beetle and the dung ball. It is typical Fabre, making the discovery of insects and insect behavior into an adventure.

It happened like this. There were five or six of us: myself, the oldest, officially their master but even more their friend and comrade; they, lads with warm hearts and joyous imaginations, overflowing with that youthful vitality which makes us so enthusiastic and so eager for knowledge. We started off one morning down a path fringed with dwarf elder and hawthorn, whose clustering blossoms were already a paradise for the Rose-chafers [beetles] ecstatically drinking in their bitter perfumes. We talked as we went. We were going to see whether the Sacred Beetle had yet made his appearance on the sandy plateau of Les Angles [a village near Avignon], whether he was rolling that pellet of dung in which ancient Egypt beheld an image of the world. . . .

But let us . . . clamber up the bluff to the plateau above us. Up there, sheep are grazing and horses being exercised for the approaching races, while all are distributing manna to the enraptured Dung-beetles.

Here are the scavengers at work, the Beetles whose proud mission it is to purge the soil of its filth. One would never weary of admiring the variety of tools wherewith they are supplied, whether for shifting, cutting up and shaping the matter or for excavating deep burrows in which they will seclude themselves with their booty. This equipment resembles a technical museum where every digging-implement is represented. It includes things that seem copied from those appertaining to human industry and others of so original a type that they might well serve us as models for new inventions.

The Spanish *Copris* carries on his forehead a powerful pointed horn, curved backwards, like the long blade of a mattock. . . . All are supplied with a shovel, that is to say, they have a broad, flat head with a sharp edge; all use a rake, that is to say, they collect materials with their toothed forelegs.

As some sort of compensation for their unsavoury task, several of them give out a powerful scent of musk, while their bellies shine like polished metal. The Mimic *Geotrupes* has gleams of copper and gold beneath violet. But generally their colouring is black. The Dung-beetles in gorgeous raiment, those veritable living gems, belong to the tropics. Upper Egypt can show us under its Camel-dung a Beetle rivaling the emerald's brilliant green; Guiana, Brazil and Senegambia boast of *Copres* that are a metallic red, rich as copper and rubybright. The Dung-beetles of our climes cannot flaunt such jewelry, but they are no less remarkable for their habits.

What excitement over a single patch of Cow-dung! Never did adventurers hurrying from the four corners of the earth display such eagerness in working a Californian claim. Before the sun becomes too hot, they are there in their hundreds, large and small, of every sort, shape and size, hastening to carve themselves a slice of the common cake. There are some that labour in the open air and scrape the surface; there are others that dig themselves galleries in the thick of the heap, in search of choice veins; some work the lower stratum and bury their spoil without delay in the ground just below; others again, the smallest, keep on one side and crumble a morsel that has slipped their way during the mighty excavations of their more powerful fellows. Some, newcomers and doubtless the hungriest, consume their meal on the spot; but the greater number dream of accumulating stocks that will allow them to spend long days in affluence, down in some safe retreat. A nice, fresh patch of dung is not found just when you want it, in the barren plains overgrown with thyme; a windfall of this sort is as manna from the sky; only fortune's favourites receive so fair a portion. Wherefore the riches of to-day are prudently hoarded for the morrow. The stercoraceous scent has carried the glad tidings half a mile around; and all have hastened up to get a store of provisions. A few laggards are still arriving, on the wing or on foot.

Who is this that comes trotting towards the heap, fearing lest he reach it too late? His long legs move with awkward jerks, as though driven by some mechanism within his belly; his little red antennae unfurl their fan, a sign of anxious greed. He is coming, he has come, not without sending a few banqueters sprawling. It is the Sacred Beetle, clad all in black, the biggest and most famous of our Dung-beetles. Behold him at table, beside his fellow-guests, each of whom is giving the last touches to his ball with the flat of his broad fore-legs or else enriching it with yet one more layer before retiring to enjoy the fruit of his labours in peace. Let us follow the construction of the famous ball in all its phases.

The clypeus, or shield, that is the edge of the broad, flat head, is notched with six angular teeth arranged in a semicircle. This constitutes the tool for digging and cutting up, the rake that lifts and casts aside the unnutritious vegetable fibres, goes for something better, scrapes and collects it. A choice is thus made, for these connoisseurs differentiate between one thing and another, making a rough selection when the Beetle is occupied with his own provender, but an extremely scrupulous one when it is a matter of constructing the maternal ball, which has a central cavity in which the egg will hatch. Then every scrap of fibre is conscientiously rejected and only the stercoral quintessence is gathered as the material for building the inner layer of the cell. The young larva, on issuing from the egg, thus finds in the very walls of its lodging a food of special delicacy which strengthens its digestion and enables it afterwards to attack the coarse outer layers.

Where his own needs are concerned, the Beetle is less particular and contents himself with a very general sorting. The notched shield then does its scooping and digging, its casting aside and scraping together more or less at random. The fore-legs play a mighty part in the work. They are flat, bow-shaped, supplied with powerful nerves and armed on the outside with five strong teeth. If a vigorous effort be needed to remove an obstacle or to force a way through the thickest part of the heap, the Dung-beetle makes use of his elbows, that is to say, he flings his toothed legs to right and left and clears a semicircular space with an energetic sweep. Room once made, a different kind of work is found for these same limbs: they collect armfuls of the stuff raked together by the shield and push it under the insect's belly, between the four hinder legs. These are formed for the turner's trade. They are long and slender, especially the last pair, slightly bowed and finished with a very sharp claw. They are at once recognised as compasses, capable of embracing a globular body in their curved branches and of verifying and correcting its shape. Their function is, in fact, to fashion the ball.

Armful by armful, the material is heaped up under the belly, between the four legs, which, by a slight pressure, impart their own curve to it and give it a preliminary outline. Then, every now and again, the rough-hewn pill is set spinning between the four branches of the double pair of spherical compasses; it turns under the Dung-beetle's belly until it is rolled into a perfect ball. Should the surface layer lack plasticity and threaten to peel off, should some too-stringy part refuse to yield to the action of the lathe, the fore-legs touch up the faulty places; their broad paddles pat the ball to give consistency to the new layer and to work the recalcitrant bits into the mass.

Under a hot sun, when time presses, one stands amazed at the turner's feverish activity. And so the work proceeds apace: what a moment ago was a tiny pellet is now a ball the size of a walnut; soon it will be the size of an apple. I have seen some gluttons manufacture a ball the size of a man's fist. This indeed means food in the larder for days to come!

The Beetle has his provisions. The next thing is to withdraw from the fray and transport the victuals to a suitable place. Here the Scarab's most striking characteristics begin to show themselves. Straightway he begins his journey; he clasps his sphere with his two long hind-legs, whose terminal claws, planted in the mass, serve as pivots; he obtains a purchase with the middle pair of legs; and, with his toothed fore-arms, pressing in turn upon the ground, to do duty as levers, he proceeds with his load, he himself moving backwards, body bent, head down and hind-quarters in the air. The rear legs, the principal factor in the mechanism, are in continual movement backwards and forwards, shifting the claws to change the axis of rotation, to keep the load balanced and to push it along by alternate thrusts to right and left. In this way the ball finds itself touching the ground by turns with every point of its surface, a process which perfects its shape and gives an even consistency to its outer layer by means of pressure uniformly distributed.

And now to work with a will! The thing moves, it begins to roll; we shall get there, though not without difficulty. Here is a first awkward place: the Beetle is wending his way athwart a slope and the heavy mass tends to follow the incline; the insect, however, for reasons best known to itself, prefers to cut across this natural road, a bold project which may be brought to naught by a false step or by a grain of sand that disturbs the balance of the load. The false step is made: down goes the ball to the bottom of the valley; and the insect, toppled over by the shock, is lying on its back, kicking. It is soon up again and hastens to harness itself once more to its load. The machine works better than ever. But look out, you dunderhead! Follow the dip of the valley—that will save labour and mishaps; the road is good and level; your ball will roll quite easily. Not a bit of it! The Beetle prepares once again to mount the slope that has already been his undoing. Perhaps it suits him to return to the heights. Against that I have nothing to say: the Scarab's judgment is better than mine as to the advisability of keeping to lofty regions; he can see farther than I can in these matters. But at least take this path, which will lead you up by a gentle incline! Certainly not! Let him find himself near some very steep slope, impossible to climb, and that is the very path which the obstinate fellow will choose. Now begins a Sisyphean labour. The ball, that enormous burden, is painfully hoisted, step by step, with infinite precautions, to a certain height, always backwards. We wonder by what miracle of statics a mass of this size can be kept upon the slope. Oh! An ill-advised movement frustrates all this toil: the ball rolls down, dragging the Beetle with it. Once more the heights are scaled and another fall is the sequel. The attempt is renewed, with greater skill this time at the difficult points; a wretched grass-root, the cause of the previous falls, is carefully got over. We are almost there; but steady now, steady! It is a dangerous ascent and the merest trifle may yet ruin everything. For see, a leg slips on a smooth bit of gravel! Down come ball and Beetle, all mixed up together. And the insect begins over again, with indefatigable obstinacy. Ten times, twenty times, he will attempt the hopeless ascent, until his persistence vanquishes all obstacles, or until, wisely recognizing the futility of his efforts, he adopts the level road.

Fabre, J.-Henri. 1918. *The Sacred Beetle and Others.* New York: Dodd, Mead & Co. (Translation by Alexander Teixeira de Mattos.)

Dung beetles: the sacred *Scarabaeus* of Egypt, by Mm. E. Blanchard (*The Insect World* by Louis Figuier, Chapman and Hall, London, 1869).

Enjoying Insects in the Home Garden

Howard Ensign Evans

Howard E. Evans (b. 1922) is a brilliant scientist who has explored seminal questions about the evolution of complex behaviors in wasps. One of the finest popular writers about entomology, past or present, he is the author of, among other books, Wasp Farm, The Pleasures of Entomology, *and* Life on a Little Known Planet*—which he dedicated to "the book lice and silverfish that share my study with me. May they find it digestible!" In this selection Evans turns conventional gardening wisdom on its head by asking, Why not treat insects as ornamentals?*

The growing of vegetables in one's backyard has much to recommend it. Some thirty-eight million Americans have such gardens, altogether occupying nearly two million acres. The flavor and food value of freshly picked vegetables so far exceed those of the store-bought equivalent that gardeners look forward eagerly to each year's harvest—and spend cold winter days planning next year's. (I always order my seeds from the catalog on the day the IRS forms arrive; it takes away some of the pain.)

It is claimed that despite the high cost of food these days, one doesn't really save money by having a home garden—if one counts the cost of one's own labor. But who wants to count that when gardening is such good exercise and so full of challenges and rewards? What will the summer's weather be like? Drought? Hail? Early frost? Will we have enough sweet corn to invite friends for a corn roast? How many canning jars will be needed?

Then, of course, there are insects to think about. Will flea beetles be abundant this year? Will leaf miners decimate the beet greens? Will the cabbage butterflies find the broccoli? (Of course, they always do.) Shall I ring the garden with marigolds to keep the grasshoppers out? (Doesn't work; they love marigolds.) Shall I sacrifice my ideals and stock up on insecticides? Or shall I plant a little more than we need and simply enjoy the insects, the rabbits, the birds?

There is much to be said for the last suggestion. Our garden is not far outside our picture windows, and as we drink our coffee we can watch the robins and grackles slipping into our strawberry patch and emerging with their beaks smeared with red. Now and then a squirrel stares into the patch, then dives in, emerges with a berry, and takes off for the nearest tree. Like the rabbits that nibble the peas, he has learned that we may pursue him. It is fun for all; of course we never catch anything.

Each year I look forward to the insect inhabitants of the garden. I don't begrudge them their share: they amuse us, inform us, and often stimulate our sense of beauty. If they take what I consider more than their share (a fairly rare event) I don't mind being a bit brutal. No animal should multiply so as to destroy its environment (though, being a member of the species *Homo sapiens,* I may have no right to say that).

As I write this I am rearing some zebra caterpillars that were skeletonizing the leaves of our roses. (We do make space for flowers, which after all are food for the spirit.) The zebra caterpillar is very beautiful indeed: its head is orange, its body yellow, with three pairs of black, longitudinal stripes, each pair separated by a white streak. In color it rivals the orioles that are nesting in our cottonwoods. One does, of course, need a microscope to fully appreciate the beauty of zebra caterpillars, but that is an essential item in the household of everyone who admires insects. The caterpillars don't remind me much of zebras. I prefer the scientific name, *Ceramica picta,* which I would translate from the Greek and Latin to mean "a painted earthen vessel."

It has been a good year for asparagus beetles. I am glad; we had a good harvest, and the beetles are most decorative on the bushy flowering stalks. There are two kinds, and we have them both. One, officially called *the* asparagus beetle, has a black head, antennae, and legs, with an orange collar and blue-black wing covers bordered with orange and bearing six light yellow spots—as elaborate a color pattern as one could design. The other, called the spotted asparagus beetle, is wholly orange except for black "polka dots" on its wing covers. The two kinds have similar defensive behavior: if the bush is shaken they simply release their grip and drop to the ground, a response called "thanatosis" in the scientific literature, where suggestive terms such as "playing dead" are frowned upon. If captured, both emit high-pitched squeaks, presumably addressed to birds and other predators, which may drop them in surprise.

Neither species feeds on anything other than asparagus, and hardly anything else in the insect world will eat asparagus. Is there some chemical in the plant that repels herbivores? How does it happen that two related species of beetles can live together on the same plant at the same time—a seeming contradiction to the rule that complete competitors cannot coexist? What is the significance of the fact that the eggs of the asparagus beetle are laid in rows in an erect position, while those of the spotted asparagus beetle are laid singly, flat against the stems? How did two related species happen to evolve such brilliant and such different color patterns?

It appears that the larvae of the asparagus beetle feed freely on the foliage, while those of its spotted cousin live primarily within the seeds. Thus to a certain extent they share the plants and are not complete competitors. Perhaps, with some research, I could answer the other questions. But these are difficult questions, without simple answers. And there are so many other unanswered problems within a few yards of our back door! . . .

Parsleyworms (the larvae of black swallowtail butterflies) feed on carrot tops, parsley, dill, and parsnips, all members of the carrot family (Umbelliferae).

Parsleyworms are most welcome denizens of our garden. They are elegant creatures, transversely banded with green and black and ringed with orange spots. When disturbed, they erect a pair of orange "horns," actually eversible glands that secrete a repellent substance that smells like rancid butter—in fact, it is butyric acid, the essence of rancid butter. It is an enjoyable experience to collect the mature larvae or chrysalids and rear the butterflies indoors, where one can follow the expansion of their glossy, spangled wings. We always release them outdoors, of course, where they can find mates and plants on which to lay their eggs. Their larvae are seldom abundant; it would take a lot of them, all season, to do as much damage to our carrots as the local rabbits can do in one evening.

Parsleyworms are fastidious feeders, requiring food containing certain essential oils, such as methyl chavicol, which is found principally in members of the parsley family. They will even attempt to eat paper if it is soaked in these oils. On members of the parsley family, they can feed without serious competition from other insects, since parsleyworms have evolved the ability to thrive in the presence of psoralens, substances that deter feeding by most other insects by binding DNA in the presence of ultraviolet light. G. Wayne Ivie and his colleagues in the U.S. Department of Agriculture have recently fed tissues treated with carbon-14-labeled psoralens to fall armyworms (which are very general feeders) and to parsleyworms. They found that parsleyworms rapidly detoxify the poisons in their midgut, so that they do not enter the body fluids to any great extent; within 1.5 hours, 50 percent of the carbon-14 passes out with the feces. By contrast, fall armyworms accumulate more psoralens in the body tissues, and within 1.5 hours only 1 percent of the carbon-14 has appeared in the feces. So it is not surprising that, despite the appetite of armyworms for plants of many kinds, they do not flourish on parsley and related plants.

It is interesting that some members of the parsley and carrot family are relatively unpalatable to parsleyworms. Paul Feeny and his colleagues at

Cornell University have shown that cow parsnip and angelica have evolved certain modifications of the form of the psoralen molecule that cause a reduction of the growth rate and fecundity of parsleyworms. They cite this as an example of coevolution: parsleyworms first evolved a means of overcoming the effects of psoralens, in members of the carrot family, and later certain members of this family evolved a modification of the molecule that parsleyworms could not handle. Will parsleyworms further evolve the ability to overcome this novel plant defense? In a few hundred years, perhaps a few thousand, the answer should be apparent.

On the other hand, some plants of the parsley family that grow as wildflowers in woodlands lack psoralens, which are ineffective as deterrents in the absence of plenty of light. Unlike most umbellifers, these plants are attacked by a variety of generalist feeders.

Parsleyworms, cabbageworms, and similar insects were the subject of a 1964 article that has become something of a classic of entomology: "Butterflies and Plants: a Study of Coevolution." The authors were Paul Ehrlich, of Stanford University, and Peter Raven, of the Missouri Botanical Garden. The science of plant-insect relationships has since blossomed into a major field, the subject of several books and innumerable scientific articles. Once these relationships are better understood, it may be possible to breed varieties of crop plants that either lack the chemical cues required by pest species or that have repellent or toxic properties with respect to these insects. To some extent this is already being done. For example, plant breeders have for some years been developing kinds of wheat resistant to the notorious Hessian fly, a stem-infesting insect reputed to have been brought to this country in straw bedding used by Hessian mercenaries employed during the Revolutionary War. Resistant varieties provide a theoretically perfect insect control, requiring no intervention with insecticides. Unfortunately, insects can sometimes overcome, in a few generations of evolution, resistant factors that have been bred into these stocks. This has happened with the Hessian fly, requiring a continuing program of plant breeding.

That, incidentally, is one of the strongest reasons for preserving as much of the wild plant world as possible: the genetic material needed for the breeding of resistant stocks may lie in wild relatives of wheat, corn, and other crops. The rapid extinction of species and locally adapted populations that is occurring as a result of widespread habitat destruction is not pleasant to contemplate as we look forward to feeding the crowded world of the future.

Many of the seeds available to home gardeners are, in fact, those of varieties that have been developed for resistance to various diseases and insects. However, neither our basic knowledge of this complex subject nor our technology has advanced to the point that we need fear that next year we will miss the cabbageworms, the parsleyworms, the asparagus beetles, and all the other insects that provide half the joy of gardening. . . .

For gardeners everywhere, the late Cynthia Westcott wrote *The Gardener's Bug Book*, an encyclopedic treatment of all the "bugs" one might possibly encounter. Although ostensibly addressed even to "the organic gardener who shuns all chemicals and [to] the wildlife enthusiast who is worried about them," two early chapters are devoted to chemicals and how to use them. After listing alphabetically many hundreds of pests ("and a few friends"), she provides an alphabetical list of host plants with the important pests of each. It is an intriguing list, with all kinds of insects I could never hope to lure into my backyard, surely not the rednecked peanutworm or the longtailed banyan mealybug. But I am disappointed never to have attracted the harlequin cabbage bug or the imbricated snout beetle, both of which sound exciting. But no matter; I shall still look forward each year to my old friends, all of whom have their own stories to tell.

Evans, Howard E. 1985. *The Pleasures of Entomology.* Washington, D.C.: Smithsonian Institution Press.

The Ways of a Mud Dauber

George D. Shafer

The standard in scientific writing is supposed to be one of objective detachment. This is especially true of writing about animal behavior, which for years has conformed to the knee-jerk behaviorist dictum of avoiding anthropomorphism at all costs. George Shafer didn't have to worry about this dictum, however, because he wasn't a behaviorist (he had made his reputation as a physiologist) and, by the time he wrote the words below, he was a Stanford professor emeritus studying wasps just for fun.

The result is what most readers will agree is an acceptable level of anthropomorphic affection for his little subjects. The story of Shafer's pet wasps, begun in this selection, concludes with "Little Crumple-Wing," in Chapter 10 (page 299).

Shortly after receiving the title of Emeritus Professor, I was out early on the day that a new year of University work was to open. This was the day when, for many years, I had gone to office and classroom for another term of strenuous work with young men and women—University students—and I had liked it. How much I had liked it was never so fully realized as on this morning. Long before, the resolution had been made to let this day come and go as a vacation day, with hardly a passing thought. But here I was, up early in spite of my resolution, and entertaining the uncomfortable feeling that there was no place to go.

And yet there was a place to go. Instead of going to hear the voices of students I could go and listen to the hum of my bees. Experimenting with bees was one of my hobbies. The abandoned University beehouse had been turned over to me, and a room had been built where I might carry on experiments of my own choosing. The air was cool that fall morning, but the sun was out and the bees were out. The music of their wings was in the air. Aware of all this upon arriving at the bee yard, but without stopping to enjoy it, I walked aimlessly past the row of beehives and entered the old beehouse. Across the room the sun was streaming through a window. I stepped over and stood in its warmth, half insensible to my surroundings. Suddenly I became aware of a gentle, low whir of wings passing my head. They were not the wings of a bee; but they made a most pleasing sound. They brought me around, alert, at once. Their owner alighted on the window sill. She was a beautiful mud-dauber wasp, her regal, richly black body circled with golden-yellow stripes—our largest species of thread-waisted wasps, *Sceliphron cementarium.* She walked leisurely and yet with a businesslike air along the window ledge, twitching her neat, slender wings as is the habit of all mud daubers. She seemed to know just what she wanted to do. She selected a particularly warm spot in the sunlight, lay down on her stomach, stretched her forefeet out in front of her like a kitten, and began to enjoy a sun bath. Every move she made, it seemed to me, was eloquent of assurance and purpose. She had a certain likable air of self-sufficiency. She was only a solitary wasp; her universe was big, but she had a place in it and seemed to know it. She accepted the things that nature provided and used them. More than that, she aroused my admiration.

Every morning, in the early part of that new University year, this mud dauber was at the window enjoying the warmth of the sun, and I came to the window regularly to see her. I began to feel that this little insect was much more than a mere automaton. Somehow she showed personality and confidence, and she inspired me to confidence in my own judgment. Yes (I confess it), a mud dauber was helping to set me right with the world again. Besides the mud daubers, there were many other creature-people at the old bee yard, and now they all took on new interest. Of course, there were the bees—twenty colonies of them, with their thousands of individuals. At one corner of the yard, in a hole in the ground, was a big nest of yellow jackets. Individuals were numerous there too. At several places within my reach the paper wasps *(Polybia flavitarsus)* had nests, and some of their nests were inhabited not alone by wasp larvae but also by the larvae of an interesting moth. Little lizards looked sideways at me as they sat ready to catch bees near the entrances to my hives. An old gopher snake came often to the beehouse that fall looking for mice. All these, and several other creatures, were either inhabitants or visitors of my bee yard. Most were welcome, some not so welcome—but I became interested in them all. They all seemed worthy of study, worthy of acquaintance; and if I was curious about them, many of them seemed just as curious about me. They were not human beings, but often they showed traits and emotions similar to those of humans. Evidences of timidity and aggressiveness, of anger and joy, I observed daily.

While the weather remained favorable that fall, I could step into the beehouse any warm afternoon and listen to mud daubers here and there on the rafters above my head, singing as they spread their pellets of mud to build a new cell or to plaster the surface of a nest almost completed. "Buzt, buz-z-z, buz, buzz-z-z, buzt," they sang as they shaped the mud to their purpose. It was a joyful song. I liked it and was drawn more and more to a study of the songster. Most of the songs came from the large, yellow-banded species that I liked best. There were some specimens of the steel-blue species *(Chalybion coerulium);* and I found a few nests, under the covers of my beehives, of a very small species. These nests were odd, and of interest, but I had little opportunity to become acquainted with the owners, which probably belonged to the genus *Eumenes.* Around the beehouse, *Sceliphron cementarium*—the one with the golden bands—was most plentiful. It was with this species, during the next five years, that I became best acquainted. It is my feeling that, with a few individuals of this species, I even gained a real, though necessarily brief, friendship. The relationship seemed like a friendship and, always, it seemed too brief.

It was not long that first fall until noticeably fewer mud daubers could be seen. In the afternoons the number of nest-building songs fell off. Within a few weeks they had ceased altogether, and in the mornings there were no more visits at the sunny window. The adult mud daubers had completed their work. Their life span was over; but their nests were hidden away on the rafters of the beehouse, and the young larvae of a new generation of wasps were in the sealed cells of the nests.

A little later in the fall, I gathered some of the nests in the beehouse and examined the contents of their cells. At that time the cavity of each mud cell was occupied by a brown cocoon within which a larva lay limp and at rest. In this condition, I knew, they were to remain until spring. Then they would pupate, and after a while, adults would emerge from the pupae; mating would occur; and, in due time, each female would start to build the cells of a new nest for herself. Having built the first cell to full length, she would commence stocking it with spiders, captured and paralyzed by her sting. On one of these paralyzed captives, as it was stored in the cell, an egg would be laid. When the cell had been fully stocked, it would be sealed with a plug of mud before the next cell of the nest was begun. The stored spiders would serve as food for her young—the tiny white larva that would come from the egg in every cell. Each larva would eat and grow until its food supply was constituted. Soon after that, it would split a cocoon about itself, within its mud cell, and be ready to spend a winter there at rest, just as had its ancestors before it. This much—the main features of the life cycle of mud daubers—was known. But while studying larvae removed from their cocoons that fall, I was impressed by the number and size of peculiar little white pellets which were easily visible through the body wall. These were conspicuous enough to be readily seen in photographs. . . .

It was interest in these strikingly unusual pellets that started me on an investigation carried on intermittently for five years.

The period when one can observe adult mud daubers and study their larvae, during the active growing stage, is comparatively short. The earliest adults I have had at my beehouse came out May 12, and the very latest I have observed a living adult was October 8. Although they were present in this period of nearly five months, they were numerous during scarcely three months; and individual adult females live but three months at the most—at the greatest age limit. It is during about this same time that the active stage of larval life must be studied. This is the time, also, when honeybees, yellow jackets, and paper wasps are most active. However, in the five seasons of my association with *S. cementarium,* I not only learned about the little white pellets in the larvae—what they are, why they are present, when and how they are removed from the body of the insect—but in that connection there was revealed also a most astonishing method of sanitation possessed by the larva. Moreover, during the detailed study, the mud dauber revealed some new and interesting secrets about her nest-building, egg-laying, and working habits. Finally, an intimate acquaintance with a few individual wasps brought to light certain responses and reactions tending to convince me that adult females of this species possess a nervous system which, though tiny in size, enables them to remember, to learn, and to show individuality.

Shafer, George D. 1949. *The Ways of a Mud Dauber.* Stanford, Cal.: Stanford University Press.

Ode to the Cricket

William Cowper

The British poet William Cowper (1731–1800) wrote this ode, which hints at the close rapport possible between humans and insects. Jiminy Cricket, the 1940 Walt Disney motion picture star, was the four-legged cricket as conscience to Pinocchio, but crickets remain underappreciated. Cowper's verses, below, seem to sing with the rhythm of crickets.

Little inmate, full of mirth,
Chirping on my kitchen hearth,
Whereso'er be thine abode
Always harbinger of good,
Pay me for thy warm retreat
With a song more soft and sweet;
In return thou shalt receive
Such a strain as I can give.

Thus thy praise shall be expressed
Inoffensive, welcome guest! . . .
Frisking thus before the fire,
Thou hast all thine heart's desire . . .
Wretched man, whose years are spent
In repining discontent,
Lives not, aged though he be,
Half a span, compared to thee.

Cowper, William. "Ode to the Cricket."

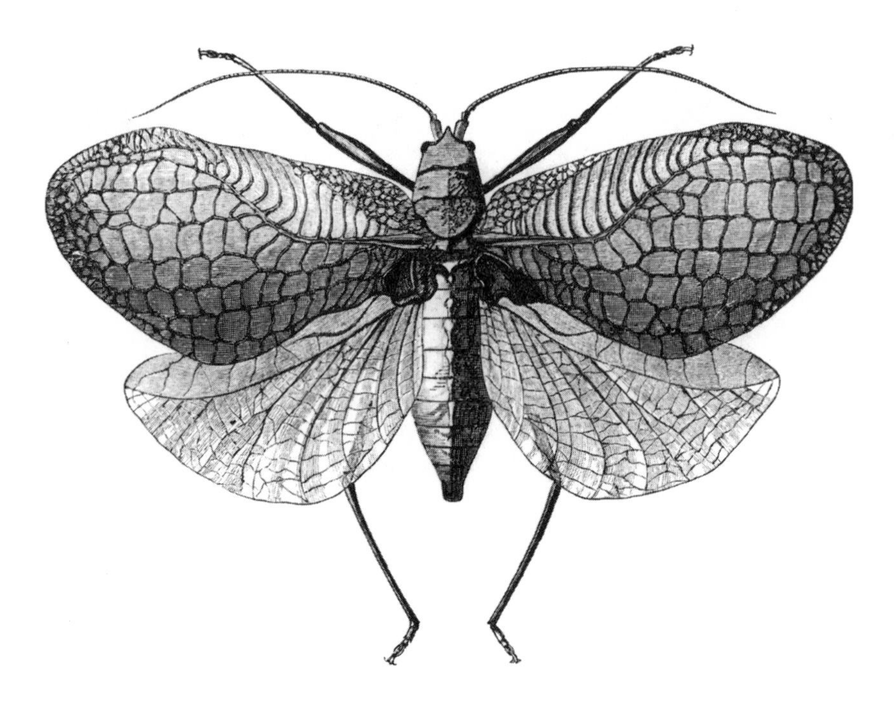

Musical cricket: the lobes of the wing-cases are transformed into a musical instrument. Illustration by E. W. Robinson (*The Naturalist on the River Amazons* by Henry Walter Bates, Murray, London, 1892).

Manna from Heaven

Exodus, The Bible

In Exodus 16, Moses and Aaron lead the people of Israel into the desert where they are near starving and beginning to wonder about the glory of the Lord and why they left Egypt in the first place. And this time, insects come to the rescue, saving their lives. Of course, the "manna," which the Bible says looked like fine frost on the ground and tasted like honey, was thought to be a miracle from God, but it is generally believed today that it was insect excrement known as "honeydew." It came from sap-feeding scale insects, or coccids (Coccidae). These scale insects feed on the phloem sap of plants, and the sugar-rich liquid travels through the gut and out the anus. A single insect can process and expel an extraordinary 2 percent to more than 100 percent of its weight per hour, depending on the species. The insects flick the excretions away with their hind legs, using the propulsive power of their contracting rectum, sometimes even by collapsing their abdomen and expelling the air to get the honeydew out of their way. Typically, it lands on the surrounding ground or vegetation. Arabs in the desert still gather this material, which they call "man." It is apparently very nutritious and, in a desert environment, akin to "manna from heaven"!

13 And it came to pass, that at even the quails came up, and covered the camp: and in the morning the dew lay round about the host.

14 And when the dew that lay was gone up, behold, upon the face of wilderness there lay a small round thing, as small as the hoar frost on the ground.

15 And when the children of Israel saw it, they said one to another, It is manna: for they wist not what it was. And Moses said unto them, This is the bread which the Lord hath given you to eat.

16 This is the thing which the Lord hath commanded, Gather of it every man according to his eating, an omer [one-tenth of a bushel] for every

man, according to the number of your persons; take ye every man for them which are in his tents.

17 And the children of Israel did so, and gathered, some more, some less.

18 And when they did mete it with an omer, he that gathered much had nothing over, and he that gathered little had no lack; they gathered every man according to his eating.

19 And Moses said, Let no man leave of it till the morning.

20 Notwithstanding they hearkened not unto Moses; but some of them left of it until the morning, and it bred worms, and stank: and Moses was wroth with them.

21 And they gathered it every morning, every man according to his eating: and when the sun waxed hot, it melted.

22 And it came to pass, that on the sixth day they gathered twice as much bread, two omers for one man: and all the rulers of the congregation came and told Moses.

23 And he said unto them, This is that which the Lord hath said, To-morrow is the rest of the holy sabbath unto the Lord: bake that which ye will bake today, and seethe that ye will seethe; and that which remaineth over lay up for you to be kept until the morning.

24 And they laid it up till the morning, as Moses bade: and it did not stink, neither was there any worm therein.

25 And Moses said, Eat that today; for today is a sabbath unto the Lord: today ye shall not find it in the field.

26 Six days ye shall gather it; but on the seventh day, which is the sabbath, in it there shall be none.

27 And it came to pass, that there went out some of the people on the seventh day for to gather, and they found none.

28 And the Lord said unto Moses, How long refuse ye to keep my commandments and my laws?

29 See, for that the Lord hath given you the sabbath, therefore he giveth you on the sixth day the bread of two days: abide ye every man in his place, let no man go out of his place on the seventh day.

30 So the people rested on the seventh day.

31 And the house of Israel called the name thereof Manna: and it was like coriander seed, white; and the taste of it was like wafers made with honey.

32 And Moses said, This is the thing which the Lord commandeth, Fill an omer of it to be kept for your generations; that they may see the bread wherewith I have fed you in the wilderness, when I brought you forth from the land of Egypt.

33 And Moses said unto Aaron, Take a pot, and put an omer full of manna therein, and lay it up before the Lord, to be kept for your generations.
34 As the Lord commanded Moses, so Aaron laid it up before the Testimony, to be kept.
35 And the children of Israel did eat manna forty years, until they came to a land inhabited: they did eat manna, until they came unto the borders of the land of Canaan.

Exodus 16:13–35. Authorized King James Version (translated into English 1603–1611).

Things Clean and Unclean

Leviticus, The Bible

Insects have long served as human food, and it is only cultural prejudice that largely excludes them from the twentieth-century Western larder. Leviticus 11 has one of the earliest references to "entomophagy" (eating of insects), in which commonly consumed winged insects such as locusts, crickets, and grasshoppers are distinguished from crawling insects, which are identified as "unclean," possibly because a few of them serve to transmit human disease.

Other insects, such as ants, are neither suggested nor rejected as food in the Bible, though the Chinese and other ancient cultures ate ants and used them as medicine. A story recently circulated on the Internet tells of a very worried and upset woman who called a poison control center because she had caught her little daughter eating ants. A medical student doing a rotation in toxicology was said to have quickly reassured her that the ants were not harmful and there would be no need to bring her daughter to the hospital. The woman calmed down, and at the end of the conversation happened to mention that she gave her daughter some ant poison to eat in order to kill the ants. The medical student told her to bring her daughter to the ER right away.

1 And the Lord spake unto Moses and to Aaron, saying unto them,

2 Speak unto the children of Israel, saying, These are the beasts which ye shall eat among all the beasts that are on the earth. . . .

22 Even these of them ye may eat; the locust after his kind, and the bald locust after his kind, and the beetle after his kind, and the grasshopper after his kind.

23 But all other flying creeping things . . . shall be an abomination unto you. . . .

41 And every creeping thing that creepeth upon the earth shall be an abomination; it shall not be eaten.

42 Whatsoever goeth upon the belly, and whatsoever goeth upon all four, or whatsoever hath more feet among all creeping things that creep upon the earth, them ye shall not eat; for they are an abomination.

43 Ye shall not make yourselves abominable with any creeping thing that creepeth, neither shall ye make yourselves unclean with them, that ye should be defiled thereby.

44 For I am the Lord your God. . . .

Leviticus 11:1–2, 22–23, 41–44. Authorized King James Version (translated into English 1603–1611).

The Culinary Marvels of Insect Life

Edward Step

In many human cultures, insects provide essential nutrition to humans foraging in marginal environments, but certain species are sought-after gourmet items that demand top dollar in Oriental markets and upper-crust Mexico City restaurants. Among the few exceptions to this widespread entomophagy are the European-derived cultures, which are increasingly imposing their misplaced squeamishness onto indigenous peoples, with malnutrition as the unfortunate result. As one Mexican agronomist, quoted in the Food Insects Newsletter, *observed, "More Mexicans would be eating bugs were it not for decades of ad campaigns by international companies pushing white bread and Spam."*

After all, doesn't it make more sense to eat locusts (as generations of Africans and Native Americans have done) than to dump pesticides on them? From all reports, they are quite tasty, similar to fried shrimp. In fact, pound for pound, insect pests are often more nutritious than the crops they eat!

A visitor from one of the other planets would probably be surprised to learn that though the civilized races of the earth indulge in the eating of live oysters and some other strange foods, they abstain as a rule from the eating of insects. Why it should be so is rather difficult to understand, when one considers the things we do eat, and the fact that uncivilized and semi-civilized peoples retain the primitive habit. The abstention from insects as food has been brought about, no doubt, by fashion, for that it is not merely culture and civilization that produce an abhorrence of such food is proved by the fact that the cultured Greeks and Romans found nothing disgusting in it. They ate their cossus, their cicadas, and their locusts; and even their poets praised such fare.

We are all familiar with the story of John the Baptist subsisting upon locusts and wild honey, and though controversialists have sought to show that the locusts in question were the seed-pods of a small tree, there can be no doubt that they were the insects still eaten by the Arabs and other races of countries where the swarms of migratory locusts visit. Hasselquist found at Mecca, when corn was scarce, the Arabs ground locusts in their hand-mills or pounded them in stone mortars to make a substitute for flour. Moistened with water and worked into a sort of dough, this was made into cakes and baked. He adds that they also eat locusts without waiting for the excuse of a famine, boiling them in water, and then stewing with butter "into a kind of fricassee of no bad flavour." Sparrman, too, tells us of the Hottentots that they rejoice when swarms of locusts visit their country, though it means the destruction of all verdure. They feast upon the locusts, and make a coffee-coloured soup of their eggs, getting visibly fatter on such nourishing diet. Forskål, a Swede, says, there is no great relish in this diet, and that if indulged in too freely it thickens the blood, and becomes injurious to those of melancholic temperament. For all that, in Calcutta the natives still regard a swarm of locusts as a providential event, and their dried bodies form an ingredient in the preparation of curries.

In parts of South Africa it is not only the migratory locust that is eaten, but other large grasshoppers, including one that we figure that is coloured black and red. This is a livery that is regarded as being outward evidence of unwholesomeness; but we are told that in spite of it the thick leaping legs are eaten as a relish with "mealies." The entire insect is fried, and the legs are detached for food as they have a salt flavour. Our informant says the custom of eating insects is dying out and is now only practised by the poorest of the natives, the reason being that meat and salt are now so easily obtainable, and most of the natives are in a position to buy them. On the other hand, Mr. W. L. Distant, author of *The Naturalist in the Transvaal,* informs the present writer that he never heard of the natives eating this particular black and red species, not does he think it likely that they do so.

Caterpillars of several kinds are eaten by the African Bushmen, the Australian Blackfellow, and the Chinese. The latter also utilize the chrysalids of the silkworm after the silk has been unwound from the cocoon. Indian rearers of the Tussar silk-moth make a similar use of the chrysalids. More important as a food is the white ant. Sparrman tells us that the Hottentots eat them boiled and raw, and Smeathman says that the Africans among whom he explored parch them in iron pots over a gentle fire, much as coffee is roasted. In that state they are delicious food; and they eat them by handfuls. He thought them delicate, nourishing, and wholesome, being sweeter than the big grub of the palm weevil, and tasting like sugared cream or sweet almond paste. The palm weevil grub, the gru-gru of the West Indies, is also eaten on the Amazons, according to Wallace. In Nyassaland they make a paste of a species of mayfly and gnats, and eat it under the name of "kungu".

Step, Edward. 1916. *Marvels of Insect Life.* New York: Robert M. McBride & Co.

Insect Extravaganza

Jim Mertins

The Food Insects Newsletter, *a labor of love from the Department of Entomology of the University of Wisconsin, always has a few useful recipes. The following excerpt has all that and more under the apt headline "Insect Extravaganza at Iowa State University."*

The Insect Horror Film Festival which unfolded at Iowa State University on September 6–8 had more to offer than bug horror movies. A horror movie was indeed shown, as advertised, at 8:00 pm each evening. But the doors opened at 7:30 pm and those who were attracted by the prospect of seeing bug violence found that they had inadvertently exposed themselves to some additional education about insects. The Festival was sponsored by the Entomology Club, the Student Union Board and the ISU Committee on Lectures.

Before each film, at 7:30 pm:

- Praying Mantis Give-away to first 50 audience members.
- Insect Tasting Event including corn borer bread.
- Insect Petting Zoo with hissing cockroaches, drone bees, tarantula, giant walking stick, giant water beetle and dragonfly nymph.
- Concert of recorded insect mating calls.
- Two short bug cartoons.
- Introduction by Dr. John Obrycki of the ISU Entomology Department who discussed how insects are used to develop moods in films and also the accuracy of the portrayal of insects and scientists in film. . . .

Among the recipes featured at the Insect Tasting Event:

Corn Borer Corn Bread: Use favorite corn bread recipe and substitute 1/2 cup ground dry roasted corn borer larvae for 1/2 cup corn meal. Any larvae without hairs or bright colors can be substituted for corn borer larvae.

Chocolate Covered Crickets: 2 squares of semi-sweet chocolate and 25 dry roasted crickets and/or grasshoppers with legs and wings plucked. Melt chocolate as directed on the box. Dip insects in chocolate, place on wax paper and refrigerate until party.

Insect preparation: To clean insects, place in a colander or fine mesh strainer, rinse and pat dry. Dry roast in a 200° oven, until crispy. They can then be ground into flour, cut into pieces or used whole.

The entomology students serving the insects attempted to make them irresistible by displaying them on silver trays with white tablecloths and candlelight. Cynthia Lidtke, an entomology student helping with the festival, said she hopes movie-goers will come away a little more appreciative of insects. "I hope they won't just step on them any more, and say, 'Well, there's another one.' "

Mertins, Jim. 1990. "Insect Extravaganza at Iowa State University." *The Food Insects Newsletter* III(3):2.

Why Not Eat Insects?

Vincent M. Holt

The nineteenth-century food writer Simmonds suggested that the scientific name for locusts, Gryllus, *was tantamount to an invitation to cook them. Vincent Holt—author of the little-known 1885 treatise* Why Not Eat Insects?*—took things a step further by creating a complete insect menu, rendered first in French, then in English. Despite Mr. Holt's unfailing enthusiasm, most will probably prefer the French version. "Of course these menus are unnaturally crowded with insect items," he writes, "but they are merely intended to show how such dishes may be usefully introduced into the chief courses of an ordinary dinner."*

FRENCH

Menu

Potage aux Limaces à la Chinoise
Morue bouillie à l'Anglaise, Sauce aux Limaçons

Larves de Guêpes frites au Rayon
Phalènes à l'Hottentot
Boeuf aux Chenilles
Petites Carottes, Sauce blanche aux Rougets

Crême de Groseilles aux Nemates
Larves de Hanneton Grillées
Cerfs Volants à la Gru Gru

ENGLISH

MENU

Slug soup
Boiled Cod with Snail Sauce

Wasp Grubs fried in the Comb
Moths sautés in Butter
Braized Beef with Caterpillars
New Carrots with Wireworm Sauce

Gooseberry Cream with Sawflies
Devilled Chafer Grubs
Stag Beetle Larvae on Toast

Holt, Vincent. 1885. *Why Not Eat Insects?* (Reprinted 1978, Faringdon, England: E. W. Classey.)

Sugaring for Moths

W. J. Holland

W. J. Holland (1848–1932) was a lepidopterist (moth and butterfly specialist) at the Carnegie Museum in Pittsburgh, Pennsylvania, around the turn of the century. His The Moth Book: A Guide to the Moths of North America *is especially appealing for its pleasant literary style and copious use of quotations from prose and poetry. This excerpt conveys the sheer joy of encountering, through a particularly entomological collecting ritual, these rare and mysterious denizens of the natural world.*

The day has been hot and sultry. The sun has set behind great banks of clouds which are piling up on the northwestern horizon. Now that the light is beginning to fade, the great masses of cumulus, which are slowly gathering and rising higher toward the zenith, are lit up by pale flashes of sheet-lightning. As yet the storm is too far off to permit us to hear the boom of the thunder, but about ten or eleven o'clock to-night we shall probably experience all the splendor of a dashing thundershower.

Along the fringe of woodland which skirts the back pastures is a path which we long have known. Here stand long ranks of ancient beeches; sugar maples, which in fall are glorious in robes of yellow and scarlet; ash trees, the tall gray trunks of which carry skyward huge masses of light pinnated foliage; walnuts and butternuts, oaks, and tulip-poplars. On either side of the path in luxuriant profusion are saplings, sprung from the monarchs of the forest, young elm trees planted by the winds, broad-leaved papaws, round-topped hawthorns, viburnums, spreading dogwoods, and here and there in moist places clumps of willows. Where the path runs down by the creek, sycamores spread their gaunt white branches toward the sky, and drink moisture from the shallow reaches of the stream, in which duckweed, arrow-weed, and sweet pond-lilies bloom.

The woodland is the haunt of many a joyous thing, which frequents the glades and hovers over the flowers. To-night the lightning in the air, the suggestion of a coming storm which lurks in the atmosphere, will send a thrill through all the swarms, which have been hidden through the day on moss-grown trunks, or among the leaves, and they will rise, as the dusk gathers, in troops about the pathway. It is just the night upon which to take a collecting trip, resorting to the well-known method of "sugaring."

Here we have a bucket and a clean whitewash brush. We have put into the bucket four pounds of cheap sugar. Now we will pour in a bottle of stale beer and a little rum. We have stirred the mixture well. In our pockets are our cyanide jars. Here are the dark lanterns. Before the darkness falls, while yet there is light enough to see our way along the path, we will pass from tree to tree and apply the brush charged with the sweet semi-intoxicating mixture to the trunks of the trees.

The task is accomplished! Forty trees and ten stumps have been baptized with sugar-sweetened beer. Let us wash our sticky fingers in the brook and dry them with our handkerchiefs. Let us sit down on the grass beneath this tree and puff a good Havana. It is growing darker. The bats are circling overhead. A screech-owl is uttering a plaintive lament, perhaps mourning the absence of the moon, which to-night will not appear. The frogs are croaking in the pond. The fireflies soar upward and flash in sparkling multitudes where the grass grows rank near the water.

Now let us light our lamps and put a drop or two of chloroform into our cyanide jars, just enough to slightly dampen the paper which holds the lumps of cyanide in place. We will retrace our steps along the path and visit each moistened spot upon the tree-trunks.

Here is the last tree which we sugared. There in the light of the lantern we see the shining drops of our mixture clinging to the mosses and slowly trickling downward toward the ground. Turn the light of the lantern full upon the spot, advancing cautiously, so as not to break the dry twigs under foot or rustle the leaves. Ha! Thus far nothing but the black ants which tenant the hollows of the gnarled old tree appear to have recognized the offering which we have made. But they are regaling themselves in swarms about the spot. Look at them! Scores of them, hundreds of them are congregating about the place, and seem to be drinking with as much enjoyment as a company of Germans on a picnic in the wilds of Hoboken.

Let us stealthily approach the next tree. It is a beech. What is there? Oho! my beauty! Just above the moistened patch upon the bark is a great *Catocala.* The gray upper wings are spread, revealing the lower wings gloriously banded with black and crimson. In the yellow light of the lantern the wings appear even more brilliant than they do in sunlight. How the eyes glow like spots of fire! The moth is wary. He has just alighted; he has not yet drunk deep. Move cautiously! Keep the light of the lantern steadily upon him. Uncover your poisoning jar. Approach. Hold the jar just a little under the moth, for he will drop downward on the first rush to get away. Clap the jar over him! There! you have done it! You have him securely. He flutters for a moment, but the chloroform acts quickly and the flutterings cease. Put that jar into one pocket and take out another. Now let us go to the next tree. It is an old walnut. The trunk is rough, seamed, and full of knotted excrescences. See what a company has gathered! There are a dozen moths, large and small, busily at work tippling. Begin with those which are nearest to the ground. When I was young my grandfather taught me that in shooting wild turkeys resting in a tree, it is always best to shoot the lowest fowl first, and then the next. If you shoot the gobbler which perches highest, as he comes tumbling down through the flock, he will startle them all, and they will fly away together; but if you take those which are roosting, well down among the branches, those above will simply raise their heads and stare about for a moment to find out the source of their peril, and you can bag three or four before the rest make up their minds to fly. I follow the same plan with my moths, unless, perchance, the topmost moth is some unusual rarity, worth all that suck the sweets below him.

Bravo! You have learned the lesson well. You succeeded admirably in bottling those *Taraches* which were sucking the moisture at the lower edge of the sweetened patch. There above them is a fine specimen of *Strenoloma lunilinea.* Aha! You have him. Now take that *Catocala.* It is *amasia,* a charming little species. Above him is a specimen of *cara,* one of the largest and most superb of the genus. Well done! You have him, too. Now wait a moment! Have your captives ceased their struggles in your jar? Yes; they seem to be thoroughly stunned. Transfer them to the other jar for the cyanide to do its work. Look at your lantern. Is the wick trimmed? Come on then.

Let us go to the next tree. This is an ash. The moist spot shows faintly upon the silvery-gray bark of the tree. Look sharply! Here below are a few *Geometers* daintily sipping the sweets. There is a little *Eustixis pupula,* with its silvery-white wings dotted with points of black. There is a specimen of *Harrisimemna,* the one with the coppery-brown spots on the fore wings. A good catch!

Stop! Hold still! Ha! I thought he would alight. That is *Catocala coccinata*—a fine moth—not overly common, and the specimen is perfect.

Well, let us try another tree. Here they are holding a general assembly. Look! See them fairly swarming about the spot. A dozen have found good places; two or three are fluttering about trying to alight. The ants have found the place as well as the moths. They are squabbling with each other. The moths do not like the ants. I do not blame them. I would not care to sit down at a banquet and have ants crawling all over the repast. There is a specimen of *Catocala relicta,* the hind wings white, banded with black. How beautiful simple colors are when set in sharp contrast and arranged in graceful lines! There is a specimen of *Catocala neogama,* which was originally described by Abbot from Georgia. It is not uncommon. There is a good *Mamestra,* and there *Pvrophila pyramidoides.* The latter is a common species; we shall find scores of them before we get through. Do not bother with those specimens of *Agrotis Ypsilon;* there are choicer things to be had. It is a waste of time to take them to-night. Let them drink themselves drunk, when the flying squirrels will come and catch them. Do you see that flying squirrel there peeping around the trunk of the tree? Flying squirrels eat insects. I have seen them do it at night, and they have robbed me of many a fine specimen.

Off now to the next tree!

And so we go from tree to tree. The lightning in the west grows more vivid. Hark! I hear the thunder. It is half-past nine. The storm will be here by ten. The leaves are beginning to rustle in the tree-tops. The first pulse of the tornado is beginning to be felt. Now the wind is rising. Boom! Boom! The storm is drawing nearer. We are on our second round and are coming up the path near the pasture-gate. Our collecting jars are full. We have taken more than a hundred specimens representing thirty species. Not a bad night's work. Hurry up! Here are the draw-bars. Are you through? Put out the light in your lantern. Come quickly after me. I know the path. Here is the back garden gate. It is beginning to rain. We shall have to run if we wish to avoid a wetting. Ah! here are the steps of the veranda. Come up!

My! what a flash and a crash that was! Look back and see how the big trees are bowing their heads as the wind reaches them, and the lightning silhouettes them against the gray veil of the rain. We may be glad we are out of the storm, with a good roof overhead. To-morrow morning the sun will rise bright and clear, and we shall have work enough to fill all the morning hours in setting the captures we have made.

Holland, W. J. 1903. *The Moth Book.* New York: Doubleday, Page and Company. (Reprinted 1968, New York: Dover.)

Bug Off!

Dave Barry

Popular American humorist Dave Barry (b. 1947) holds forth on the choice of a U.S. Official National Insect.

I am sick and tired of our so-called representatives in Washington being influenced by powerful special-interest groups on crucial federal issues. As you have no doubt gathered, I am referring to the current effort to name an Official National Insect.

This effort, which I am not making up, was alertly brought to my attention by Rick Guldan, who's on the staff of U.S. Representative James Hansen of Utah, at least until this column gets published. Rick sent me a letter that was mailed to congresspersons by the Entomological Society of America. (An "entomologist" is defined by Webster's as "a person who studies entomology.") The letter urges Representative Hansen to support House Joint Resolution 411, which would "designate the monarch butterfly as our national insect." The letter gives a number of reasons, including that "the durability of this insect and its travels into the unknown emulate the rugged pioneer spirit and freedom upon which this nation was settled."

The letter is accompanied by a glossy political-campaign-style brochure with color photographs showing the monarch butterfly at work, at play, relaxing with its family, etc. There's also a list entitled "Organizations Supporting the Monarch Butterfly," including the Friends of the Monarchs, the National Pest Control Association, the Southern Maryland Rock and Mineral Club, and the Saginaw County Mosquito Abatement Commission.

Needless to say I am strongly in favor of having an official national insect. If history teaches us one lesson, it is that a nation that has no national insect is a nation that probably also does not celebrate Soybean Awareness Month. I also have no problem with the monarch butterfly *per se.* ("Per se" is Greek for "unless it lays eggs in my salad.") Butterflies are nice to have around, whereas with a lot of other insects, if they get anywhere near you, your immediate reaction, as an ecologically aware human being, is to whomp them with a hardcover work of fiction at least the size of *Moby Dick.*

But what bothers me is the way the Entomological Society is trying to slide this thing through Congress without considering the views of the average citizen who does not have the clout or social standing to belong to powerful elite "insider" organizations such as the Saginaw County Mosquito Abatement Commission. Before Congress makes a decision of this magnitude, we, the public, should get a chance to vote on the national insect. We might feel that, in these times of world tension, we don't want to be represented by some cute little flitting critter. We might want something that commands respect, especially in light of the fact that the Soviet Union recently selected as its national insect the Chernobyl Glowing Beetle, which grows to a length of 17 feet and can mate in midair with military aircraft.

Fortunately, we Americans have some pretty darned impressive insects ourselves. In South Florida, for example, we have industrial cockroaches that have to be equipped with loud warning beepers so you can get out of their way when they back up. Or we could pick a fierce warlike insect such as the fire ant, although this could create problems during the official White House National Insect Naming Ceremony ("WASHINGTON—In a surprise development yesterday that political observers believe could affect the 1992 election campaign, President Bush was eaten").

Other strong possible candidates for National Insect include: the gnat, the imported Japanese beetle, the chigger, the praying mantis, Jiminy Cricket, the laughing mantis, the lobster, the dead bugs in your light fixture, the skeet-shooting mantis, and Senator Jesse Helms. I could go on, but my purpose here is not to name all the possibilities; my purpose is to create strife and controversy for no good reason.

And you can help. I recently acquired a highly trained, well-staffed, modern Research Department. Her name is Judi Smith, and she is severely underworked because I never need anything researched other than the question of what is the frozen-yogurt Flavor of the Day at the cafeteria.

So I'm asking you to write your preference for National Insect on a POSTAL CARD. (If you send a letter, the Research Department has been instructed to laugh in the diabolical manner of Jack Nicholson as The Joker and throw it away unopened.) Send your card to: National Insect Survey, c/o Judi Smith, The Miami Herald Tropic Magazine, 1 Herald Plaza, Miami, FL 33132.

Judi will read all the entries and gradually go insane. Then I'll let you know which insect is preferred by you, The People, and we can start putting serious pressure on Congress. If all goes well, this could wind up costing the taxpayers millions of dollars.

In closing, let me stress one thing, because I don't want to get a lot of irate condescending mail from insect experts correcting me on my facts: I am well aware that Senator Helms is, technically, a member of the arachnid family.

Barry, Dave. 1991. "Bug Off!" *Dave Barry Talks Back.* New York: Crown Publishers. (Column originally appeared in the *Miami Herald.*)

2

Plagues of Vermin: Insects Reviled

Perhaps the most prevalent view of insects in human culture—certainly so in modern times—has been one of revulsion. We have reacted with fear, hatred, and deliberate efforts at extermination toward those insects that invade our gardens and granaries or, even worse, parasitize our bodies and transmit disease. These negative attitudes about insects are as old as humans but, as noted in chapter 1, were in times past directed only at particular offending species. In recent years, however, and especially in the Western cultures, these attitudes have intensified. As we have become increasingly estranged from the natural world, we have correspondingly lost the ability to distinguish one insect from another, and, as a result, our negative emotions have become generalized to almost all insects. We shriek in horror and reach for the insecticide without regard to whether the object of our revulsion is beneficial, harmful, or simply indifferent to humankind; we certainly never stop to contemplate whether it might be beautiful in form or complex in the story of its hidden life. This twentieth-century "insect fear" is reflected in Hollywood films, cartoons, magazine articles, and various other forms of popular culture.

In the following pages we will look at the negative side of insects in human history, the stories of insects reviled.

A Treatise of Buggs

John Southall

Frank E. Lutz, former curator of entomology at the American Museum of Natural History, once suggested that an obscure historical figure named John Southall should be regarded as the patron saint of exterminators. In London, in 1730, Southall produced and sold this booklet and offered a comprehensive range of services using a secret formula purchased—for some tobacco and a few pennies—from an old African in Jamaica. Bedbugs ("buggs") were apparently rampant in eighteenth-century London, and if Southall's pamphlet is to be believed, his secret potion was one of the only reliable ways of eliminating (at least temporarily) this unpleasant scourge.

A

TREATISE

OF

BUGGS:

SHEWING

When and How they were firft brought into *England*. How they are brought into and infect Houfes.

Their Nature, feveral Foods, Times and Manner of Spawning and Propagating in this Climate.

Their great INCREASE accounted for, by Proof of the Numbers each Pair produce in a Seafon.

REASONS given why all Attempts hitherto made for their Deftruction have proved ineffectual.

VULGAR ERRORS concerning them refuted.

That from *September* to *March* is the beft Seafon for their total Deftruction, demonftrated by Reafon, and proved by Facts.

Concluding with

DIRECTIONS for fuch as have them not already, how to avoid them; and for thofe that have them, how to deftroy them.

By *JOHN SOUTHALL*,

Maker of the Nonpareil Liquor for deftroying *Buggs* and *Nits*, living at the *Green Pofts* in the *Green Walk* near *Faulcon-ftairs*, *Southwark*.

LONDON: Printed for J. ROBERTS, near the *Oxford-Arms*, in *Warwick-Lane*. M.DCC XXX.

(Price One Shilling.)

Illustration of bedbug, or nit, and original book cover (from Southall, 1730).

As Buggs have been known to be in England above sixty Years, and every Season increasing so upon us, as to become terrible to almost every Inhabitant in and about this Metropolis, it were greatly to be wished that some more learned Person than my self, studious for the Good of Human Kind, and the Improvement of natural Knowledge, would have oblig'd the Town with some Treatise, Discourse or Lecture on that nauseous venomous Insect.

But as none such have attempted it, and I have ever since my return from America made their destruction my Profession, and was at first much baffled in my Attempts for want, (as I then believ'd, and have since found) of truly knowing the Nature of those intolerable Vermin: I determin'd by all means possible to try if I could discover and find out as much of their Nature, Feeding and Breeding, as might be conducive to my being better able to destroy them.

And tho' in attempting it I must own I had a View at private Gain, as well as the publick Good; yet I hope my Design will appear laudable, and the Event answer both Ends.

Southall, John. 1730. *A Treatise of Buggs.* London: J. Roberts. (We are indebted to Frank E. Lutz, a curator of the American Museum of Natural History, who found this manuscript in the museum's collection and included part of it in his *A Lot of Insects,* published in 1941.)

DEATH-WATCH BEETLES AND THE FLYPAPER SELLERS OF LONDON

Frank Cowan

Prior to the late nineteenth century, the recording of the "natural history" of a group of plants or animals necessarily included an account of the human beliefs and superstitions about that group. But even by these standards, Frank Cowan's classic text departed from typical natural history because it focused on what would later come to be called "cultural entomology," as suggested in his subtitle: "A complete collection of the legends, superstitions, beliefs, and ominous signs connected with insects; together with their uses in medicine, art, and as food; and a summary of their remarkable injuries and appearances." As such, no rival to Cowan's work has appeared in the more than 130 years since his book was written.

Here are two vintage Frank Cowan (1844-1905) selections. His essay on death-watch beetles tells of the terror they inspired in the public as well as the annoyance felt by entomologists. Under "flies," Cowan presents a slice of life from mid-nineteenth-century London revealing the lot of the "catch-'em-alive" boys who sold flypaper on the streets—a piece that evokes Charles Dickens.

Ptinidae—Death-watch, etc.

The common name of *Death-watch,* given to the *Anobium tesselatum,* sufficiently announces the popular prejudice against this insect; and so great is this prejudice, that, as says an editor of Cuvier's works, the fate of many a nervous and superstitious patient has been accelerated by listening, in the silence and solitude of night, to this imagined knell of his approaching dissolution. The learned Sir Thomas Browne considered the superstition connected with the Death-watch of great importance, and remarks that "the man who could eradicate this error from the minds of the people would save from many a cold sweat the meticulous heads of nurses and grandmothers," for such persons are firm in the belief, that

The solemn Death-watch clicks the hour of death.

The witty Dean of St. Patrick endeavored to perform this useful task by means of ridicule. And his description, suggested, it would appear, by the old song of "A cobbler there was, and he lived in a stall," runs thus:

—A wood worm
That lies in old wood, like a hare in her form,
With teeth or with claws, it will bite, it will
scratch;
And chambermaids christen this worm a
Death-watch;
Because, like a watch, it always cries click.
Then woe be to those in the house that are sick!
For, sure as a gun, they will give up the ghost,
If the maggot cries click when it scratches the
post.
But a kettle of scalding hot water injected,
Infallibly cures the timber affected;
The omen is broken, the danger is over,
The maggot will die, and the sick will recover.

Grose, in his Antiquities, thus expresses this superstition: "The clicking of a Death-watch is an omen of the death of some one in the house wherein it is heard." Watts says: "We learn to presage approaching death in a family by ravens and little worms, which we therefore call a Death-watch." Gay, in one of his Pastorals, thus alludes to it:

When Blonzelind expired . . .
The solemn Death-watch click'd the hour she
died.

And Train,—

An' when she heard the Dead-watch tick,
She raving wild did say,
"I am thy murderer, my child;
I see thee, come away."

And Pope,—

Misers are muck-worms, silkworms beaux,
And Death watches physicians.

"It will take," says Mrs. Taylor, a writer in Harper's New Monthly Magazine, "a force unknown at the present time to physiological science to eradicate the feeling of terror and apprehension felt by almost every one on hearing this small insect." She herself, an entomologist, confesses to have been very much annoyed at times by coming in contact with this "strange nuisance;" but she was cured by an overapplication. "I went to pay a visit," says she, "to a friend in the country. The first night I fancied I should have gone mad before morning. The walls of the bed-room were papered, and from them beat, as it were, a thousand watches—tick, tick, tick! Turn which way I would, cover my head under the bedclothes to suffocation, every pulse in my body had an answering tick, tick, tick! But at last the welcome morning dawned, and early I was down in the library; even here every book, on shelf above shelf, was riotous with tick, tick, tick! At the breakfast table, beneath the plates, cups, and dishes, beat the hateful sound. In the parlor, the withdrawing-room, the kitchen, nothing but tick, tick! The house was a huge clock, with thousands of pendulums ticking from morning till night. I was careful not to allow my great discomfort to annoy others. I argued what they could tolerate, surely I could; and in a few days habit had rendered the fearful, dreaded ticking a positive necessity."

The Death-watch commences its clicking, which is nothing more than the call or signal by which the male and female are led to each other, chiefly when spring is far advanced. The sound is thus produced: Raising itself upon its hind legs, with the body somewhat inclined, it beats its head with great force and agility upon the plane of position. The prevailing number of distinct strokes which it beats in succession is from seven to nine or eleven; which circumstance thinks Mr. Shaw, may perhaps still add, in some degree, to the ominous character which it bears. These strokes follow each other quickly, and are repeated at uncertain intervals. In old houses, where these insects abound, they may be heard in warm weather during the whole day.

Baxter, in his World of Spirits . . . most sensibly observes that "there are many things that ignorance causeth multitudes to take for prodigies. I have had many discreet friends that have been affrighted with the noise called a Death-watch, whereas I have since, near three years ago, oft found by trial that it is a noise made upon paper by a little nimble, running worm, just like a louse, but whiter and quicker; and it is most usually behind a paper pasted to a wall, especially to wainsot; and it is rarely, if ever, heard but in the heat of summer." Our author, however, relapses immediately into his honest credulity, adding: "but he who can deny it to be a prodigy, which is recorded by Melchior Adamus, of a great and good man, who had a clock-watch that had layen in a chest many years unused; and when he lay dying, at eleven o'clock, of itself, in that chest, it struck eleven in the hearing of many."

In the British Apollo, 1710 . . . is the following query: "Why Death-watches, crickets, and weasels do come more common against death than at any other time? A. We look upon all such things as idle superstitions, for were anything in them, bakers, brewers, inhabitants of old houses, etc., were in a melancholy condition."

To an inquiry in the British Apollo, concerning a Death-watch, whether you suppose it to be a *living creature*, answer is given: "It is nothing but a little worm in the wood."

"How many people have I seen in the most terrible palpitations, for months together, expecting every hour the approach of some calamity, only by a little worm, which breeds in old wainscot, and, endeavoring to eat its way out, makes a noise like the movement of a watch!" Secret Memoirs of the late Mr. Duncan Campbell, 1732.

Authors were formerly not agreed concerning the insect from which this sound of terror proceeded, some attributing it to a kind of wood-louse, others to a spider.

M. Peignot mentions an instance where, in a public library that was but little frequented, *twenty-seven folio* volumes were perforated in a straight line by one and the same larva of a small insect (*Anobium pertinax* or *A. striatum*?) in such a manner that, on passing a cord through the perfectly round hole made by the insect, these twenty-seven volumes could be raised at once.

Muscidae—Flies

Mr. Henry Mayhew, in that part of his interesting work on London Labor and London Poor devoted to the London Street-folk, has given us the narratives of several "Catch-'em-Alive" sellers—a set of poor boys who sell prepared papers for the purpose of catching Flies. He discovered, as he relates, a colony of these "Catch-'em-Alive" boys residing in Pheasant-court, Gray's-inn-lane. They were playing at "pitch-and-toss" in the middle of the paved yard, and all were very willing to give him their statements; indeed, the only difficulty he had was in making his choice among the youths.

"Please, sir," said one with teeth ribbed like celery, to him, "I've been at it longer than him."

"Please, sir, he ain't been out this year with the papers," said another, who was hiding a handful of buttons behind his back.

"He's been at shoe-blacking, sir; I'm the only reg'lar fly-boy," shouted a third, eating a piece of bread as dirty as London snow.

A big lad with a dirty face, and hair like hemp, was the first of the "catch-'em-alive" boys who gave him his account of his trade. He was a swarthy featured boy, with a broad nose like a negro's, and on his temple was a big half-healed scar, which he accounted for by saying that "he had been runned over" by a cab, though, judging from the blackness of one eye, it seemed to Mr. Mayhew to have been the result of some street fight. He said:

"I'm an Irish boy, and nearly turned sixteen, and I've been silling fly-papers for between eight and nine year. I must have begun to sill them when they first come out. Another boy first tould me of them, and he'd been silling them about three weeks before me. He used to buy them of a part as lives in a back-room near Drury-lane, what buys paper and makes the catch 'em alive for himself. When they first come out they used to charge sixpence a dozen for 'em, but now they've got 'em to twopence ha'penny. When I first took to silling 'em, there was a tidy lot of boys at the business, but not so many as now, for all the boys seem at it. In our court alone I should think there was about twenty boys silling the things.

"At first, when there was a good time, we used to buy three or four gross together, but now we don't no more than half a gross. As we go along the streets we call out different cries. Some of us says, 'Fly-papers, fly-papers, ketch 'em all alive.' Others make a kind of song of it, singing out, 'Fly-paper, ketch 'em all alive, the nasty flies, tormenting the baby's eyes. Who'd be fly-blow'd, by all the nasty blue-bottles, beetles, and flies?' People likes to buy of a boy as sings out well, 'cos it makes 'em laugh.

"I don't think I sell so many in town as I do in the borders of the country, about Highbury, Croydon, and Brentford. I've got some regular customers in town about the City-prison and the Caledonian-road; and after I've served them and the town custom begins to fall off, then I goes to the country. We goes two of us together, and we takes about three gross. We keep on silling before us all the way, and we comes back the same road. Last year we sould very well in Croydon, and it was the best place for gitting the best price for them; they'd give a penny a piece for 'em there, for they didn't know nothing about them. I went off one day at ten o'clock and didn't come home till two in the morning. I sould eighteen dozen out in that d'rection the other day, and got rid of them before I had got half-way. But flies are very scarce at Croydon this year, and we haven't done so well. There ain't half as many flies this summer as last.

"Some people says the papers draws more flies than they ketches, and that when one gets in, there's twenty others will come to see him. It's according to the weather as the flies is about. If we have a fine day it fetches them out, but a cold day kills more than our papers.

"We sills the most papers to little cook-shops and sweetmeat shops. We don't sill so many at private houses. The public-houses is pretty good customers, 'cos the beer draws the flies. I sould nine dozen at one house—a school—at Highgate, the other day. I sould 'em two for three-ha' pence. That was a good hit, but then t'other days we loses. If we can make a ha'penny each we thinks we does well.

"Those that sills their papers at three a-penny buys them at St. Giles's, and pays only three ha-pence a dozen for them, but they ain't half as big and good as those we pays tuppence-ha'penny a dozen for.

"Barnet is a good place for fly-papers; there's a good lot of flies down there. There used to be a man at Barnet as made 'em, but I can't say if he do now. There's another at Brentford, so it ain't much good going that way.

"In cold weather the papers keep pretty well, and will last for months with just a little warming at the fire; for they tears on opening when they are dry. You see we always carry them with the stickey sides doubled up together like a sheet of writing-paper. In hot weather, if you keep them folded up, they lasts very well; but if you opens them, they dry up. It's easy opening them in hot weather, for they comes apart as easy as peeling a horrange. We generally carries the papers in a bundle on our arm, and we ties a paper as is loaded with flies round our cap, just to show the people the way to ketch 'em. We get a loaded paper given to us at a shop.

"When the papers come out first, we use to do very well with fly-papers; but now it's hard work to make our own money for 'em. Some days we used to make six shillings a day regular. But then we usen't to go out every day, but take a rest at home. If we do well one day, then we might stop idle another day, resting. You see, we had to do our twenty or thirty miles silling them to get that money, and then the next day we was tired.

"The silling of papers is gradual falling off. I could go out and sill twenty dozen wonst where I couldn't sill one now. I think I does a very grand day's work if I yearns a shilling. Perhaps some days I may lose by them. You see, if it's a very hot day, the papers gets dusty; and besides, the stuff gets melted and oozes out; though that don't do much harm, 'cos we gets a bit of whitening and rubs 'em over. Four years ago we might make ten shillings a day at the papers, but now, taking from one end of fly-season to the other, which is about three months, I think we makes about one shilling a day out of papers, though even that ain't quite certain. I never goes out without getting rid of mine, somehow or another, but then I am obleeged to walk quick and look about me.

"When it's a bad time for silling the papers, such as a wet, could day, then most of the fly-paper boys goes out with brushes, cleaning boots. Most of the boys is now out hopping. They goes reg'lar every year after the season is give over for flies.

"The stuff as they puts on the paper is made out of boiled oil and turpentine and resin. It's seldom as a fly lives more than five minutes after it gets on the paper, and then it's as dead as a house. The blue-bottles is tougher, but they don't last long, though they keeps on fizzing as if they was trying to make a hole in the paper. The stuff is only p'isonous for flies, though I never heard of anybody as ever eat a fly-paper."

A second lad, in conclusion, said: "There's lots of boys going selling 'ketch-'em-alive oh's' from Golden-lane, and White-chapel and the Borough. There's lots, too, comes out of Gray's-inn-lane and St. Giles's. Near every boy who has nothing to do goes out with fly-papers. Perhaps it ain't that the flies is falled off that we don't sill so many papers now, but because there's so many boys at it."

A third, of the lot the most intelligent and gentle in his demeanor, though the smallest in stature, said:

"I've been longer at it than the last boy, though I'm only getting on for thirteen, and he's older than I'm; 'cos I'm little and he's big, getting a man. But I can sell them quite as well as he can, and sometimes better, for I can holler out just as loud, and I've got reg'lar places to go to. I was a very little fellow when I first went out with them, but I could sell them pretty well then, sometimes three or four dozen a day. I've got one place, in a stable, where I can sell a dozen at a time to country people.

"I calls out in the streets, and I goes into the shops, too, and calls out, 'Ketch-'em-alive, ketch-'em-alive; ketch all the nasty black-beetles, blue-bottles, and flies; ketch 'em from teasing the baby's eyes.' That's what most of us boys cries out. Some boys who is stupid only says, 'Ketch 'em alive,' but people don't buy so well from them.

"Up in St. Giles's there is a lot of fly-boys, but they're a bad set, and will fling mud at gentlemen, and some prigs the gentlemen's pockets. Sometimes, if I sell more than a big boy, he'll get mad and hit me. He'll tell me to give him a halfpenny and he won't touch me, and that if I don't he'll kill me. Some of the boys takes an open fly-paper, and makes me look another way, and then they sticks the ketch-'em-alive on my face. The stuff won't come off without soap and hot water, and it goes black, and looks like mud. One day a boy had a broken fly-paper, and I was taking a drink of water, and he come behind me and slapped it up in my face. A gentleman as saw him give him a crack with a stick and me twopence. It takes your breath away, until a man comes and takes it off. It all sticked to my hair, and I couldn't rack (comb) right for some time. . . .

"I don't like going along with other boys, they take your customers away; for perhaps they'll sell 'em at three a penny to 'em, and spoil the customers for you. I won't go with the big boy you saw, 'cos he's such a blackgeyard; when he's in the country he'll go up to a lady and say, 'Want a fly-paper, marm?' and if she says 'No,' he'll perhaps job his head in her face—butt at her like.

"When there's no flies, and the ketch-'em-alive is out, then I goes tumbling. I can turn a cat'enwheel over on one hand. I'm going to-morrow to the country, harvesting and hopping—for, as we says, 'Go out hopping, come in jumping.' We start at three o'clock to-morrow, and we shall get about twelve o'clock at night at Dead Man's Barn. It was left for poor people to sleep in, and a man was buried there in a corner. The man had got six farms of hops; and if his son hadn't buried him there, he wouldn't have had none of the riches.

"The greatest number of fly-papers I've sold in a day is about eight dozen. I never sells no more than that; I wish I could. People won't buy 'em now. When I'm at it I makes, taking one day with another, about ten shillings a week. You see, if I sold eight dozen, I'd make four shillings. I sell 'em at a penny each, at two for three-ha'pence, and three for twopence. When they gets stale I sells 'em for three a penny. I always begin by asking a penny each, and perhaps they'll say, 'Give me two for three ha'pence?' I'll say, 'Can't, ma'am,' and then they pulls out a purse full of money and gives a penny.

"The police is very kind to us, and don't interfere with us. If they see another boy hitting us they'll take off their belts and hit 'em. Sometimes I've sold a ketch-'em-alive to a policeman; he'll fold it up and put it into his pocket to take home with him. Perhaps he's got a kid, and the flies teazes its eyes.

"Some ladies like to buy fly-cages better than ketch-'em-alive's, because sometimes when they're putting 'em up they falls in their faces, and then they screams."

The history of the manufacture of Fly-papers was thus given to Mr. Mayhew by a manufacturer, whom he found in a small attic-room near Drury-lane: "The first man as was the inventor of these fly-papers kept a barber's shop in St. Andrew-street, Seven Dials, of the name of Greenwood or Greenfinch, I forget which. I expect he diskivered it by accident, using varnish and stuff, for stale varnish has nearly the same effect as our composition. He made 'em and sold 'em at first at threepence and fourpence a piece. Then it got down to a penny. He sold the receipt to some other parties, and then it got out through their having to employ men to help 'em. I worked for a party as made 'em, and then I set to work making 'em for myself, and afterwards hawking them. They was a greater novelty then than they are now, and sold pretty well. Then men in the streets, who had nothing to do, used to ask me where I bought 'em, and then I used to give 'em my own address, and they'd come and find me."

Cowan, Frank. 1865. *Curious Facts in the History of Insects Including Spiders and Scorpions: A Complete Collection of the Legends, Superstitions, Beliefs, and Ominous Signs Connected with Insects; Together with Their Uses in Medicine, Art, and as Food; and a Summary of their Remarkable Injuries and Appearances.* Philadelphia: J. B. Lippincott & Co.

Insecticides

Tim Hunkin

For fourteen years (1973–1987), Tim Hunkin produced The Rudiments of Wisdom, *a weekly cartoon in the* London Sunday Observer *that became a "must read" for its arcane yet authoritative exploration of every aspect of life. Later, the columns were collected into a fat reference book subtitled "the antidote to boring reference books." Hunkin (he often styles himself using only his surname) has investigated ants, bees, locusts, wood lice, and other insects and related topics. This strip on insecticides—a sample of his Sunday-best—gives a wide-ranging view of one of the more fundamental relationships between humans and insects.*

Hunkin, Tim. 1988. *Almost Everything There Is to Know by Hunkin.* London: Pyramid Books.

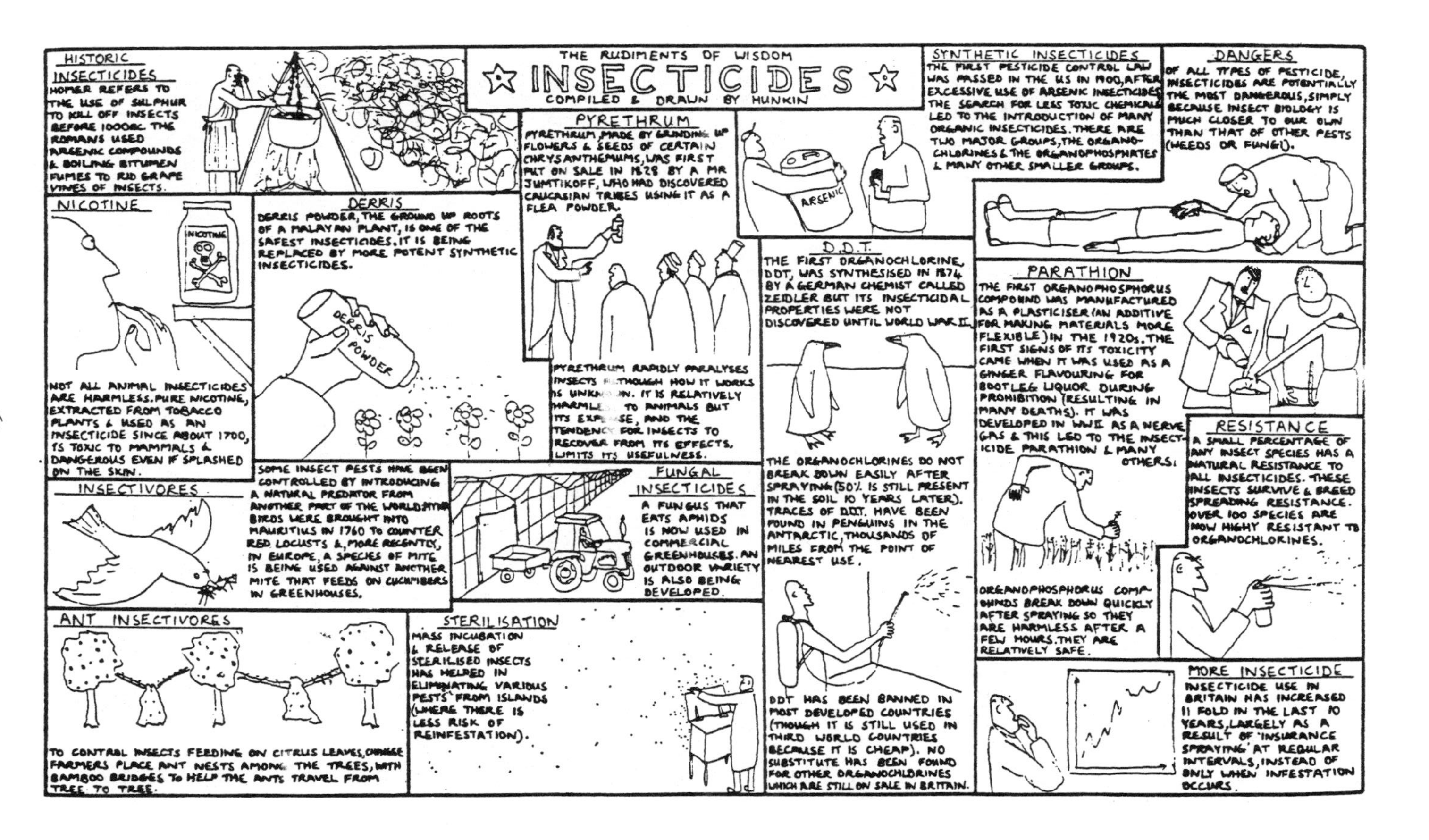
THE RUDIMENTS OF WISDOM
INSECTICIDES
COMPILED & DRAWN BY HUNKIN
HISTORIC INSECTICIDES
HOMER REFERS TO THE USE OF SULPHUR TO KILL OFF INSECTS BEFORE 1000BC. THE ROMANS USED ARSENIC COMPOUNDS & BOILING BITUMEN FUMES TO RID GRAPE VINES OF INSECTS.
PYRETHRUM
PYRETHRUM, MADE BY GRINDING UP FLOWERS & SEEDS OF CERTAIN CHRYSANTHEMUMS, WAS FIRST PUT ON SALE IN 1828 BY A MR JUMTIKOFF, WHO HAD DISCOVERED CAUCASIAN TRIBES USING IT AS A FLEA POWDER.
PYRETHRUM RAPIDLY PARALYSES INSECTS ALTHOUGH HOW IT WORKS IS UNKNOWN. IT IS RELATIVELY HARMLESS TO ANIMALS BUT ITS EXPENSE, AND THE TENDENCY FOR INSECTS TO RECOVER FROM ITS EFFECTS, LIMITS ITS USEFULNESS.
ARSENIC
SYNTHETIC INSECTICIDES
THE FIRST PESTICIDE CONTROL LAW WAS PASSED IN THE US IN 1900, AFTER EXCESSIVE USE OF ARSENIC INSECTICIDES. THE SEARCH FOR LESS TOXIC CHEMICALS LED TO THE INTRODUCTION OF MANY ORGANIC INSECTICIDES. THERE ARE TWO MAJOR GROUPS, THE ORGANOCHLORINES & THE ORGANOPHOSPHATES & MANY OTHER SMALLER GROUPS.
DANGERS
OF ALL TYPES OF PESTICIDE, INSECTICIDES ARE POTENTIALLY THE MOST DANGEROUS, SIMPLY BECAUSE INSECT BIOLOGY IS MUCH CLOSER TO OUR OWN THAN THAT OF OTHER PESTS (WEEDS OR FUNGI).
NICOTINE
NICOTINE
NOT ALL ANIMAL INSECTICIDES ARE HARMLESS. PURE NICOTINE, EXTRACTED FROM TOBACCO PLANTS & USED AS AN INSECTICIDE SINCE ABOUT 1700, IS TOXIC TO MAMMALS & DANGEROUS EVEN IF SPLASHED ON THE SKIN.
DERRIS
DERRIS POWDER, THE GROUND UP ROOTS OF A MALAYAN PLANT, IS ONE OF THE SAFEST INSECTICIDES. IT IS BEING REPLACED BY MORE POTENT SYNTHETIC INSECTICIDES.
DERRIS POWDER
D.D.T.
THE FIRST ORGANOCHLORINE, DDT, WAS SYNTHESISED IN 1874 BY A GERMAN CHEMIST CALLED ZEIDLER BUT ITS INSECTICIDAL PROPERTIES WERE NOT DISCOVERED UNTIL WORLD WAR II.
THE ORGANOCHLORINES DO NOT BREAK DOWN EASILY AFTER SPRAYING (50% IS STILL PRESENT IN THE SOIL 10 YEARS LATER). TRACES OF DDT. HAVE BEEN FOUND IN PENGUINS IN THE ANTARCTIC, THOUSANDS OF MILES FROM THE POINT OF NEAREST USE.
DDT HAS BEEN BANNED IN MOST DEVELOPED COUNTRIES (THOUGH IT IS STILL USED IN THIRD WORLD COUNTRIES BECAUSE IT IS CHEAP). NO SUBSTITUTE HAS BEEN FOUND FOR OTHER ORGANOCHLORINES WHICH ARE STILL ON SALE IN BRITAIN.
PARATHION
THE FIRST ORGANOPHOSPHORUS COMPOUND WAS MANUFACTURED AS A PLASTICISER (AN ADDITIVE FOR MAKING MATERIALS MORE FLEXIBLE) IN THE 1920s. THE FIRST SIGNS OF ITS TOXICITY CAME WHEN IT WAS USED AS A GINGER FLAVOURING FOR BOOTLEG LIQUOR DURING PROHIBITION (RESULTING IN MANY DEATHS). IT WAS DEVELOPED IN WWII AS A NERVE GAS & THIS LED TO THE INSECTICIDE PARATHION & MANY OTHERS.
ORGANOPHOSPHORUS COMPOUNDS BREAK DOWN QUICKLY AFTER SPRAYING SO THEY ARE HARMLESS AFTER A FEW HOURS. THEY ARE RELATIVELY SAFE.
RESISTANCE
A SMALL PERCENTAGE OF ANY INSECT SPECIES HAS A NATURAL RESISTANCE TO ALL INSECTICIDES. THESE INSECTS SURVIVE & BREED SPREADING RESISTANCE. OVER 100 SPECIES ARE NOW HIGHLY RESISTANT TO ORGANOCHLORINES.
INSECTIVORES
SOME INSECT PESTS HAVE BEEN CONTROLLED BY INTRODUCING A NATURAL PREDATOR FROM ANOTHER PART OF THE WORLD; MYNA BIRDS WERE BROUGHT INTO MAURITIUS IN 1760 TO COUNTER RED LOCUSTS &, MORE RECENTLY, IN EUROPE, A SPECIES OF MITE IS BEING USED AGAINST ANOTHER MITE THAT FEEDS ON CUCUMBERS IN GREENHOUSES.
FUNGAL INSECTICIDES
A FUNGUS THAT EATS APHIDS IS NOW USED IN COMMERCIAL GREENHOUSES. AN OUTDOOR VARIETY IS ALSO BEING DEVELOPED.
ANT INSECTIVORES
TO CONTROL INSECTS FEEDING ON CITRUS LEAVES, CHINESE FARMERS PLACE ANT NESTS AMONG THE TREES, WITH BAMBOO BRIDGES TO HELP THE ANTS TRAVEL FROM TREE TO TREE.
STERILISATION
MASS INCUBATION & RELEASE OF STERILISED INSECTS HAS HELPED IN ELIMINATING VARIOUS PESTS FROM ISLANDS (WHERE THERE IS LESS RISK OF REINFESTATION).
MORE INSECTICIDE
INSECTICIDE USE IN BRITAIN HAS INCREASED 11 FOLD IN THE LAST 10 YEARS, LARGELY AS A RESULT OF 'INSURANCE SPRAYING' AT REGULAR INTERVALS, INSTEAD OF ONLY WHEN INFESTATION OCCURS.

Bee Bites

Roger B. Swain

Some insects bite (sink their teeth into you) and others sting (poke you with a barb from their rear end); the vast majority do neither. Most nonentomologists don't know the difference (which is it that mosquitoes do?). When Roger Swain makes polite dinner conversation about bee stings, he educates us all in the process. Entomologist Swain is a popular gardening and science writer who appears weekly on the PBS TV show The Victory Garden.

At dinner last evening, the woman across from me was describing how swollen her arm had been after three bee bites. "Mosquitoes bite," I corrected her, "honey-bees sting." I was tempted to add something about asses and elbows, but manners prevailed. My distinction was dismissed with a toss of bracelets, and she went on to explain that her doctor had warned her that she was allergic to bees.

Rebuffed, I returned to dissecting my pork chop. Eleven-year-olds can describe precisely the armament on an F-14 Tomcat. ("The world's deadliest combat aircraft. A 20-mm rotary cannon with 675 rounds. External load of 14,500 pounds of bombs and combinations of Phoenix, Sparrow, and Sidewinder air-to-air missiles.") Yet most adults have difficulty distinguishing the front end of an insect from its rear. Why is information about human weaponry collected with the same enthusiasm as baseball cards when information about insect weaponry is so overlooked? As a deterrent, at least as a deterrent to going outdoors, insects are far more effective than F-14s.

People remain unenlightened about insects for two reasons. Although most of them have run into bees, wasps, or ants, they are usually too excited by the encounter to take notes. When a yellow jacket is sucked in the car's window vent, the driver is apt to be too busy running off the road to notice whether he is being bitten or stung. These distinctions are better made either before or after the encounter, but then the curious run into a lack of information, or at least a lack of accessible information. Robert Snodgrass has written a fine book called *The Anatomy of the Honey Bee*, but it is 334 pages long and not likely to be a best seller or even an editor's choice.

The woman across from me newly convalescent from an encounter with *Apis mellifera*, was clearly in need of more information. It would amuse her and fascinate her to learn more about honeybee stings. She would welcome the instruction and find the knowledge useful. I waited until dessert was served, and then I began.

"You know," I said, pointing my spoon at her, "only females sting. It's because the sting of a bee or wasp or an ant is a modified ovipositor. The ovipositor, which in most insects is used to pierce and deposit eggs inside something, has evolved into a device for injecting venom. Since only females lay eggs, only females sting. The bad news is that in colonies of bees, wasps, and ants, almost all the members are females."

Afraid that I was beginning to sound like a misogynist, I smiled warmly and hurried on.

"The sting of a honeybee is a shaft with a bulbous base. The shaft is made of three separate pieces: a stylet and two lancets. The three pieces are neatly fitted together to form a central tube called a poison canal. Venom is manufactured by two glands that empty into a poison sac, which in turn empties into the bulb at the base of the sting."

I looked around for a napkin on which to draw a diagram. Finding only linen, I had to forgo the visual aids.

"When a honeybee stings you, she bends her abdomen sharply downward and extrudes the sting that is ordinarily concealed in a sting chamber. With a quick jab, she sticks the tip of the sting into you. The two lancets have barbed tips and, furthermore, are capable of being moved independently. Muscles attached to the sting contract in rapid alternation advancing the lancets, the barbs on one securing it while the other is pushed in deeper. Meanwhile, the same muscles that are embedding the sting are also pumping venom down the poison canal where it escapes through a cleft near the tip of the sting shaft."

The woman across from me was looking pained. Her banana pudding lay untouched.

"What did you do when you were stung?" I inquired.

"I screamed," she replied, "and swatted at it. When I opened my eyes it was gone."

Beside me a freckle-faced child with gaps in her teeth chimed in: "The sting wasn't. Bees lose their stings and then they die." She looked at me for confirmation.

"That's right." I beamed at her, delighted to have a partner in my educational endeavor. "Although wasps can sting you over and over, a honeybee can only sting you once. Your soft skin grips the barbs on the lancets too tightly for the bee to withdraw them. If the bee were to sting another insect, she wouldn't lose her sting, but if she stings a soft-skinned animal like you, the entire sting apparatus rips right out of her. It isn't very strongly attached in the first place, and as she flies off, she leaves her sting behind, often with a few attached innards."

At the mention of innards, other people seemed to become interested. At least they put down their spoons.

"The most remarkable thing about the honeybee sting is that it's like an automatic hypodermic syringe. Even after the bee has flown away, the muscles attached to the sting keep working it in deeper and deeper and keep pumping in venom. If you remove the sting from yourself in a few seconds, you'll cut way down on the amount of venom that's injected. It's no good to grab the sting between your fingers, because you'll just squeeze the rest of the venom into you. It would be like grabbing the top of an eyedropper. Instead, take your fingernail and scrape the sting out. If your fingernails are too short, use a penknife blade or a nail file. You'll find that the sting comes out easily.

"From the bee's point of view, having a sting that tears out, even if it is ultimately fatal, is a neat way to guarantee that more venom gets into the victim. If she had to hang around, she might be brushed off prematurely. In general, of course, bees don't sting people. You have to be disturbing the hive to get attacked, or else step on or bump into an individual bee somewhere. Even if you throw rocks at a beehive, less than half of one percent of the bees in the hive are likely to sting. Alarmed bees look for movement and color, and at close range, odor. Dark clothing, hair, and leather elicit stinging. Away from the hive, brightly colored clothing and sweet perfumes may attract a bee, and chemicals in a person's breath may cause her to sting.

"I don't suppose you smelled your arm after the bee stung you?" I asked the woman across the table. From her expression, I gathered she hadn't.

"When the sting tears out of the bee, an alarm pheromone is released, a chemical that signals other bees to come and attack. If there are other bees in the vicinity, one sting is likely to lead to others. Guess what the chemical is?" I brandished the remains of my banana pudding. "It's isoamyl acetate. You can smell it anytime; just go to the garden and mash a bee and sniff its abdomen. Straight banana oil."

I paused to finish the pudding in my cup and noticed that the others had resumed eating. But I could see they were doing so cautiously.

"Now the venom that's being pumped into you is remarkable stuff," I continued. "There's not much of it, less than 0.3 milligrams, and most of that is water, but the rest is a mix of at least 10 ingredients. There are some low-molecular-weight agents like histamine, dopamine, and noradrenaline. There are also some high-molecular-weight toxins like the hemolyzing melittin, a neurotoxic apamin, mast-cell-degranulating peptide, and minimine, and there are some enzymes like phospholipase A and B, and hyaluronidase. The various ways in which these compounds affect a person's body are still being worked out, but there's hope that some of them will prove to be medically useful.

"The idea that bee venom might be beneficial goes back at least to the ancient Greeks. Since then a lot of people have used bee venom to treat various ailments, most importantly arthritis."

"That's right!" came a voice from the end of the table. "As a child, I used to watch my aunt go into her garden, catch a bee, and let it sting her arm. She said it helped her arthritis."

"Well, there are a lot of people experimenting with bee venom as a treatment for arthritis. It's called apitherapy. Up in Middlebury, Vermont, a beekeeper named Charles Mraz administers bee venom to people suffering from arthritis. There was an article about it a couple of years ago in *Country Journal.* According to the article, he isn't accused of practicing medicine without a license, because he uses bees to administer the venom and he doesn't charge patients. Now I'm not saying that bee venom will work for everyone. There are certainly enough beekeepers who have arthritis in spite of being repeatedly stung. Also the U.S. medical establishment is skeptical about bee venom as a treatment for arthritis. Nevertheless, doctors acknowledge that there haven't been sufficient studies. In Europe especially, they are continuing to look into bee-venom therapy.

"The trouble with most doctors, in my opinion, is that they spend too much time worrying people about the harmful effects of bee venom. Now I'm not suggesting that you're not allergic to bees," I said, pointing my spoon across the table again, "it's just that nearly everyone is being told they are. About the only people who aren't allergic these days are beekeepers. When I get stung, it still burns at the moment the sting goes in, but there is no swelling afterward. In the spring, when I haven't been stung all winter, there is a little swelling, but by the end of the summer I'm immune.

"Most people when they get stung, say, by walking barefoot in clover, have a localized reaction. There's a swelling and soreness near the sting. That's not an allergy.

"What everyone should worry about are those very few cases in which a sting is followed by shock, unconsciousness, and death.

"But such a serious reaction is incredibly rare. In the whole United States there are only forty deaths a year from bee, wasp, yellow jacket, and hornet stings combined. That makes these insects slightly more dangerous than snakes, which kill fifteen people per year, but virtually harmless compared to cars, which kill fifty thousand.

"At Harvard there is an allergist named Dr. Howard Rubenstein who thinks doctors are scaring people unnecessarily about bee stings. He claims that there is no evidence to support the belief that there is a predictable progression from simple localized reaction after one sting, through generalized itching and hives, breathing difficulties, and abdominal cramps, to death after subsequent stings. In fact, there are many cases in which people have died from a bee sting without apparently ever having been stung before. There are also people who have had difficulty breathing after one sting and had a much milder reaction after the next."

I had hoped that the woman across from me would be relieved by this news. She didn't seem to be.

"Among the causes of the forty deaths that occur each year, Dr. Rubenstein suspects that, in addition to some genuine allergic responses, there are also heart attacks, toxic reactions to the venom itself, and fear. Some people, convinced by their doctors that they are allergic, probably just die of fright.

"Now, fear is a perfectly real thing. If you're afraid of bees (for whatever reason), you should carry a bee kit with you that has antihistamine in it and epinephrine hydrochloride. But as Dr. Rubenstein says, most people need to be protected not so much from bees as from alarmist propaganda."

The woman across from me got up from the table. "That's fascinating," she said. "I simply had no idea bees could be so interesting. How did you ever manage to learn all that? I'll be lucky to remember half of it if I'm ever bitten again."

Swain, Roger B. 1983. *Field Days: Journal of an Itinerant Biologist.* New York: Scribner's. Reprinted by permission of Don Congdon Associates, Inc.

A Pain Scale for Bee, Wasp, and Ant Stings

Christopher K. Starr

This delightful scientific paper is entirely serious. Yet one can sense the author grinning as he writes, tongue firmly in cheek, pushing the bounds of the "objective" scientific standard. As the author points out, "The question of objectivity is entirely irrelevant in evaluating any pain scale, as we are concerned with how it feels.*" It is less easy to imagine the author grinning during the research phase of his project, however. Chris Starr (b. 1949) is a professor at the University of the British West Indies in Trinidad and is an expert on social insects. His pain scale begins to answer the age-old question parents pose to children after they've been stung by a bee or wasp: How badly does it hurt?*

Venom injection (stinging) is an important defense tactic among various animal taxa, the most obvious of which is the aculeate hymenoptera (bees, wasps, and ants). It would be useful, then, to be able to compare the defensive power of stinging by different species or colonies. The components of such a comparison would be the number of potential defenders in the colony or aggregation, their readiness to attack, and the effectiveness of a single sting. The first of these is easily and often known. No standard measure has yet been derived for the second, but there is good reason to believe it correlates positively with the first, especially within species. That means that individuals of larger colonies appear to need less provocation to attack. This paper deals with the third component.

By "stinger" is meant here the venom-injection apparatus, while "sting" refers to the event. Research into the stinger and its venom has shown impressive progress along three lines: a) morphology of the stinger, b) venom chemistry and c) the toxic effects of venom.

At the same time, our knowledge of the pain caused by venom is still at the anecdotal stage. This is consistent with the fact that recent progress in the psychophysics of pain has been based on temperature, pressure, and electric shock, but not chemically-induced pain.

Yet in the evolution of defense against large, primarily vertebrate predators, pain must be the key factor in sting effectiveness, much more important than toxicity or paralyzing power. Given the extensive toxicological literature, it would be convenient if toxicity were a good index of painfulness. This appears not to be so. Schmidt and Blum and Schmidt *et al.* give examples of wasps with very painful stings yet only slightly toxic venoms. Even if we find that more toxic venoms are *usually* more painful, as appears reasonable, it will be the exceptions which are of special interest.

The Hymenoptera literature contains many brief descriptions of stings received in the course of field research. A common standard is lacking, though, so that it is often difficult to infer which of two species-stings mentioned in different reports is the more painful. The intention of this paper is to provide just such a standard, with the hope that over time a systematized body of comparative observations will accumulate.

The Pain Scale

Pain is the body's alarm system in the face of injury, so that it is not surprising that its perception is graded into relatively few intensity levels. Humans can distinguish about 570 levels of light intensity, from barely perceptible to dazzling and about 90 levels of warmth below the pain threshold. But for pain induced by pressure or pricking, Hardy *et al.* put the number of distinguishable levels at just 22.

The complete fineness of discrimination is not commonly used in experimental pain studies. Chapman gives as the most common method of assessment a scale from 0 to 10, in which 1 represents the lower threshold of perception and 10 the upper threshold. The McGill Pain Questionnaire has 6 levels over the same range, and that of Lutterbeck and Triay just 4. The scale described below has 5 levels, from 0 to 4. It is very close to the McGill scale, the only substantial difference being that McGill levels 4 and 5 are approximately equal to my level 4. Schmidt *et al.* use a pain scale of 1–4 for stinging hymenoptera, though without defining the levels. Inasmuch as all of the species they rank can penetrate human skin, and as their rankings agree very well with my own (Table 1) and with those given by Schmidt using the present scale, it seems that the rankings 1–4 of Schmidt *et al.* and this paper are virtually identical.

0. No pain
1. Pain so slight as to constitute no real deterrent
2. Painful

3. Sharply and seriously painful
4. Traumatically painful

Rank 0 is common, as many species with a functional stinger are too small or weak to penetrate human skin. Rank 1 lies in that area in which the sting is clearly perceived (pain above threshold), yet most people would not say it "hurts." Stings of rank 4 are often medically serious events, producing strong physical reactions and durable pain even in persons without a history of acute reaction to stings (numerous pers. comm.), but attention is given here only to short-term pain, within a few seconds of the sting.

The distinction between ranks 2 and 3 may often be unclear. The intention is to distinguish between the great mass of painful stings (2) and those which stand out as clearly more painful than, for example, most honey bee stings, though not of traumatic intensity (3). One possibly useful characterization is that rank-3 stings, just from the pain itself and apart from any surprise or fear, produce loud cries, groans and/or long preoccupation. The examples in Table 1 will add to this distinction.

Table 1. Examples of ranked Hymenoptera stings. Each of these is based on at least two stings, and as far as I know all follow the restrictions recommended in the text. Those marked with an asterisk are based on induced stings, explained in the text. Where a species has two rankings, this represents variation in stings, rather than uncertainty. Rankings of social species are all based on workers or the subcaste most commonly encountered outside the nest.

Family	Species	Common Name	Rank	Source
Mutillidae	*Dasymutilla klugii*	[Velvet ant]	3	d
	Dasymutilla small sp.	[Velvet ant]	1–2	d
	Pepsis formosa pationii	[Spider wasp]	4	d
	Monobia quadridens	[Twig-nesting wasp]	2*	a
	Eustenogaster luzonensis	[Hover wasp]	3	a
Vespidae:				
Polistinae	*Polistes annularis*	[Paper wasp]	3	a,c
	Polybia diguetana	[Paper wasp]	0–1	a
Vespidae:				
Vespinae	*Dolichovespula maculata*	[Bald-faced hornet]	2	c,d
	Vespa mandarinia	[Hornet]	2	d
	Vespula maculiforns	[Yellow jacket]	2	a,c
	Vespula squamosa	[Yellow jacket]	2	a,c,d

Formicidae:				
Myrmeciinae	*Myrmecia pyriformis*	[Bulldog ant]	2–3	d
	Dinoponera gigantea	[Dinosaur ant]	1–2	c,d
	Ectatomma tuberculatum	[Kelep ant]	2	c
	Odontomachus sp.	[Trap-jaw ant]	2*	a
	Paraponera clavata	[Bullet ant or bala]	4	b,c,d
Formicidae:				
Dorylinae	*Eciton burchelli*	[Army ant]	1–2	a,c,d
Formicidae:				
Myrmicinae	*Monomorium pharaonis*	[Pharaoh's ant]	0	a
	Pogonomyrmex spp.	[Harvester ant]	3	d
	Solenopsis geminata	[Native fire ant]	low 2	a
	Solenopsis invicta	[Imported fire ant]	1–2	d
Formicidae:				
Formicinae	*Oecophylla smaragdina* (bite, with spraying formic acid into the wound)	[Weaver ant]	2	a
Anthophoridae	*Centris pallida*	[Digger bee]	1–2	d
	Xylocopa virginica	[Carpenter bee]	1–2	c,d
Apidae	*Apis cerana*	[Indian honey bee]	2	a
	Apis mellifera	[Honey bee]	2	a,d
	Bombus sonorus	[Bumble bee]	2	d

a = Personal observation
b = Numerous personal communications, Costa Rica
c = Schmidt *et al.* 1984
d = Schmidt, in press
e = J. O. Schmidt, personal communication

I suggest that for experienced observers it will often be useful to distinguish between low-2 and high-2 stings. This should he done with caution, only when a sting seems clearly at the lower or upper end of rank 2.

In order for the pain scale to have its intended reliability, certain constraints on use are necessary. I suggest the following:

1. Reports should be made only by adult observers in good health.
2. Disregard all stings accompanied by allergic reactions.
3. Reports should not come from observers who are rarely stung. This is to avoid mixing pain with novelty.

4. Reports should be based only on events in which a very small number of stings are received at once.
5. A ranking should never be based on just one sting. Although individual social wasps probably sting rarely (I suspect that most never do), so that significant day-to-day variation in venom volume is unlikely, uncompleted or grazing stings are not uncommon. It is not known to what extent the regular use of the stinger by solitary wasps causes variation in venom delivery.
6. Reports on stings received through free attack by the insect (volunteer stings) are preferable to those deliberately induced by holding her between the fingers or against the skin (induced stings). We are not always so fortunate, though, as to be attacked by those species of special interest. Induced stings can contribute useful data if used with caution. Species which fail to penetrate the skin in an induced situation can sometimes sting under their own power, as with *Polybia diguetana* (personal observation); no rank 0 should be based on induced stings. Care must also be taken that induced stings are solid and direct. In addition, reports based on induced stings should be identified as such.

Discussion

Given a hymenopteran defender against a vertebrate intruder, the two biologically relevant questions about the sting are: a) Does it have immediate value, by way of turning back that particular intrusion? and b) Does it have long-term value, by way of the intruder's memory? Any method of ranking sting pain from different species will contribute to answering these questions.

The goal in this regard must be a standardized, exact clinical method, such as those used in comparing venom toxicities. I have elsewhere suggested an approach to this, but we are still a long way from having a tested method of this type. Until we do, a non-clinical scale such as described above seems the best hope for progress.

The question of objectivity is entirely irrelevant in evaluating any pain scale, as we are concerned with how it *feels*. On the other hand, the question of reliability is central. If perception variation between different humans is so great that same species reports from different people would show no strong positive correlation, then the scale is worse than valueless. There is good reason to expect such correlation. Although it has yet to be shown for specifically chemically-induced pain, the reliability of reported pain between individual humans and within individuals at different times is in general much greater than biologists would tend to expect. This result is summarized in Wolff and Wolff's statement that all healthy human beings have approximately the same capacity to feel pain. The expected reliability of pain-scale data is well within the norms of present-day pain research.

A second assumption is that other vertebrate species would each rank stings in the same *sequence*, i.e., that given two stings, they would respond similarly to the question "Which is more painful?" This reasonable working assumption is completely untested at present.

Clearly, though, different species must have differing thresholds for slight, serious and traumatic pain. To a small mammal our ranks 2, 3 and 4 might well all be so painful as to have identical biological meaning, while our rank 1 could represent a much wider pain spectrum than it does for us. This serious limitation of the pain scale cannot be overcome. It underlines the fact that stings of the same rank do not make up a natural universal grouping, but simply indicate the limits of our own resolution.

To what extent can the pain scale meaningfully rank non-hymenopteran stings, or the pain from defensive tactics other than stinging? To each of these the answers is: to only a very limited extent. The venoms of some taxa, such as snakes, and tiffids, produce pain largely as an irrelevant-by-product. In addition, it must be asked whether the animal is normally able to manipulate venom into a large aggressor. The venom apparatus of spiders, for example, must nearly always be useless against a vertebrate attacker. Some few other groups, such as scorpions and centipedes, appear to use pain as a deterrent, and it may sometimes be of interest to rank them.

For non-stinging defense, such as the purely mechanical use of sharp structures, the pain scale seems applicable only in a very few cases.

It should sometimes be meaningful, for example, to compare the bite-pain of some ant and termite workers with the sting-pain of some other ant workers or the jab-pain from the pseudostinger-genitalia of many aculeate males. The action of *Oecophylla smaragdina* major workers in spraying formic acid directly into a bite wound is very like a sting in form and effect.

Starr, Christopher K. 1985. "A Simple Pain Scale for Field Comparison of Hymenopteran Stings." *J. Entomol. Sci.* 20, no. 2:225–31.

Fancy Footwork

David George Gordon

Many university biology departments have cockroach problems that would surprise even a New York City apartment dweller. That's because the cockroach is one of the most studied of invertebrates, a veteran model for neurological and physiological study. And cockroaches escape. So, even in biology departments where they haven't had a cockroach researcher in years, the subjects live on.

Although cockroaches have six legs, they usually move only three at a time. The first and third legs on one side of the body and the second leg on the other side remain stationary, forming a tripod to support the cockroach's body. Freed from their supporting role, the three other legs move forward. Having taken one step, these legs form a fresh tripod, so the legs that were stationary can now take a step.

This sequence of steps ensures that a cockroach's center of gravity is always within the support area provided by its legs. Thus, the evenly distributed cockroach can stop at any time in the walking pattern without toppling over—something that horses, humans, and many other animals in motion can't do.

A walking cockroach can break into a run simply by picking up the pace of its movements. When brownbanded cockroaches and a few other species are really cruising, they spread their wings and shift their body weight to the rear. With their wings angled upward like a Stealth bomber's, these creatures attain peak speeds by running on their two hind legs.

Workers at the University of California's Berkeley campus have clocked sprinting American cockroach specimens at speeds of around fifty-nine inches per second. This pencils out to roughly 3.4 miles per hour. It may not seem like much to most people, who can easily walk at more than twice this speed. However, the roach's feat must be considered on a proportional scale.

American cockroaches can cover fifty body lengths per second. That's about ten times the number a human can cover in the same amount of time, and more than eight times the number that a horse can cover. It's nearly three times the relative speed of a cheetah, the fastest animal on land, which can reach a peak speed of forty-five miles per hour.

Considerably smaller than their American relatives, German cockroaches travel at a proportionately slower pace. One of these runners is having a great day if it can go faster than a foot per second.

A Day at the Races

Each spring, thoroughbred racing cockroaches go for the gold (or, in this case, fifty dollars and a trophy shaped like a garbage can) at The All-American Trot. The Kentucky Derby of the six-legged set, this exciting event is hosted by Purdue University's Department of Entomology. Since 1991, students and staff of this department have supplied the custom-built circular track and recruited the racers from their own research stocks.

"These puppies were born to run," claims Arwin Provonsha, curator of insect collections and announcer for the cockroach races. His contestants have names like "Fluttering Antenna," "Hot to Trot," and "Plain Disgusting." They wear racing colors, artfully applied in acrylic paint on each animal's back. Pedigrees ("Seattle Sewer" by "Sewer Sam," out of "Septic System Sally") are displayed on a big board.

Spectators—more than seven thousand in 1995—are encouraged to place their bets for what Provonsha calls a "two-furshort" (as opposed to a two furlong) race. He asserts that the event has been sanctioned by the Indiana Roach Racing Commission, and that "betting is permitted under their auspices."

Provonsha's pedigreed racers are kept in the dark until the starter's gun. Then they're turned loose into the bright light of day, and, if necessary, prodded into action. Sometimes, the cockroaches run in the same direction.

After continued prodding, the roaches head into the far turn and make a dash for the finish line. Occasionally there's a photo finish, with a *Periplaneta* winning the trophy by a labial palp.

Following the All-American Trot, the action shifts to the three-foot-long straight track, scene of the exhibition tractor pulls. For this event, three Madagascan hissing cockroaches lug miniature green-and-yellow John Deere tractors. The first to tow its load across the finish line—a distance of three feet—is the winner.

Both tractor pull and race are staged within the confines of what Purdue entomologists have dubbed Roachill Downs. Designed and constructed by Provonsha, this unusual backdrop features a grandstand full of dead roaches in sunglasses and baseball caps, waving pennants, drinking soda pop, and eating what their outfitter calls "Green Gunk on a Stick." Other dried specimens pose in lawn chairs, go for grasshopper rides, or wait in a long line outside a port-a-potty.

Harborage Horrors

In the autumn of 1983, a middle-aged woman checked into the emergency room at Charity Hospital in New Orleans, complaining of ear discomfort. Help was summoned. Resident physician Kevin O'Toole peered into the woman's ear canal, where he discovered the problem: an adult American cockroach, comfortably ensconced in this warm and snug space.

Initially, O'Toole didn't think much of his discovery. "At Charity Hospital where I was participating in a trauma training program," he told *Omni Magazine,* "we found that problem almost everyday, particularly in people from lower socioeconomic groups."

It wasn't until he'd found a second adult American cockroach, living in the woman's other ear canal, that O'Toole grasped the significance of this event. "We recognized immediately that fate had granted us the opportunity for an elegant comparative therapeutic trial," O'Toole and his associates, P. M. Paris and R. D. Stewart, later reported in the *New England Journal of Medicine.*

The emergency-room team put mineral oil in one car canal and sprayed the second with 2-percent solution of the anesthetic lidocaine. The roach in mineral oil "succumbed after a valiant but futile struggle, but its removal required much dexterity on the part of the house officer." The other animal, sprayed with lidocaine, leapt from the opposite ear "at a convulsive rate of speed," making its way across the floor in a desperate attempt to escape. "A fleet-footed intern promptly applied an equally time-tested remedy," dispatching the insect with what O'Toole, Paris, and Stewart laconically described as "the simple crush method."

Gordon, David George. 1996. *The Compleat Cockroach: A Comprehensive Guide to the Most Despised (and Least Understood) Creature on Earth.* Berkeley, Cal.: Ten Speed Press.

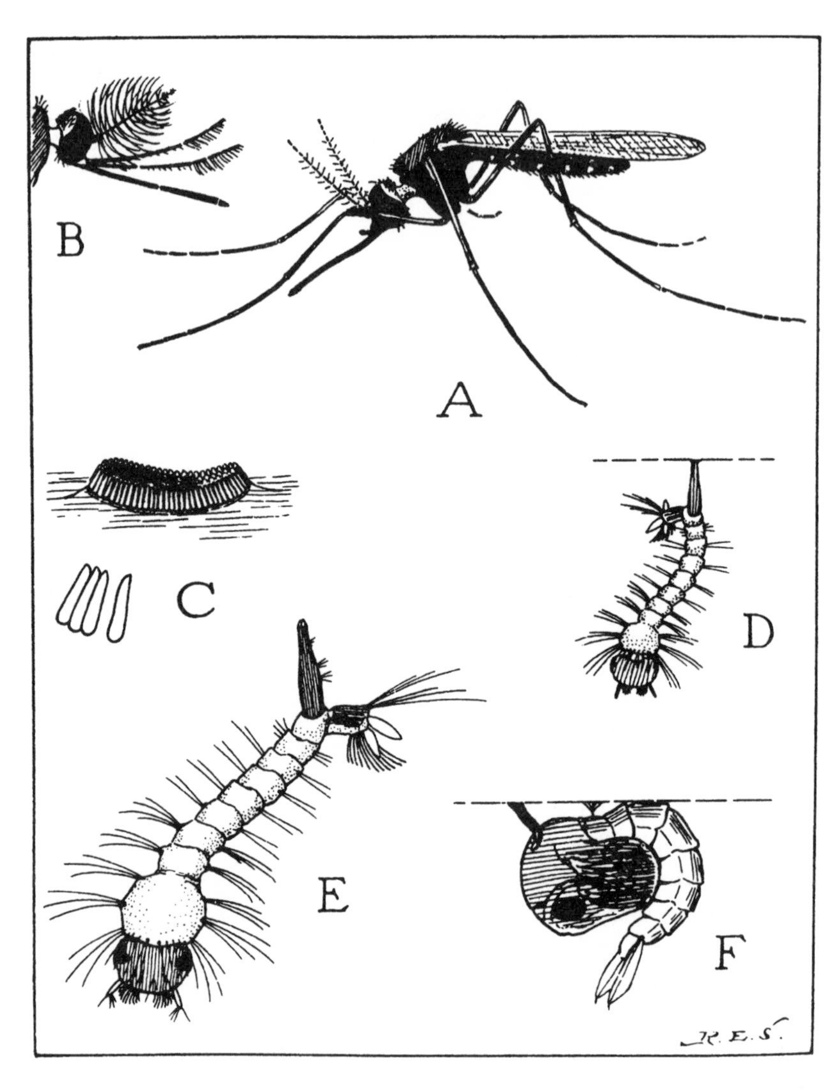

Life stages of a mosquito: (A) adult female, (B) head of adult male, (C) a floating egg raft with four eggs shown enlarged, (D) a young larva suspended near the water's surface, (E) full-grown larva, (F) the pupa resting against the surface film of the water. Illustration by R. E. Snodgrass (*Insects. Their Ways and Means of Living* by Robert Evans Snodgrass, Smithsonian Institution, Washington, DC, 1930).

Sympathy for the Devil

David Quammen

Whether an organism is pestiferous or beneficial is in the eye of the beholder. Step back from a human-centered viewpoint for a moment and appreciate David Quammen's novel view of mosquitoes as rain-forest protectors that serve to check human growth.

Undeniably they have a lot to answer for: malaria, yellow fever, dengue, encephalitis, filariasis, and the ominous tiny whine that begins homing around your ear just after you've gotten comfortable in the sleeping bag. All these griefs and others are the handiwork of that perfidious family of biting flies known as Culicidae, the mosquitoes. They assist in the murder of millions of humans each year, carry ghastly illness to millions more, and drive not a few of the rest of us temporarily insane. They are out for blood.

Mosquitoes have been around for 50 million years, which has given them time to figure all the angles. Judged either by sheer numbers, or by the scope of their worldwide distribution, or by their resistance to enemies and natural catastrophe, they are one of the great success stories on the planet. They come in 2,700 different species. They inhabit almost every land surface, from Arctic tundra to downtown London to equatorial Brazil, from the Sahara to the Himalaya, though best of all they like tropical rainforests, where three-quarters of their species lurk. Mosquitoes and rainforests, in fact, go together like gigolos and bridge tournaments, insurance salesmen and Elks lunches, panhandlers and. . . . But more on all that in a moment.

They hatch and grow to maturity in water, any entrapment of quiet water, however transient, or funky. A soggy latrine, for instance, suits them fine. The still edge of a crystalline stream is fine. In the flooded footprint of an elephant, you might find a hundred mosquitoes. As innocent youngsters they use facial bristles resembling cranberry rakes to comb these waters for smorgasbord, but on attaining adulthood, they are out for blood.

It isn't a necessity for individual survival, quenching that blood thirst, just a prerequisite of motherhood. Male mosquitoes do not even bite. A male mosquito lives his short, gentle adult life content, like a swallowtail butterfly, to sip nectar from flowers. As with black widow spiders and mantids, it is only the female that is fearsome. Make of that what larger lessons you dare.

She relies on the blood of vertebrates—mainly warm-blooded ones but also sometimes reptiles and frogs—to finance, metabolically, the development of her eggs.

A female mosquito in a full lifetime will lay about ten separate batches of eggs, roughly 200 in a batch. That's a large order of embryonic tissue to be manufactured in one wispy body, and to manage it the female needs a rich source of protein; the sugary juice of flowers will deliver quick energy to wing muscles, but it won't help her build 2,000 new bodies. So she has evolved a hypodermic proboscis and learned how to *steal* protein in one of its richest forms, hemoglobin. In some species her first brood will develop before she has tasted blood, but after that she must have a bellyfull for each set of eggs coming to term.

When she drinks, she drinks deeply: The average blood meal amounts to 2½ times the original weight of the insect. Picture Audrey Hepburn sitting down to a steak dinner, getting up from the table weighing 380 pounds; then, for that matter, flying away. In the Canadian Arctic, where species of the genus *Aedes* emerge in savage, sky-darkening swarms like nothing seen even in the Amazon, and work under pressure of time because of the short summer season, an unprotected human could be bitten 9,000 times per minute. At that rate a large man would lose half his total blood in two hours. Arctic hares and reindeer move to higher ground, or die. And sometimes solid mats of *Aedes* will continue sucking the cool blood from a carcass.

Evidently the female tracks her way to a blood donor by flying upwind toward a source of warmer air, or air that is both warm and moist, or that contains an excess of carbon dioxide, or a combination of all three. The experts aren't sure. Perspiration, involving both higher skin temperature and released moisture, is one good way to attract her attention. In Italy it is established folk wisdom that to sleep with a pig in the bedroom is to protect oneself from malaria, presumably because the pig, operating at a higher body temperature, will be preferred by mosquitoes. And at the turn of the century, Professor Giovanni Grassi, then Italy's foremost zoologist, pointed out that garrulous people seemed to be bitten more often than those who kept their mouths shut. The experts aren't sure, but the Italians are full of ideas.

Guided by CO_2 or idle chatter or distaste for pork or whatever, a female mosquito lands on the earlobe of a human, drives her proboscis (actually a thin bundle of tools that includes two tubular stylets for carrying fluid and four serrated ones for cutting) through the skin, gropes with it until she taps a capillary, and then an elaborate interaction begins. Her saliva flows down one tube into the wound, retarding coagulation of the spilled blood and provoking an allergic reaction that will later be symptomized by itching. A suction pump in her head draws blood up the other tube, a valve closes, another pump pulls the blood back into her gut. And that alternate pumping and valving continues quickly for three orgiastic minutes, until her abdomen is stretched full like a great bloody balloon, or until a fast human hand ends her maternal career, whichever comes first.

But in the meantime, if she is an *Anopheles gambiae* in Nigeria, the protozoa that cause malaria may be streaming into the wound with her saliva, heading immediately off to set up bivouac in the human's liver. Or if she is *Aedes aegypti* in Costa Rica, she may be drooling out an advance phalanx of the yellow fever virus. If she is *Culex pipiens* in Malaysia, long tiny larvae of the filaria worm may be squirting from her snout like a stage magician's springwork snakes, dispersing to breed in the unfortunate person's lymph nodes and eventually clog them, causing elephantiasis. Definitely, this is misanthropic behavior.

No wonder, then, that in the rogues' pantheon of select creatures not only noxious in their essential character but furthermore lacking any imaginable forgiving graces, the mosquito is generally ranked beyond even the wood tick, the wolverine, or the black toy poodle. The mosquito, says common bias and on this the experts agree, is an unmitigated pain in the ass.

But I don't see it that way. To begin with, the family is not monolithic, and it does have—from even the human perspective—its beneficent representatives. In northern Canada, for instance, *Aedes nigripes* is an important pollinator of arctic orchids. In Ethiopia, *Toxorhynchites brevipalpis* as a larva preys voraciously on the larvae of other mosquitoes, malaria carriers, and then *Taxo* itself transforms to a lovely huge iridescent adult that, male or female, drinks only plant juices and would not dream of biting a human.

But even discounting these aberrations, and judging it by only its most notorious infamies, the mosquito is taking a bad rap. It has been victimized, I submit to you, by a strong case of anthropocentric press-agentry. In fact the little sucker can be viewed, with only a small bit of squinting, as one of the great ecological heroes of planet Earth. If you consider rainforest preservation.

The chief point of blame, with mosquitoes, happens also to be the chief point of merit: They make tropical rainforests, for humans, virtually uninhabitable.

Tropical rainforest constitutes by far the world's richest and most complex ecosystem, a boggling entanglement of life forms and habits and equilibriums and relationships. Those equatorial forests—mainly confined to the Amazon, the Congo basin and Southeast Asia—account for only a small fraction of the Earth's surface, but serve as home for roughly *half* of the Earth's total plant and animal species, including 2,000 kinds of mosquito. But rainforests lately, in case you haven't heard, are under siege.

They are being clear-cut for cattle ranching, mowed down with bulldozers and pulped for paper, corded into firewood, gobbled up hourly by human development on the march. The current rate of loss amounts to eight acres of rainforest gone poof since you began reading this sentence; within a generation, at that pace, the Amazon will look like New Jersey. Conservation groups are raising a clamor, a few of the equatorial governments are adopting plans for marginal preservation. But no one and no thing has done more to delay this catastrophe, over the past 10,000 years, than the mosquito.

The great episode of ecological disequilibrium we call "human history" began, so the Leakey family tell us, in equatorial Africa. Then immediately the focus of intensity shifted elsewhere. What deterred mankind, at least until this half of this century, from hacking space for his farms and his cities out of the tropical forests? Yellow fever did, and malaria, dengue, filariasis, o'nyong-nyong fever.

Clear the vegetation from the brink of a jungle waterhole, move in with tents and cattle and Jeeps, and *Anopheles gambiae,* not normally native there, will arrive within a month, bringing malaria. Cut the tall timber from five acres of rainforest, and species of infectious *Aedes*—which would otherwise live out their lives in the high forest canopy, passing yellow fever between monkeys—will literally fall on you, and begin biting before your chainsaw has cooled. Nurturing not only more species of snake and bird than anywhere else on earth, but also more forms of disease-causing microbe, and more mosquitoes to carry them, tropical forests are elaborately booby-trapped against disruption.

The native forest peoples gradually acquired some immunity to these diseases, and their nondisruptive hunting-and-gathering economies minimized their exposure to mosquitoes that favored the canopy or disturbed ground. Meanwhile the occasional white interlopers, the agents of empire, remained vulnerable. West Africa in high colonial days became known as "the white man's grave."

So as Europe was being stripped of its virgin woods, and India and China, and the North American heartland, the rainforests escaped, lasting into the late, twentieth century—with some chance at least that they may endure a bit longer. Thanks to what? Thanks to ten million generations of jungle-loving, disease-bearing, blood-sucking insects: the Culicidae, nature's Viet Cong.

And a time, says Ecclesiastes, to every purpose.

Quammen, David. 1981. "Sympathy for the Devil." *Outside.* June–July. Also published in David Quammen, *Natural Acts.* New York: Nick Lyons Books/Schocken Books, 1985.

Of Maggots and Murderers

May Berenbaum

May R. Berenbaum (b. 1953) is head of the Department of Entomology at the University of Illinois in Urbana, Illinois, and a member of the National Academy of Sciences. Aside from her entomological research, she organizes the Annual Insect Fear Film Festival at the University of Illinois and is the author of Ninety-Nine Gnats, Nits, and Nibblers; Ninety-Nine More Gnats, Nits, and Nibblers; *and* Bugs in the System, *from which this selection is taken. Berenbaum has, incidentally, become a part of popular culture in a way most entomologists never will: A 1996 episode of the popular* X-Files *television series, a story about alien killer cockroaches, featured a female entomologist named Bambi Berenbaum.*

It's bad enough knowing that insects pester humans in their daily lives but it's really adding insult to injury to think that they will continue to do so after the humans are dead and supposedly gone. Dead bodies (of any vertebrate) provide nourishment and livelihood to a large number of insects. That such is the case has been recognized since time immemorial and may in fact be at least part of the motivation behind the distinctly human practice of burying the dead with elaborate ceremony. Biblical references to carrion-feeders abound. In Isaiah (14:11), for example, it is written, "Thy pomp is brought down to the grave and the noise of thy viols, the worm is spread under thee and the worms cover thee." If that's not depressing enough, there's a passage in Job (21:26) reminding mortals that "they shall lie down alike in the dust, and the worms shall cover them."

Insects associate with carrion or corpses for a variety of reasons. Necrophagous species are those that actually consume dead flesh. Most of the truly necrophagous species are flies or beetles. Among the flies, members of the family Calliphoridae (the blow flies) and Sarcophagidae (the flesh flies) are particularly conspicuous elements of the dead-animal fauna. House flies, black scavenger flies, brine flies, coffin flies, and, particularly in later stages, cheese skippers are also associated with carrion. Among the beetles, necrophagous families include the burying beetles and the hide or skin beetles; histerids and tenebrionids are associated with later stages of decay. About the only other conspicuous members of the dead animal fauna are lepidopterans, especially tineids, or clothes moths.

From the insect perspective, there are definite advantages to utilizing carrion as a food resource. Unlike many forms of animal protein, carrion does not struggle or run away. There are, however, decided drawbacks as well, notably that carrion is neither an abundant nor predictable resource. Its suitability also changes dramatically over a relatively short time. Life cycles and behavioral traits in the carrion fauna reflect the ephemeral nature of carrion. Species associated exclusively with carrion, particularly in its early stages, tend to have a phenomenal capacity for finding dead animals. Many members of the carrion fauna are exceedingly sensitive to odors of decay and putrefaction at short range; silphids can, for example, detect the presence of skatole, a degradation product of protein, at concentrations in the air of about 9 parts per million (although the effective range over which they are sensitive is only on the order of about 1 meter). House flies and blow flies are attracted to a number of decomposition products, including acetic acid, benzoic acid, butyric acid, valeric acid, indole, acetone, phenol, *p*-cresol, and methyl disulfide.

Aside from having remarkable sensory capabilities, many members of the carrion fauna also have remarkable digestive capabilities. Larval blow flies, for example, are among only a handful of species that can metabolize collagen, the principal protein constituent of connective tissues. Immature clothes moths and hide beetles possess enzymes that can metabolize keratin, the proteinaceous component of hair and skin, which is indigestible to most organisms due to the fact that some of its amino acids are linked by usually unbreakable sulfur-containing chemical bonds.

If there is one word that characterizes the dung fauna, it's "competitive"—it's a dead-dog-eat-dead-dog world out there. Competition comes not only from other arthropods but from vertebrate carrion-feeders as well. Scavengers (such as vultures, wild dogs, or hyenas) may remove from 60 to 100% of the aboveground carcasses before they can be colonized by arthropods at all. High mortality and wide population fluctuations are thus typical of the lives of many arthropod carrion feeders. It often happens that more eggs are laid on a single carcass than can possibly reach maturity; ecologists refer to this sort of competition for food, in which, due to overcrowding, the possibility exists that all individuals fail to complete development, as scramble competition (in contrast to contest competition, after which clear winners and clear losers can be identified).

Many life-history attributes of carrion insects are designed to reduce competition both within a species and between species. Fast growth and flexibility of resource requirements are typical of carrion feeders. Blow flies, for example, can complete development in ten to fourteen days, gaining an average of 5% of their final larval weight per hour. Flexibility is key, too; they can pupate, depending on the availability of food, at a wide range of sizes (with the minimum size only 12% of the standard weight) and over a wide range of time intervals (from 50 to about 3,000 hours). In addition, some carrion feeders actually kill potential competitors. Despite the fact that legless, sightless maggots would hardly seem to be formidable enemies, flesh fly maggots routinely kill blow fly maggots in large numbers.

Also characteristic of carrion fauna is the ability to make quick and efficient use of a corpse. The ability to find carrion quickly is at a premium, since first arrivals can preempt resources and outcompete latecomers. *Sarcophaga* flesh flies practice larviposition (deposition of maggots, rather than eggs), which may provide them with a competitive edge compared to flies that must undergo an egg stage in carrion. *Lucilia* blow flies go to the extreme of laying their eggs directly in wounds of injured animals, not even waiting for them to die before moving in.

Some carrion insect behaviors are aimed at preventing access of the carrion to would-be competitors. *Dryomyza anilis* flies actively defend a body from other members of the same species that attempt to oviposit. Food relocation accomplishes the same end for *Nicrophorus* and other silphids. Silphids are known as burying, or sexton, beetles because they take the bodies of dead mammals and birds and bury them, thereby preventing potential competitors from finding, ovipositing, and feeding on the body. They are aided and abetted in their struggles by an unusual mutualistic association with a mite. The bodies of the silphid beetle *Nicrophorus* are covered with tiny mites *(Poecilochirus necrophori)* found only on the bodies of burying beetles. When a silphid locates a carcass, the mites leave the body of the beetle and move onto the carcass. There, they feed not on the dead remains but rather on the eggs and developing maggots of carrion flies that lay their eggs on carcasses at about the same time that burying beetles begin to prepare a carcass for burial. After processing and burying a carcass, burying beetles lay their eggs in the burial chamber. The larvae are fed by the parents initially, who in a tender example of parental care regurgitate partly digested dead flesh to meet the nutritional needs of their offspring. Eventually, the larvae can feed directly on the carrion. When they are ready to pupate, the adults dig an escape tunnel for them and, after pupation, the newly eclosed adults emerge and seek out fresh carrion. As they emerge, they are boarded by the mites, which accompany them to the carrion and to new sources of fly eggs and maggots. The association between the mites and the silphids appears to be mutually beneficial—the mites, which are flightless, are able to move from one fresh carcass to another, and the silphids are able to keep a carcass free from maggots for the duration of the development of their offspring.

Not every insect found in a dead body is necessarily necrophagous. A sizable community of predators and parasites has arisen that specializes on the insects that are found in carrion. Among the most important predators of carrion-feeding insects are certain staphylinids (or rove beetles). Even some of the necrophagous blow fly maggots turn carnivorous in late instars. There are also omnivorous species in carrion that include dead meat as part of a wide and varied diet. Wasps are among these species and are frequently found feeding on dead flesh. Most of the stories of augury attributed to "bees" in carcasses in all probability involved wasps such as yellowjackets rather than bees. Finally, there are adventive species, species that visit a corpse simply as a bit of topographical relief in their environment. Springtails and other litter dwellers, including mites and millipedes, are often found on or around dead bodies, not so much as a result of deliberate colonization but rather just as a result of their soil-dwelling, debris-feeding habits.

Like dung, dead bodies are habitats that are extremely unpredictable in distribution and extremely variable in terms of size, condition, and general composition. The changes in physical features of a corpse over time, however, are so predictable that distinct waves of colonization by stage-specific species have been described (although there is some controversy as to whether these represent distinct stages or a continuum). The period immediately after death is associated with chemical changes originating within the body, such as the release of body fluids and gases. Bluebottle and house flies visit and lay eggs at this stage. They arrive so quickly and develop so fast that for centuries people believed that maggots appeared spontaneously when meat began to decay. This notion of spontaneous generation was extremely difficult to dispel. Francesco Redi of seventeenth-century Italy finally performed a series of classic experiments, however, that demonstrated that maggots appear in meat only if flies are allowed to lay eggs on the meat first.

Shortly after death, fermentative changes attributable to microbes within the body result in detectable odor emission, and flesh flies move in. Wasps are also frequent visitors of carcasses at this stage. When fats begin to break down, volatile fatty acids are released. These attract hide beetles, *Aglossa* (a pyralid moth), and other species that can feed on decomposing fats. This stage lasts from three to six months. Caseic fermentation—

protein breakdown—follows and attracts a number of species specialized at metabolizing protein-breakdown products. In this group is the cheese skipper, *Piophila casei.*

Fermentative changes are accompanied by evaporation of many body fluids. Carcasses in later stages attract silphid and hister beetles, as well as some flies. Eventually, fluids evaporate completely and the body becomes desiccated. At this stage, hide and skin beetles move in as well as clothes moths and their relatives. These are species that can break down and metabolize collagen, the major protein constituent of connective tissue. Mites also colonize at this stage. Finally, only the real dregs remain. Such things as fecal material from the resident insects are consumed by scavengers such as spider beetles *(Ptinus brunneus)* and other such unfussy types.

The progression of species in a decomposing corpse is so precise and predictable that it is actually admissible as evidence in court. Forensic entomology makes use of life-history information and species identification to estimate time of death (postmortem interval), usually of crime victims, so entomologists can be called into court to testify as expert witnesses. Probably the earliest recorded incident in which insects helped to identify a murderer was the one described in a thirteenth-century Chinese manuscript, in which a murder by sickle was under investigation. All of the local farmers were called in and asked to lay their sickles on the ground. Flies were attracted to land on only one sickle; faced with such damning testimony, the guilty party promptly confessed. The modern use of forensic entomology for crime solving dates back to 1855, in which the knowledge of the insect fauna of human corpses was used to solve the murder of a child. A plasterer working on a mantelpiece in 1850 discovered the body of a child while he was working and Dr. M. Bergeret of Jura was called in to conduct the autopsy. He estimated from the presence of mite eggs and certain life stages of a blow fly *(Sarcophaga carnaria)* that the occupants of the house in 1848 were the likely perpetrators.

By the end of the century, a number of landmark papers had been published, including classics by J.-P. Mégnin (*La Faune des Tombeaux* in 1887 and *La Faune des Cadavres* in 1894) and the report in 1898 by M.G. Motter, "A contribution to the study of the grave. A study of one hundred and fifty disinterments, with some additional experimental observations." Since that time, entomological evidence has been used in countless murder cases and has played a critical role in obtaining convictions.

Berenbaum, May R. 1995. *Bugs in the System: Insects and Their Impact on Human Affairs.* Reading, Mass.: Addison-Wesley.

3

To Conquer the Earth: Insects Take Over

Many Hollywood scripts and science fiction tales suggest a world swarming with insects and threatening to overrun and conquer humans. Indeed, it could be argued that insects already control Earth, or vast portions of it. From earliest writers to present-day scientists, there have been widespread reports of various swarms, emergences, infestations, and invasions of many different insect species—from locusts in Egypt to red velvet mites in the western United States, to fire ants in tropical Brazil. Harvard entomologist E. O. Wilson, the first to report that imported fire ants had invaded his native Alabama (some decades before they conquered the American South), calls ants simply "the little creatures who run the world." The future of Earth, if Jonathan Schell is right, cannot be much different; if nuclear weapons destroy Earth, it will become a "Republic of Insects and Grass."

Disturbing the Composure of an Entomologist's Mind

Charles Darwin

Darwin's story of his five years' travel around the world as a young man makes stirring reading. In this selection, he reflects on the insect diversity of the Neotropics; in fact, Darwin (1809-1882) would be shocked by current estimates of the number of insect species, which range from 5 million to 30 million. In 1917, William Beebe said, "Yet another continent of life remains to be discovered, not upon the Earth, but one to two hundred feet above it." But it took Terry Erwin from the U.S. National Museum to propose in 1982 that there may be 30 million insect species, based on extrapolations from, in part, the great numbers of carabid beetles in the tropical forest canopy. Carabids are the same family of beetles that Darwin, who looked only on the ground, mistakenly speculated were rare in the tropics.

During our stay at Brazil I made a large collection of insects. A few general observations on the comparative importance of the different orders may be interesting to the English entomologist. The large and brilliantly-coloured Lepidoptera bespeak the zone they inhabit, far more plainly than any other race of animals. I allude only to the butterflies; for the moths, contrary to what might have been expected from the rankness of the vegetation, certainly appeared in much fewer numbers than in our own temperate regions. I was much surprised at the habits of *Papilio feronia.* This butterfly is not uncommon, and generally frequents the orange-groves. Although a high flier, yet it very frequently alights on the trunks of trees. On these occasions its head is invariably placed downwards; and its wings are expanded in a horizontal plane, instead of being folded vertically, as is commonly the case. This is the only butterfly which I have ever seen, that uses its legs for running. Not being aware of this fact, the insect, more than once, as I cautiously approached with my forceps, shuffled on one side just as the instrument was on the point of closing, and thus escaped. But a far more singular fact is the power which this species possesses of making a noise.* Several times when a pair, probably male and female, were chasing each other in an irregular course, they passed within a few yards of me; and I distinctly heard a clicking noise, similar to that produced by a toothed wheel passing under a spring catch. The noise was continued at short intervals, and could be distinguished at about twenty yards' distance: I am certain there is no error in the observation.

I was disappointed in the general aspect of the Coleoptera. The number of minute and obscurely-coloured beetles is exceedingly great.† The cabinets of Europe can, as yet, boast only of the larger species from tropical climates. It is sufficient to disturb the composure of an entomologist's mind, to look forward to the future dimensions of a complete catalogue. The carnivorous beetles, or Carabidae, appear in extremely few numbers within the tropics: this is the more remarkable when compared to the case of the carnivorous quadrupeds, which are so abundant in hot countries. I was struck with this observation both on entering Brazil, and when I saw the many elegant and active forms of the Harpalidae re-appearing on the temperate plains of La Plata. Do the very numerous spiders and rapacious Hymenoptera supply the place of the carnivorous beetles? The carrion-feeders and Brachelytera are very uncommon; on the other hand, the Rhyncophora and Chrysomelidae, all of which depend on the vegetable world for subsistence, are present in astonishing numbers. I do not here refer to the number of different species, but to that of the individual insects; for on this it is that the most striking character in the entomology of different countries depends. The orders Orthoptera and Hemiptera are particularly numerous; as likewise is the stinging division of the Hymenoptera; the bees, perhaps, being excepted. A person, on first entering a trop-

* Mr. Doubleday has lately described (before the Entomological Society, March 3rd, 1845) a peculiar structure in the wings of this butterfly, which seems to be the means of its making its noise. He says, "It is remarkable for having a sort of drum at the base of the fore wings, between the costal nervure and the subcostal. These two nervures, moreover, have a peculiar screw-like diaphragm or vessel in the interior." I find in Langsdorff's travels (in the years 1803–7, p. 74) it is said, that in the island of St. Catherine's on the coast of Brazil, a butterfly called *Februa Hoffmanseggi*, makes a noise, when flying away, like a rattle.

† I may mention, as a common instance of one day's (June 23rd) collecting, when I was not attending particularly to the Coleoptera, that I caught sixty-eight species of that order. Among these, there were only two of the Carabidae, four Brachelytra, fifteen Rhyncophora, and fourteen of the Chrysomelidae. Thirty-seven species of Arachnidae, which I brought home, will be sufficient to prove that I was not paying overmuch attention to the generally favoured order of Coleoptera.

ical forest, is astonished at the labours of the ants: well-beaten paths branch off in every direction, on which an army of never-failing foragers may be seen, some going forth, and others returning, burdened with pieces of green leaf, often larger than their own bodies.

A small dark-coloured ant sometimes migrates in countless numbers. One day, at Bahia, my attention was drawn by observing many spiders, cockroaches, and other insects, and some lizards, rushing in the greatest agitation across a bare piece of ground. A little way behind, every stalk and leaf was blackened by a small ant. The swarm having crossed the bare space, divided itself, and descended an old wall. By this means many insects were fairly enclosed; and the efforts which the poor little creatures made to extricate themselves from such a death were wonderful. When the ants came to the road they changed their course, and in narrow files reascended the wall. Having placed a small stone so as to intercept one of the lines, the whole body attacked it, and then immediately retired. Shortly afterwards another body came to the charge, and again having failed to make any impression, this line of march was entirely given up. By going an inch round, the file might have avoided the stone, and this doubtless would have happened, if it had been originally there: but having been attacked, the lion-hearted little warriors scorned the idea of yielding.

Darwin, Charles. 1839. *Journal of Researches into the Geology and Natural History of the Various Countries Visited by H.M.S. Beagle.* London: Henry Colburn.

Locusts in the Land of Egypt

Exodus, The Bible

Locust plagues and swarms provide some of the most dramatic examples of mass insect movements in the world. One of the more vivid, lyrical scenes in the Bible is the story in Exodus of Moses and the Seven Plagues of Egypt; it produced the great Afro-American spiritual "Let My People Go" and became a metaphor for the struggle of many peoples to escape slavery. As part of the effort to make the Egyptian pharaoh realize that Moses and his people were serious, God produced a number of signs and wonders, triggered by Moses and Aaron, and chief among these were swarms of gnats, followed by flies, and finally, worst of all, locusts that came and ate "every plant in the land." That insect plagues figured so prominently is indicative of how insects were viewed—as pests. Eventually, mass insect movement usually produces human movement!

12 And the Lord said unto Moses, Stretch out thine hand over the land of Egypt for the locusts, that they may come up upon the land of Egypt, and eat every herb of the land, even all that the hail hath left.

13 And Moses stretched forth his rod over the land of Egypt, and the Lord brought an east wind upon the land all that day, and all that night; and when it was morning, the east wind brought the locusts.

14 And the locusts went up over all the land of Egypt, and rested in all the coasts of Egypt: very grievous were they; before them there were no such locusts as they, neither after them shall be such.

15 For they covered the face of the whole earth, so that the land was darkened; and they did eat every herb of the land, and all the fruit of the trees which the hail had left: and there remained not any green thing in the trees, or in the herbs of the field, through all the land of Egypt.

16 Then Pharaoh called for Moses and Aaron in haste; and he said, I have sinned against the Lord your God, and against you.

17 Now therefore forgive, I pray thee, my sin only this once, and entreat the Lord your God, that he may take away from me this death only.

18 And he went out from Pharaoh, and entreated the Lord.

19 And the Lord turned a mighty strong west wind, which took away the locusts, and cast them into the Red sea; there remained not one locust in all the coasts of Egypt.

20 But the Lord hardened Pharaoh's heart, so that he would not let the children of Israel go.

Exodus 10:12–20. Authorized King James Version (translated into English 1603–1611).

Giant Red Velvet Mites

Irwin M. Newell and Lloyd Tevis Jr.

It's August 3, 1943. Entomologist Robert E. Beer of the University of Kansas is flying over the Arizona desert when he spots a vast reddish bloom on the ground below. Puzzled (August is the wrong time of year for desert wildflowers), he lands and inspects on the ground. The seething red carpet is an explosion of giant red velvet mites! Are they taking over the world?

Sound like the notes for a Hollywood science fiction script? In fact, it's a sober scientific paper from a 1960 issue of the Annals of the Entomological Society of America. *No one had ever before reported such an astonishing outbreak of mites, let alone of this species, unknown to science but called* angelitos *(little angels) by the locals. Since then, one other, similar outbreak in a closely related species has come to light, this time on the other side of the globe in a desert in India.*

The emergence was sighted from the air at an altitude of approximately 1,500 ft. near the town of Picacho, Arizona, about 48 miles northwest of Tucson. At first it was thought that it might be an unusual bloom of some desert lichen, but an inspection of the area from the ground revealed that it was in reality a heavy emergence of giant red velvet mites. An estimated 40 or 50 mites were found in each 4-inch square (16 square inches). This very heavy emergence was observed over an area of approximately 2 acres, beyond which the mites decreased in numbers. Very few mites were found outside of an area of approximately 5 acres. The population was not actually sampled, but if we accept these figures at their face value, the number of mites involved was of the order of 3 to 5 million. The eruption of the mites occurred almost immediately after a heavy but brief summer rainstorm, on August 3, 1943. The storm was of the type so characteristic of the summer rainy season in this part of the country, with a column of rain progressing across the desert. These storms are ordinarily of small diameter, but the rainfall within them is intense.

Newell, Irwin M., and Lloyd Tevis Jr. 1960. "*Angelothrombium pandorae* N.G., N. Sp. (Acari, Trombidiidae), and Notes on the Biology of the Giant Red Velvet Mites." *Annals of the Entomological Society of America,* May, 53(3):293–304.

Weird Tales illustration by Jayem Wilcox. From 1923 to 1954, *Weird Tales* was the premier magazine of horror and fantasy fiction, giving rise to such diverse authors as Tennessee Williams, Ray Bradbury, H. P. Lovecraft, Robert Bloch, and Robert E. Howard. Jayem Wilcox provided the lurid "giant insect" illustrations for a several-part serial that ran in the magazine in 1934. The story, written by David H. Keller, M.D., featured an unemployed Ivy League entomologist (shown in the illustration) and told of the realistic behavior of giant lion-size wasps—"The Solitary Hunters."

Them!

George Worthing Yates and Ted Sherdeman

From science fiction screenplays and novels to scientific papers, writers have portrayed insects as seeking to control or already being in possession of Earth. Certainly, insects have colonized Earth in a thorough and dominating way, they show resistance to radiation, and they have reproductive systems that make them extremely difficult to eradicate. On the other hand, there's not much chance that they may one day mutate and increase in size by more than a thousand times as a result of exposure to fallout from nuclear weapons testing. For one thing, an ant or a cockroach scaled up to the size of a diesel truck would have trouble standing up, much less wreaking havoc.

One of the first great insect movie thrillers was the 1954 Warner Brothers film entitled Them! *Aside from the rather sizeable matter of giant ants, the behavior and statements about the ants are, for the most part, accurate. Among other things, this must mark the only time the word "myrmecologist" (one who studies ants) has ever been used in a mainstream Hollywood film, and used several times. The American screenwriter Ted Sherdeman and the originator of the story, George Worthing Yates, gave director Gordon Douglas a well-researched script with lines that are delivered with wonderful understatement by actors James Whitmore, James Arness, and others. Two of the main characters are a father-daughter team of entomologists, the Doctors Medford. The father, played by Edmund Gwen, is a U.S. Department of Agriculture myrmecologist from Washington, D.C., and the daughter (Joan Weldon), an attractive, intelligent Katharine Hepburn–type. They arrive in New Mexico to follow up on a giant footprint they were sent by the FBI. Several people have been mysteriously attacked and killed—some have crushed chests or heads and all are full of formic acid. The local police and the FBI have been on the case and are working on the theory that a homicidal maniac is on the loose, but they still don't have a clue why so much sugar has gone missing.*

First policeman *(pointing to footprint):* How come the FBI sent this to you?

Dr. Medford *(curtly):* They weren't able to identify it. *(pursuing his own train of thought:)* Tell me in what area was the atomic bomb exploded, I mean the first one back in 1945.

First policeman: It was right here in this general area *(pointing to map),* and 12 miles down is the Johnson's Store *(where the first murder took place).* White Sands.

Dr. Medford: That's 1945, 9 years ago. Yes, genetically, that's certainly possible.

FBI Man *(losing his temper):* Now there's no need to play footsy with us; if you know what it is, you tell us. We're assigned to this case too.

Dr. Medford: We can't say anything till we're absolutely sure.

They turn to a young girl who has been rendered silent after her family was killed. Dr. Medford tries wafting formic acid under the little girl's nose to test her reaction.

Dr. Medford: It may provide the jolt she needs.

Little girl stares straight ahead.

Little girl *(as she smells the acid):* Them! Them! Them!

She starts screaming and screaming.

Dr. Medford: May we visit the desert now, gentlemen?

First policeman: It's getting late, Doctor.

Dr. Medford: Later than you think!

Scene changes to the desert. They arrive in the middle of a sandstorm, just before twilight.

Dr. Medford: Quite a breeze!

First policeman: The print was found right over there.

Dr. Medford: Has there been a report of a strange mound? Something with a cone shaped structure, something recently formed?

Policeman: No sir.

Dr. Pat Medford: Rather slim pickings for food, Dad. They turn carnivorous for lack of a vegetable diet.

Dr. Medford: I believe you're right.

FBI Man: What would turn carnivorous?

Dr. Pat Medford: My father will tell you.

FBI Man: When?

Dr. Pat Medford: When he's positive.

FBI Man: Well I've got a job to do and I've got enough mystery on my hands anyway without that old man, your father.

Dr. Pat Medford: Well that old man, as you started to call him, is one of the world's greatest myrmecologists.

FBI Man: Myr-me-col-o-gist, you see that's what I mean, that's the trouble with you people, why don't we all talk English, then we'd have a basis for an understanding?

Dr. Medford: Pat, Pat, come over here! It's the same kind of print. *(waits for her to run over)* Look at the print, it's gigantic! Over 12 centimeters.

Dr. Pat Medford: That would make the entire . . .

Dr. Medford: About two and a half meters long. Over eight feet! See if there are more. This is monstrous.

First policeman: So is the disappearance and murder of 5 people, doctor.

Dr. Medford: Yes. The direction of that print would indicate that it's going that way! Perhaps, perhaps we should visit the store. There may be more there now.

The policeman and the younger Dr. Medford help him up.

FBI Man: Now look here, Dr. Medford. Before we go visiting any place, I want to know exactly what this IT is.

Dr. Medford: Gentlemen, I understand your impatience. I know you two are interested in solving what is essentially a local crime, but please believe me, I am not being coy with you. No, if I'm wrong in my assumptions, then no harm has been done, but if I am correct, and the mounting evidence only fortifies my theory, then something incredible has happened in this desert and no one of us will dare risk revealing it, because none of us can risk a nationwide panic.

FBI Man: Panic?

Dr. Medford nods while his daughter goes out looking, and finds another print.

A strange sound steadily gets louder, and all of them look up.

Suddenly from behind Dr. Pat Medford, there appear two antennae waving only a few feet away from her. A pair of huge mandibles are snapping, closing in on her. A giant ant!

Pat screams, and leaps away as the ant comes closer and the sound becomes deafening.

The FBI Man leaps into action, shooting his gun. The policeman joins him.

DR. MEDFORD: Get the antennae, get the antennae.

More gunfire, and finally one of the antennae is hit and hangs limp.

DR. MEDFORD: Get the other antenna; get the other antenna, he's helpless without them.

Finally, the policemen retrieves a repeating gun and the other antenna is knocked off and the animal is killed. The eight-foot-long animal slumps to the ground.

In this entertaining film, things soon escalate and the battle of the species is waged in the sewers of Los Angeles. In this case, for this film, the humans win, but we are left with the impression that it is just a skirmish in a war for dominance of Earth that may well intensify.

A footnote: At the time Them! *was made, there really was a U.S. Department of Agriculture myrmecologist at the National Museum of Natural History in Washington, D.C., named Dr. Marion R. Smith. The renowned myrmecologist William L. Brown Jr. once reminisced about how he and other "ant men" had mercilessly kidded Smith about his "portrayal" in that movie, and especially badgered him to introduce them to the gorgeous daughter that he'd been hiding all these years.*

Yates, George Worthing, and Ted Sherdeman. 1954. *Them!* Directed by Gordon Douglas and produced by David Weisbart. Warner Bros. Pictures, Burbank, Cal.

Insects from Mars

In 1962, the Topps Bubble Gum Company produced a series of trading cards depicting, in fifty-five grisly installments, the invasion of Earth by Martians. Marketed in the same way as baseball cards (five cards and a stick of bubble gum in a 5-cent pack), each card featured a painted scene, usually of an imaginative method by which Martians dispensed with humans. Beginning with card #27, entitled "The Giant Flies," things took a decidedly entomological turn: Employing a "growth ray," the Martians unleashed giant insects upon the already beleaguered Earthlings. Parents' reactions to the graphically depicted results were part of the reason these cards were yanked from their test markets and never widely distributed. Today, they are among the most valuable of all collectible trading cards.

Mars Attacks trading cards. Reprinted with permission from Topps USA and Topps UK.

A Republic of Insects and Grass

Jonathan Schell

The New Yorker *writer Jonathan Schell (b. 1943) created a stir with this extended piece that later appeared in book form, jarring many people out of their early 1980s complacency about "the bomb" and its effects on Earth. The image of the United States, and for that matter Earth, becoming a republic of insects and grass remains a sobering one.*

Over the years, agencies and departments of the government have sponsored numerous research projects in which a large variety of plants and animals were irradiated in order to ascertain the lethal or sterilizing dose for each. These findings permit the prediction of many gross ecological consequences of a nuclear attack. According to "Survival of Food Crops and Livestock in the Event of Nuclear War," the proceedings of the 1970 symposium at Brookhaven National Laboratory, the lethal doses for most mammals lie between a few hundred rads and a thousand rads of gamma radiation; a rad—for "roentgen absorbed dose"—is a roentgen of radiation that has been absorbed by an organism, and is roughly equal to a rem. For example, the lethal doses of gamma radiation for animals in pasture, where fallout would be descending on them directly and they would be eating fallout that had fallen on the grass, and would thus suffer from doses of beta radiation as well, would be one hundred and eighty rads for cattle; two hundred and forty rads for sheep; five hundred and fifty rads for swine; three hundred and fifty rads for horses; and eight hundred rads for poultry. In a ten-thousand-megaton attack, which would create levels of radiation around the country averaging more than ten thousand rads, most of the mammals of the United States would be killed off. The lethal doses for birds are in roughly the same range as those for mammals, and birds, too, would be killed off. Fish are killed at doses of between one thousand one hundred rads and about five thousand six hundred rads, but their fate is less predictable. On the one hand, water is a shield from radiation, and would afford some protection; on the other hand, fallout might concentrate in bodies of water as it ran off from the land. (Because radiation causes no pain, animals, wandering at will through the environment, would not avoid it.) The one class of animals containing a number of species quite likely to survive, at least in the short run, is the insect class, for which in most known cases the lethal doses lie between about two thousand rads and about a hundred thousand rads. Insects, therefore, would be destroyed selectively. Unfortunately for the rest of the environment, many of the phytophagous species—insects that feed directly on vegetation—which "include some of the most ravaging species on earth" (according to Dr. Vernon M. Stern, an entomologist at the University of California at Riverside, writing in "Survival of Food Crops"), have very high tolerances, and so could be expected to survive disproportionately, and then to multiply greatly in the aftermath of an attack. The demise of their natural predators the birds would enhance their success. . . .

In sum, a full-scale nuclear attack on the United States would devastate the natural environment on a scale unknown since early geological times, when, in response to natural catastrophes whose nature has not been determined, sudden mass extinctions of species and whole ecosystems occurred all over the earth. How far this "gross simplification" of the environment would go once virtually all animal life and the greater part of plant life had been destroyed and what patterns the surviving remnants of life would arrange themselves into over the long run are imponderables; but it appears that at the outset the United States would be a republic of insects and grass.

Schell, Jonathan. 1982. *The Fate of the Earth.* New York: Knopf.

Insects Take Over

Gary Larson

A world survey of entomology labs would, not surprisingly, reveal any number of shared features. One of the more unexpected of these would be the presence of at least one favorite Gary Larson (b. 1950) cartoon clipped from a newspaper and tacked onto a bulletin board, wall, or door. No other cartoonist has utilized sophisticated biological themes to the extent that Larson has. Two occurrences demonstrate the affection with which he is held by biologists: His work was featured in a one-man show at the California Academy of Sciences in San Francisco, and none other than über-entomologist E. O. Wilson wrote the introduction to Larson's most recent book, There's a Hair in My Dirt.

The End

W. J. Holland

This excerpt is from the last page of a 1903 book celebrating the moth.

When the moon shall have faded out from the sky, and the sun shall shine at noonday a dull cherry-red, and the seas shall be frozen over, and the ice-cap shall have crept downward to the equator from either pole, and no keels shall cut the waters, nor wheels turn in mills, when all cities shall have long been dead and crumbled into dust, and all life shall be on the very last verge of extinction on this globe; then, on a bit of lichen, growing on the bald rocks beside the eternal snows of Panama, shall be seated a tiny insect, preening its antennae in the glow of the worn-out sun, representing the sole survival of animal life on this our earth,—a melancholy "bug."

Holland, W. J. 1903. *The Moth Book.* New York: Doubleday, Page and Company. (Reprinted, 1968, New York: Dover.)

4

A Cast of Millions on a Fantastic Journey: Mass Movement

Some insects may spend their entire lives in the same piece of fruit, but others, like the monarch butterfly, must traverse continental distances to migrate from summer feeding grounds to winter hibernation quarters. The biological equipment that enables such movement is astonishing; for instance, bees have magnetic compasses built into their brains and male moths have radarlike antennae that can sense a single molecule's worth of difference in the density of a cloud of pheromonal perfume emanating from a female miles away. From army ant raids to fly swarms, this chapter explores the mysteries of mass migration.

Insects at Sea

Charles Darwin

The theory of evolution by natural selection owes much to the five years of natural history observation that the young Charles Darwin spent aboard the HMS Beagle. *Among many other things, Darwin was awed by repeated encounters with insects in the open sea, far from any land or fresh water. As Darwin knew, with the possible exception of a species of water strider, there are no truly marine insects, probably because all the available niches had long ago been colonized by their marine counterparts, the crustaceans. But the insects observed by Darwin were strays, providing potential insight into the dispersal of terrestrial organisms between widely separated bodies of land.*

December 6th.—The *Beagle* sailed from the Rio Plata, never again to enter its muddy stream. Our course was directed to Port Desire, on the coast of Patagonia. Before proceeding any further, I will here put together a few observations made at sea.

Several times when the ship has been some miles off the mouth of the Plata, and at other times when off the shores of Northern Patagonia, we have been surrounded by insects. One evening, when we were about ten miles from the Bay of San Blas, vast numbers of butterflies, in bands or flocks of countless myriads, extended as far as the eye could range. Even by the aid of a telescope it was not possible to see a space free from butterflies. The seamen cried out "it was snowing butterflies," and such in fact was the appearance. More species than one were present, but the main part belonged to a kind very similar to, but not identical with, the common English *Colias edusa.* Some moths and hymenoptera accompanied the butterflies; and a fine beetle *(Calosoma)* flew on board. Other instances are known of this beetle having been caught far out at sea; and this is the more remarkable, as the greater number of the Carabidae seldom or never take wing. The day had been fine and calm, and the one previous to it equally so, with light and variable airs. Hence we cannot suppose that the insects were blown off the land, but we must conclude that they voluntarily took flight. The great bands of the *Colias* seem at first to afford an instance like those on record of the migrations of another butterfly, *Vanessa canlui;* but the presence of other insects makes the case distinct, and even less intelligible. Before sunset a strong breeze sprung up from the north, and this must have caused tens of thousands of the butterflies and other insects to have perished.

On another occasion, when seventeen miles off Cape Corrientes, I had a net overboard to catch pelagic animals. Upon drawing it up, to my surprise I found a considerable number of beetles in it, and although in the open sea, they did not appear much injured by the salt water. I lost some of the specimens, but those which I preserved belonged to the genera *Colymbetes, Hydroporus, Hydrobius* (two species), *Notaphus, Cynucus, Adimonia,* and *Scarabaeus.* At first I thought that these insects had been blown from the shore; but upon reflecting that out of the eight species four were aquatic, and two others partly so in their habits, it appeared to me most probable that they were floated into the sea by a small stream which drains a lake near Cape Corrientes. On any supposition it is an interesting circumstance to find live insects swimming in the open ocean seventeen miles from the nearest point of land. There are several accounts of insects having been blown off the Patagonian shore. Captain Cook observed it, as did more lately Captain King in the *Adventure.* The cause probably is due to the want of shelter, both of trees and hills, so that an insect on the wing, with an off-shore breeze, would be very apt to be blown out to sea. The most remarkable instance I have known of an insect being caught far from the land, was that of a large grasshopper *(Acrydium),* which flew on board, when the *Beagle* was to windward of the Cape de Verd Islands, and when the nearest point of land, not directly opposed to the trade-wind, was Cape Blanco on the coast of Africa, 370 miles distant.

Darwin, Charles. 1839. *Journal of Researches into the Geology and Natural History of the Various Countries Visited by H.M.S. Beagle.* London: Henry Colburn.

Army Ants

Thomas Belt

The son of a seedsman and canvas-twine manufacturer, Thomas Belt (1832–1878) showed a youthful interest in natural history. In the early 1850s he sought his fortune in the gold rush to Australia, where he acquired the knowledge to write his 1861 book, Mineral Veins: An Enquiry into Their Origin, *which established him as an expert mining engineer. Belt spent most of the remaining seventeen years of his life in the eager employ of mining companies, prospecting in Siberia, southern Russia, England, and North and Central America.*

Always devoting his leisure hours to the study of natural history, Belt published scientific papers on a wide range of subjects, including whirlwinds, entomology, paleontology, and especially geology. In 1868, Belt began a four-year contract superintending the operations of the Chontales Gold Mining Company in Nicaragua. Among his important discoveries in the Central American rain forest, he was the first to note that the leafcutter, or parasol, ants use their leaf cuttings not *to thatch the roof of the colony (as Henry W. Bates had conjectured) or for food (as others believed), but as compost for growing a fungus. Leafcutter ants are underground mushroom farmers!*

Thomas Belt is best remembered for his 1874 classic of natural history, The Naturalist in Nicaragua, *which has inspired generations of tropical biologists. Here, from that book, is a lesser known passage about army ants, full of all the colorful detail that Belt had observed.*

As I returned to the boat, I crossed a column of the army or foraging ants, many of them dragging along the legs and mangled bodies of insects that they had captured in their foray. I afterwards often encountered these ants in the forests, and it may be convenient to place together all the facts I learnt respecting them.

The *Ecitons,* or foraging ants, are very numerous throughout Central America. Whilst the leaf-cutting ants are entirely vegetable feeders, the foraging ants are hunters, and live solely on insects or other prey; and it is a curious analogy that, like the hunting races of mankind, they have to change their hunting-grounds when one is exhausted, and move on to another. In Nicaragua they are generally called "Army Ants." One of the smaller species *(Eciton predator)* used occasionally to visit our house, swarm over the floors and walls, searching every cranny, and driving out the cockroaches and spiders, many of which were caught, pulled, or bitten to pieces and carried off. The individuals of this species are of various sizes; the smallest measuring one and a quarter lines, and the largest three lines, or a quarter of an inch.

I saw many large armies of this, or a closely allied species, in the forest. My attention was generally first called to them by the twittering of some small birds, belonging to several different species, that follow the ants in the woods. On approaching to ascertain the cause of this disturbance, a dense body of the ants, three or four yards wide, and so numerous as to blacken the ground, would be seen moving rapidly in one direction, examining every cranny, and underneath every fallen leaf. On the flanks, and in advance of the main body, smaller columns would be pushed out. These smaller columns would generally first flush the cockroaches, grasshoppers, and spiders. The pursued insects would rapidly make off, but many, in their confusion and terror, would bound right into the midst of the main body of ants. A grasshopper, finding itself in the midst of its enemies, would give vigorous leaps, with perhaps two or three of the ants clinging to its legs. Then it would stop a moment to rest, and that moment would be fatal, for the tiny foes would swarm over the prey, and after a few more ineffectual struggles it would succumb to its fate, and soon be bitten to pieces and carried off to the rear. The greatest catch of the ants was, however, when they got amongst some fallen brushwood. The cockroaches, spiders, and other insects, instead of running right away, would ascend the fallen branches and remain there, whilst the host of ants were occupying all the ground below. By-and-by up would come some of the ants, following every branch, and driving before them their prey to the ends of the small twigs, when nothing remained for them but to leap, and they would alight in the very throng of their foes, with the result of being certainly caught and pulled to pieces. Many of the spiders would escape by hanging suspended by a thread of silk from the branches, safe from the foes that swarmed both above and below.

I noticed that spiders were generally most intelligent in escaping, and did not, like the cockroaches and other insects, take shelter in the first hiding-place they found, only to be driven out again, or perhaps caught by the advancing army of ants. I have often seen large spiders making off many yards in advance, and apparently determined to put a good distance between themselves and their foe. I once saw one of the false spiders, or harvest-men (Phalangidae), standing in the midst of an army of ants, and with the greatest circumspection and coolness lifting, one after the other, its long legs, which supported its body above their reach. Sometimes as many as five out of its eight legs would be lifted at once, and whenever an ant approached one of those on which it stood, there was always a clear space within reach to put down another, so as to be able to hold up the threatened one out of danger.

I was much more surprised with the behaviour of a green, leaf-like locust. This insect stood immovably amongst a host of ants, many of which ran over its legs, without ever discovering there was food within their reach. So fixed was its instinctive knowledge that its safety depended on its immovability, that it allowed me to pick it up and replace it amongst the ants without making a single effort to escape. This species closely resembles a green leaf, and the other senses, which in the *Ecitons* appear to be more acute than that of sight, must have been completely deceived. It might easily have escaped from the ants by using its wings, but it would only have fallen into as great a danger, for the numerous birds that accompany the army ants are ever on the outlook for any insect that may fly up, and the heavy flying locusts, grasshoppers, and cockroaches have no chance of escape. Several species of ant-thrushes always accompany the army ants in the forest. They do not, however, feed on the ants, but on the insects they disturb. Besides the ant-thrushes, trogons, creepers, and a variety of other birds, are often seen on the branches of trees above where an ant army is foraging below, pursuing and catching the insects that fly up.

The insects caught by the ants are dismembered, and their too bulky bodies bitten to pieces and carried off to the rear. Behind the army there are always small columns engaged on this duty. I have followed up these columns often; generally they led to dense masses of impenetrable brushwood, but twice they led me to cracks in the ground, down which the ants dragged their prey. These habitations are only temporary, for in a few days not an ant would be seen in the neighbourhood; all would have moved off to fresh hunting-grounds.

Another much larger species of foraging ant *(Eciton hamata)* hunts sometimes in dense armies, sometimes in columns, according to the prey it may be after. When in columns, I found that it was generally, if not always, in search of the nests of another ant (*Hypoclinea* sp.), which rear their young in holes in rotten trunks of fallen timber, and are very common in cleared places. The *Ecitons* hunt about in columns, which branch off in various directions. When a fallen log is reached, the column spreads out over it, searching through all the holes and cracks. The workers are of various sizes, and the smallest are here of use, for they squeeze themselves into the narrowest holes, and search out their prey in the furthest ramifications of the nests. When a nest of the *Hypoclinea* is attacked, the ants rush out, carrying the larvae and pupae in their jaws, only to be immediately despoiled of them by the *Ecitons,* which are running about in every direction with great swiftness. Whenever they come across a *Hypoclinea* carrying a larva or pupa, they capture the burden so quickly, that I could never ascertain exactly how it was done.

As soon as an *Eciton* gets hold of its prey, it rushes off back along the advancing column, which is composed of two sets, one hurrying forward, the other returning laden with their booty, but all and always in the greatest haste and apparent hurry. About the nest which they are harrying, everything is confusion, *Ecitons* run here and there and everywhere in the greatest haste and disorder; but the result of all this apparent confusion is that scarcely a single *Hypoclinea* gets away with a pupa or larva. I never saw the *Ecitons* injure the *Hypoclineas* themselves, they were always contented with despoiling them of their young. The ant that is attacked is a very cowardly species, and never shows fight. I often found it running about sipping at the glands of leaves, or milking aphides, leaf-hoppers, or scale-insects that it found unattended by other ants. On the approach of another, though of a much smaller, species, it would immediately run away. Probably this cowardly and unantly disposition has caused it to become the prey of the *Eciton.* At any rate, I never saw the *Ecitons* attack the nest of other species.

The moving columns of *Ecitons* are composed almost entirely of workers of different sizes, but at intervals of two or three yards there are larger and lighter-coloured individuals that will often stop, and sometimes run a little backward, halting and touching some of the ants with their antennae. They look like officers giving orders and directing the march of the column. . . .

This well-observed early account of army ants is carried further in an explanation of their peculiar "architecture" by William Beebe on *p. 166.* *Both of these accounts were sources for the famous 1940 short story "Leiningen versus the Ants" by Carl Stephenson, in which the army ants became decidedly more lethal and devouring with an advancing horde 2 miles wide by 10 miles long! Hollywood, too, got in on the act with an adaptation of the story in the 1954* The Naked Jungle, *starring Charlton Heston as the Brazilian plantation owner. The ants in the film, however, were not army ants but the presumably more menacingly photogenic* Camponotus *carpenter ants.*

Belt, Thomas. 1874. *The Naturalist in Nicaragua.* London: John Murray.

The Spring-Cleaning. Hordes of foraging ants move through a house in Trinidad, driving out and seizing cockroaches, wasps, and other vermin. Illustration by Theo. Carreras (*Marvels of Insect Life,* edited by Edward Step, Robert M. McBride & Co., New York, 1916).

Caterpillars on the Line

George John Romanes

George Romanes (1848–1894) was a contemporary and friend of Charles Darwin, who applied a Darwinian approach to the study of behavior. In particular, he assumed a continuity between human and animal intelligence, and he worked largely by assembling verbal and written accounts of behavioral observations. Romanes' work was eclipsed and, indeed, scorned by subsequent behaviorist doctrine, which criticized the assumption of intelligence in animals as anthropocentric (rather than objective) and Romanes' methods as anecdotal (rather than experimental).

Perhaps only now is our culture coming to grips with the truly Darwinian notion that mental phenomena in humans do not necessarily qualitatively differ from those in our closest relatives, and thus that Romanes' ideas are worthy of reexamination. Here he recounts three phenomenal cases of insect mass migration.

Taking the animal kingdom from below upwards, the first animals that can properly be said to present the instincts of migration are to be found in the group Articulata. I think it is sufficient to refer to "Animal Intelligence" for the facts concerning the migrations of Crabs and Caterpillars, though as regards the latter I may add the following remarkable account, which I quote from the "Colonies and India."

"To say that a train had been stopped by caterpillars would sound like a Yankee yarn, yet such a thing (according to the *Rangitikei Advocate*) actually took place on the local railway a few days ago. In the neighbourhood of Turakina, New Zealand, an army of caterpillars, hundreds of thousands strong, was marching across the line, bound for a new field of oats, when the train came along. Thousands of the creeping vermin were crushed by the wheels of the engine, and suddenly the train came to a dead stop. On examination it was found that the wheels of the engine had become so greasy that they kept on revolving without advancing—they could not grip the rails. The guard and the engine-driver procured sand and strewed it on the rails, and the train made a fresh start, but it was found that during the stoppage

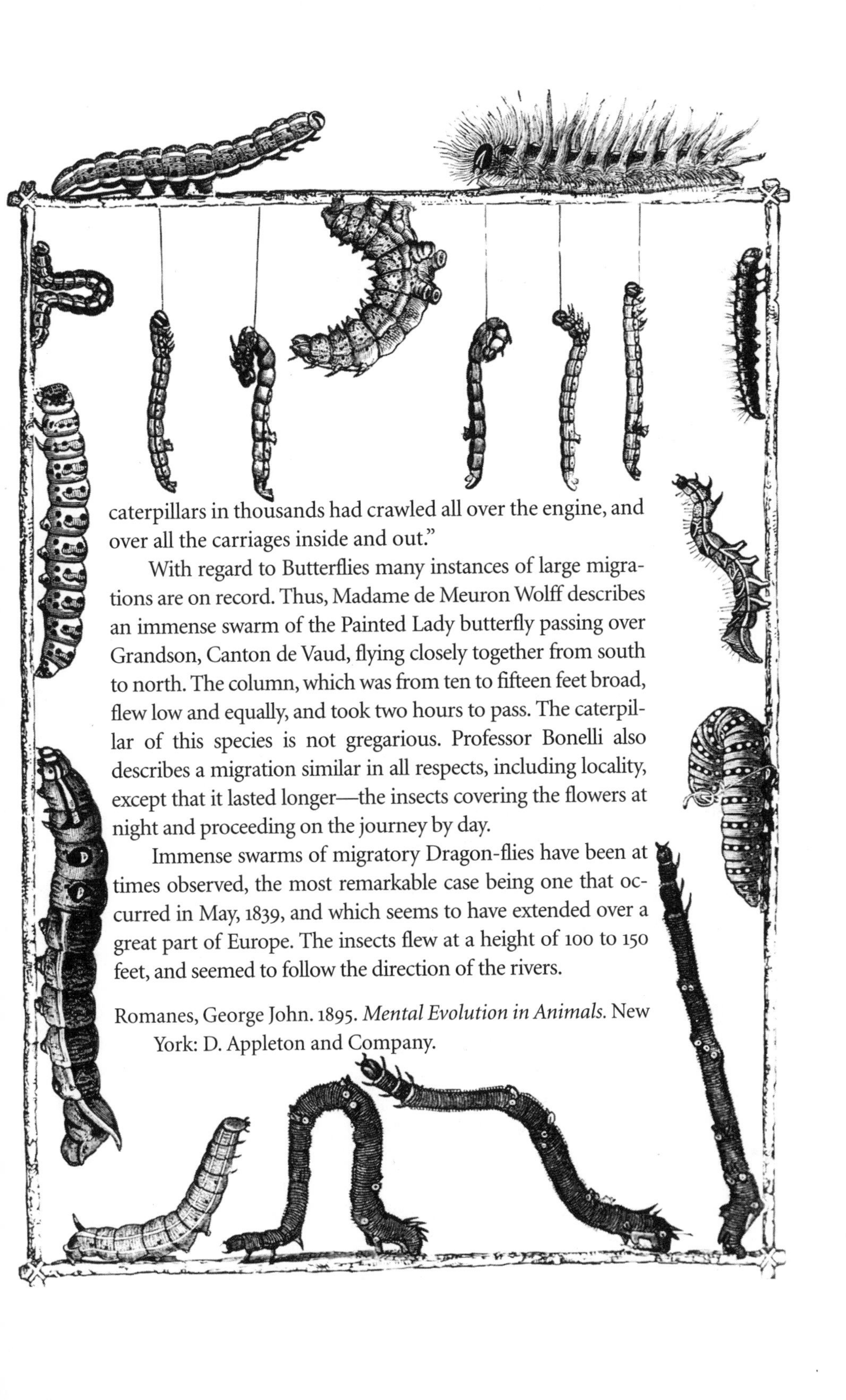

caterpillars in thousands had crawled all over the engine, and over all the carriages inside and out."

With regard to Butterflies many instances of large migrations are on record. Thus, Madame de Meuron Wolff describes an immense swarm of the Painted Lady butterfly passing over Grandson, Canton de Vaud, flying closely together from south to north. The column, which was from ten to fifteen feet broad, flew low and equally, and took two hours to pass. The caterpillar of this species is not gregarious. Professor Bonelli also describes a migration similar in all respects, including locality, except that it lasted longer—the insects covering the flowers at night and proceeding on the journey by day.

Immense swarms of migratory Dragon-flies have been at times observed, the most remarkable case being one that occurred in May, 1839, and which seems to have extended over a great part of Europe. The insects flew at a height of 100 to 150 feet, and seemed to follow the direction of the rivers.

Romanes, George John. 1895. *Mental Evolution in Animals.* New York: D. Appleton and Company.

Swarms of Flies

Harold Oldroyd

Harold Oldroyd (1913–1978) was for many years a scientist and specialist on flies working at the British Museum in London. His book The Natural History of Flies *achieves the formidable goal of making accessible to the lay reader the biology of a complex and diverse group of organisms (150,000 species and counting). Here he investigates the various reasons why flies swarm.*

Everybody talks about 'swarms' of flies, but it is only recently that serious attempts have been made to find out just what we mean when we use this expression. A swarm of bees is a corporate body, grouped round its queen, and maintaining its integrity as it moves or rests. Can we say the same about a swarm of midges or mosquitoes, about the flies that buzz round a grazing animal, or 'swarm' into the loft of a house? And what about the aggregations of larvae of flies, the Army worms, the leather-jackets and blow-fly maggots that come out of the soil on wet nights, or the clusters of Bibionid larvae in the soil?

This chapter should really have been headed 'aggregations' of flies but that is an ugly and arid word. Here German is more helpful than English, offering us *der Schwarm* for a true swarm, *der Hauf* for a mere massing of individuals, and *das Gewimmel* for a crowd in the sense of a milling throng. The last of the three seems best to fit most assemblies of flies, which we shall continue to refer to as swarms for simplicity.

The subject of aerial swarming of flies has been discussed by a number of recent authors. . . . Haddow and Corbet give an admirable summary of the whole topic, and end their paper with a quotation from Nielsen, warning us against fruitlessly speculating about the reasons why flies do certain things when we ought to be finding out much more about what it is that they actually do. I do not intend to try to guess why flies band together as adults or as larvae, but only to compare the manner in which the various assemblies come into being.

The common factor in all these assemblies is that they consist of a number of individuals doing the same thing at the same time. There is no sense of a community, in which various individuals do different things that total up to a community life. Aerial swarms take up a more or less fixed position in relation to a stationary object, and Downes says: 'It is clear . . . that no gregarious factor is necessarily involved in swarm formation, but only the individual's reactions to a common marker.' Other authors have emphasized that a 'swarm' may be as small as one individual. All through this book we have seen how nearly all families of flies have members that sometimes hover. The habit is more common in some families than in others, and generally more typical of males than of females. Hovering necessarily means staying in one place relative to the ground, and an aerial swarm comes into being when a number of flies hover in company.

Downes discusses at some length the use of a marker, a conspicuous fixed object, usually below the hovering flies, but sometimes above. The lesser house-fly, *Fannia canicularis*, seems to use an overhead marker when it circles below the electric light fitting indoors, returning again and again to the same spot in the air. Out of doors it behaves in the same way, and you will often see other small flies hovering under trees. McAlpine describes the flight of a species of *Dasiops*, an acalyptrate fly of the family Lonchaeidae, which 'dance and hover in loose swarms under mesquite trees, a few to up to a hundred, almost entirely males . . . the swarms were usually two to five feet above the ground, and within them the individuals moved slowly back and forth, and up and down, through shafts of sunlight that penetrated the leafy canopy. In this way their glistening bodies easily caught one's attention.' . . .

Downes was discussing only the more lightly built and fragile midges and gnats, most of which guided themselves by markers below rather than above. The general practice was to fly against the wind, varying the flying speed so that they alternately flew over the marker, and then allowed themselves to drift downwind of it again, much as seagulls do. According to Downes' observations, the flies vary the length of their 'run' at different heights in such a way that the apparent movement of the marker is always the same. This is only possible if the wind-speed lies between ½ mph and 9 mph. Above the upper limit the fly has too little margin of speed to make sudden accelerations, and so in a gust it is liable to lose touch with the marker; and below the lower limit the method fails because there is nothing to push against, as it were.

We can all observe this behaviour if we watch the summer swarms of non-biting midges. In the open, with a breeze of the right strength, the swarms form upon any convenient marker, even a person walking along, and will formate on him just as faithfully as on a fixed marker. If two people walk together and then separate, the swarm will often divide with them.

If the wind is too strong for them in the open, the swarms will form in the lee of objects, where the eddies provide them with just enough movement to balance themselves against. Most houses have such an eddy under the eaves, and on summer evenings, if you look out of the bedroom window you will probably see such a swarm. I shall mention these eddies again presently in connection with the small Chloropid flies that come into houses in autumn.

Swarms round high buildings such as church spires have often given rise to reports of fire. The 400-ft steeple of Salisbury Cathedral is particularly prone to this sort of false alarm.

Haddow and Corbet raised their high steel tower to 120 ft in the Mpanga Forest of Uganda, to see if they could observe the swarms they expected to find round the tops of the giant trees that emerge here and there from the general level of the forest canopy. They were rewarded for their efforts when the flies chose the tower itself as a marker, and provided the observers with a private display every night throughout the year except in stormy weather. They were concerned with swarming in the open over a marker, and contrast this with what they call 'free ground swarms at low level and not associated with a marker'. The authors do not discuss how the ground swarms keep their position, but these, like the solitary hoverers, are apparently riding the eddies, kept in station as much by the turbulence of the air as by their own efforts. They make clever use of the eddies round trees and bushes, as well as of the convection currents caused by uneven heating of the ground. This is one reason why small insects are often seen dancing in patches of sunlight, apart from the fact that the sunlight makes them conspicuous to us. . . .

Swarms over a marker necessarily stay in one place. Reports of migration or other 'purposive' movements are few. Hover-flies are among the most migratory of flies, and have been reported as moving steadily through the passes of the Pyrenees, as well as down the eastern coast of England. Occasionally they are taken on ships far out to sea. An isolated example of truly communal behaviour is that of the *Simulium* reported by Muirhead-Thompson . . . but the only flies known regularly to move *en masse* are, curiously enough, those that live on the beach, breeding in the wrack, or stranded seaweed.

Mass movements of seaweed flies a foot or two above the beach have been reported by many observers, and studied in detail by Egglishaw. There are two types of activity; a restricted mass flight just over or near the wrack, provoked by disturbance either by the rising tide or by someone walking on the weed; and a mass migration which takes place at low tide, when large numbers move in a body to new breeding-grounds. The stimuli to mass migration seem to be warmth and perhaps overcrowding, with a raised level of activity such as we see in locusts. This behaviour is characteristic of *Coelopa frigida* and *C. pilipes*, but Egglishaw also noted a mass migration of the small *Borborid Thoracochaeta zosterae*, when one sweep of his net caught 202 of them.

The direction of flight of seaweed flies seems to be generally along the coast, and generally downwind, but they are not very accurate navigators, and if they lose contact with the beach they may suddenly appear inland. Miss Dorothy Jackson records that large numbers of *Coelopa frigida* flew against the windows of her house on the Scottish coast near St. Andrews, on a mild night in January. They had risen 60 ft from sea level, and a similar swarm appeared in October round the lantern of a lighthouse in Co. Donegal, Eire.

What now of other aggregations of adult flies? Sometimes foliage is covered with what Séguy calls '*réunions sédentaires*', assemblages of adult flies, either just after emergence, or for the purpose of mating without a mating flight. Roper records such behaviour in *Scatopse picea*, and it has been often seen in the Bibionid *Dilophus febrilis*. Occasionally such swarms find their way indoors, as I have noted of the Sepsid *Themira putris*, which breeds in great numbers in sewage farms, and the Phoridae discussed by Colyer, but the notorious 'hibernating flies' that assemble in the lofts and roof-spaces of houses in autumn in temperate countries do not arrive in a mass; they assemble in the building as in a fly-trap.

Pollenia rudis, *Musca autumnalis* and *Dasyphora cyanella* are the usual loft-flies, though sometimes *Musca domestica*, *Calliphora* and even the drone-fly, *Eristalis tenax*, crowd into winter shelter. *Phormia terrae-novae* and *Limnophora humilis* also occasionally do so. *Pollenia rudis* is so confirmed in this habit that it is known as the cluster-fly. The first three species are all to be found throughout the summer in the fields, and the movement towards shelter seems to be initiated by a sudden drop in temperature on the ground at night. In England this occurs in the second half of August in an average year, and from then onwards the behaviour of the flies is characteristic. In the afternoon they settle on the upper parts of walls and on roofs, facing south or south-west, sunning themselves in the last of the warmth. As the sun sets they crawl into any crevice, and some roofs have so many that the flies simply move into the roof-space, or into crevices in the roof-lining. For a few days they come out in the daytime, and go back at night, but presently this ceases and they spend the winter inside.

Hibernation is very incomplete and easily broken by a mild spell in midwinter. In churches, where the heating often operates only at weekends, the flies may become active in a sluggish sort of way, and fall squirming on to the hair and hymnbooks of the worshippers below.

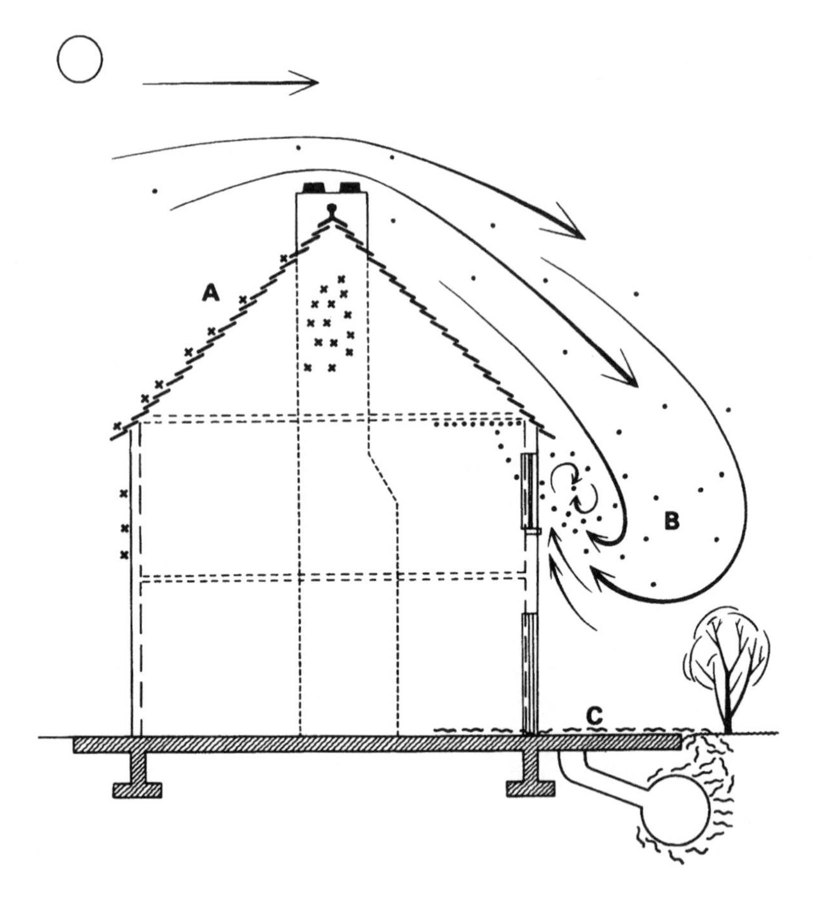

Three ways in which flies swarm into houses: (A) cluster-flies, among other fly species, settle on the warm side of the roof, crawl inside and cluster on the rafters and brickwork; (B) flies *(Thaumatomyia notata)* carried by the wind and trapped in eddies under the caves on the lee side, collect on the ceilings of upper rooms; (C) larvae of *Lucilia* flies emerge from the soil near a drain following a heavy storm and flooding (*The Natural History of Flies* by Harold Oldroyd, W. W. Norton & Co., New York, 1964).

A house or other building is therefore no more than a large flytrap. It is found that the same building is infested year after year, while the house next door may be immune. The 'choice' of a house depends on unalterable physical factors of situation and aspect, and probably on the shape of the roof, and at present there is no known remedy for these visitations except to move.

Oldroyd, Harold. 1964. *The Natural History of Flies.* New York: W. W. Norton & Co.

More Flies at Teatime

Vincent G. Dethier

Why do flies fly? In the course of trying to answer this question, insect behaviorist Vincent Dethier (1915-1993) found out something about the movement of people as well as of flies.

The charwoman who comes in to clean our laboratory at night is neither brighter nor duller than the average American, and she has a rather confused idea as to what goes on during the day. Since we who work there are obviously scientists, we must, in her mind, have something to do with rockets and space, and since at the same time we hold the title "Doctor," we devote ourselves to curing the bodily ills of mankind. The fact that we choose to experiment on flies does not bother the dear lady one whit. With blind faith she trusts our judgment implicitly in all things. As I said, she is a person of average intelligence. But recently she has come to look upon us with a rueful suspicion that we may be in league with the devil. Her disillusionment came about as a result of something the flies told me about her working habits.

To appreciate this situation you must know something about the events leading up to the revelation. By this time we knew a great deal about the fly, his sense of taste, his preferences, his hungers, his thirst, his satiation. Meditating one day on these facts I got to thinking that being hungry served no useful purpose unless hunger also stimulated one to search for food. A hungry fly should be more restless than a happily fed fly. The technical problem boiled down to this: how does one measure restlessness in a fly? There is always the method of direct observation, of course; but this is incredibly tedious and time consuming. Insects in general seem to spend enormous lengths of time waiting. I've often wondered what goes on in the microscopic brain during these periods. When not waiting, insects wander aimlessly. And yet, strangely enough, this aimless wandering is always an indirect way to something important. It is by aimless wandering that an insect efficiently covers vast quarters of space to find food, or a mate, or a place to lay eggs. People, by contrast, generally move directly from one place to another, usually to do something unimportant.

An efficient way to measure restlessness in a fly is to persuade it to operate some counting device, to operate a treadmill, or to use a pedometer. It should be feasible to do this since insects are notoriously strong. Fleas, grasshoppers, and springtails (those little snow fleas that spring mighty distances, usually into maple sirup buckets hanging on trees, by tucking their tails under their stomachs and slapping them against the ground) can jump distances equal to twenty or more times their length. The latest broadjump record for man, about twenty-seven feet, is only about four and one half times his own height. Cicada-killer wasps are able to fly with a captured cicada weighing easily five times their own weight.

This appearance of strength is an illusion, however, made possible in part by the small weight of the insect with reference to gravity. The absolute strength of an insect such as a fly is so small that the operation of most counting devices is a physical impossibility. Something very delicate had to be designed for our tests. The most delicate mechanical devices are balances, so we set about to design a balance that a fly could operate, one that would tell us how often he walked, jumped, or flew.

After many trials and many failures we finally succeeded in constructing a cylindrical cage of sheer nylon about one and one half inches long, mounted on a thin balsa wood frame, and balanced at the center on jeweled bearings from a watch. A needle-thin strip of brass ran the length of the floor of the cage. From it at each end hung a short piece of wire, wire so fine that even a Swiss watch-maker would not be able to drill a hole through it. Beneath each wire, mounted on the stand that balanced the cage, was a minute pot of mercury. Fortunately for us, flies do not run around a cage from top to bottom; instead they pace from end to end like proper zoo animals. Accordingly, the cage teeters like a seesaw. Everytime it tilted, one of the two wires dipped into its mercury pot closing an electrical circuit which activated a pen writing on a moving strip of paper. To give an aura of scholarship to the proceedings, someone tacked on the wall above the machine the familiar lines from the Rubáiyát:

> The Moving Finger writes; and, having writ,
> Moves on: nor all thy Piety nor Wit
> Shall lure it back to cancel half a Line,
> Nor all thy Tears wash out a Word of it.

The very first day's record revealed something we had known for a long time but also something unexpected and puzzling. Flies, like the majority of insects, are creatures of light. Light exerts such a profound effect on insects that moths have been flying into candles for centuries for the edification of poets and moralists, insects in a room congregate at windows to the dismay of housewives who strive to maintain a sterile house, and honeybees placed in a test tube with its open end away from light belie the intelligence ascribed to them by romanticists and never escape because of the irresistible attraction of light. Thus, it was hardly a surprise to find on examining the record of the seesaw each morning that it gave a faithful indication of the hour of sunrise and of sunset. When the light of the rising sun reached into the laboratory each morning, it stirred the laggard flies to activity. For the entire day each fly paced back and forth in his individual cage until at sunset the light retreated from the laboratory back somewhere to the world outside. Then activity ceased, and the recording instrument silently traced a smooth line.

This smooth line should have continued until the following morning. Instead, every night at eight o'clock the flies resumed activity for approximately thirty minutes. One night the burst occurred at nine. Then the answer to the mystery dawned on me. When the charwoman came in to clean, she turned on all of the lights and the flies became active until she left. Usually she came at eight; one night she had been an hour late. Confronted with the chiding accusation that she had been tardy one night, the puzzled woman readily admitted the fact, but because she knew that nobody had been in the building her curiosity demanded to know how I had known. I committed the very grave error of telling her the truth, or at least part of it. I told her that the flies had given her away. Perhaps I should have enlarged on this point, but she did not give me the opportunity. Crossing herself and muttering something about Beelzebub, Prince of flies, she retreated in haste and alarm.

The fact that flies are active in the daytime and inactive during the night did not particularly suit my purposes. After all, hunger does not automatically disappear at sundown and reappear in the morning (although the reverse is true in some human beings who suffer from a curious malady known as nocturnal eating). The overwhelming influence of light had to be removed. The simplest way to do this was to keep the flies either in constant darkness or in constant light. When we tried the first alternative, the flies still showed great activity during the day and rest at night *even though they were in total darkness!* The interesting fact is that flies which have been reared in total darkness for many generations and have never seen light still showed a diurnal rhythm. The rhythm, however, bears no relation to day and night. Some flies began to exercise at eight AM, some at eight PM, some at noon, some at midnight, etc., etc. But if they were all subjected to a flash of light at the same time in their dark world, they all synchronized their rhythms. They set their activity clocks. In nature it is the sunrise that naturally sets the clocks of all flies.

Flies do indeed have internal clocks. If you stop to think about if for a moment, you will realize that most animals have clocks. There is something inside of them that measures time. The clocks are not necessarily daily clocks; some are set to run on lunar schedules, some on seasonal schedules. The lunar schedules have captured the imagination of most people, perhaps because such interesting rhythms are based upon them. There is, for example, the Palolo worm that rises from the depths of the ocean at just the right phase of the moon and indulges in the most eye-popping orgy of reproduction. Then there is the reproductive cycle of humans, a twenty-eight day cycle, a lunar cycle in past evolution.

These cycles are controlled by some sort of biological clock. There is some mechanism within the animal that tells it when it is time to act. But fascinating as these cycles are, they interfere with, rather than help, our attempts to ascertain whether or not a hungry fly is a restless fly. Putting the flies in constant darkness failed. Since they are such slaves to light, constant light might keep them at such a high level of activity that differences due to internal rhythms would be swamped. And so it turned out. Under constant light we had more or less constant activity, a constant baseline which we could observe before and after feeding.

And then we found indeed that a hearty meal stopped activity. As a control experiment to show that the inactivity was truly a result of feeding and not merely a consequence of the added weight we burdened some hungry flies with little weights equaling a full meal. These only seemed to make the flies more active.

The behavior of the fly fits a widespread and age-old pattern. A hungry animal moves. An individual moves, a tribe moves, a race moves. They move until they find food or drop from starvation. But everyone knows this, you might argue. It is only common sense. True enough, but by showing that it occurs in a fly we establish that there is something biologically basic about the phenomenon and that the mechanism for doing it, whatever it is, is present in the small unintellectual body of a fly. Furthermore, until the fact was demonstrated experimentally it might have been an illusion.

I am reminded of an article that once appeared in a parody of a well-known scientific journal. The article had to do with the apparent increase of flies at tea-time. It went somewhat as follows: Either there are more flies at tea-time or there are not. If there are not, then there are no grounds for scientific investigation. If there are, it means either that the flies are flying faster and enter the visual field more often or are flying slower and so remain in the visual field longer. In either case we are dealing with an optical illusion.

We not only lost a charwoman, we destroyed an illusion.

Dethier, Vincent G. 1962. *To Know a Fly.* San Francisco: Holden-Day, Inc. (Reprinted with the permission of The McGraw-Hill Companies.)

Need Nectar, Will Travel

Stephen L. Buchmann and Gary Paul Nabhan

Through the winter of 1997–1998, millions of monarch butterflies died in the Sierra Madre mountains of Mexico, their corpses lying a foot deep in some places. Many die every year here, but this was more than the usual. The cause was a combination of unseasonal low temperatures and forest fires, exacerbated by logging that continues to encroach upon the traditional winter habitat of the monarch. In March 1998, Mexico's "Group of 100," led by Nobel laureate poet Octavio Paz and others, called for new federal protection for the area, and a promise to halt all logging before the butterflies disappear entirely.

Several years earlier, botanist Gary Nabhan and his entomologist colleague Steve Buchmann visited the Sierra Madres in the state of Michoacán to witness the return of the monarch, Danaus plexippus, *which migrates to its mass overwintering site from as far away as Canada and the United States, thousands of miles.*

A butterfly wind. We were enveloped in a butterfly wind, but we could still hear the incantations of warblers, orioles, and grosbeaks, as well as the distant echoes of marimbas. Steve and I had made a pilgrimage with friends to the Mexican wintering grounds of the monarch, one of the most active pollinators and nectar-feeders on milkweeds the world has ever known. We were astonished to find ourselves surrounded by some 12 million butterflies concentrated within just 3 acres. Many of them fluttered in the cool February air above, while others shuddered by the thousands on each nearby trunk of oyamel fir. These "sacred firs" formed a broken forest covering the slope of the Sierra El Campanario at 8,700 feet in elevation.

Not all of the monarchs around us could still fly, though, or even muster a shiver. One weakened monarch had slowly ambled around the herbs between my feet, then collapsed on the ground. It had arrived in Michoacán three months earlier, after up to a 2,000-mile flight from the north, and any fat reserves it might have had left were now nearly exhausted. The long migration and the cool winter had so depleted its strength that it would be unable to gain enough calories from floral nectar here to survive even another week. As many as 20,000 monarchs per acre may fall to the ground during the average overwintering season at the El Rosario sanctuary, prey for grosbeaks, orioles, mice, and ants.

I counted three dozen dead and dying butterflies within inches of our guide's feet, forming an orange, black, yellow, and beige butterfly mulch over the volcanic cobbles of the mountainside. An inchworm looped haltingly among this invertebrate carnage, its colors nearly the same as the now lifeless monarchs. Others remained among the living: one came to rest on Steve's head, while another attached itself to my jeans, where it shivered for awhile before returning to the quiescent masses clinging to the closest oyamel fir.

Right above us was another sight to behold. Flamboyant shingles of tens of thousands of monarchs swayed in the breeze, blending in with the boughs of the sacred fir as if they were fir needle clusters. I craned my neck back and looked straight up: there were thousands of other monarchs in each column of air that filled the spaces between the trees, for they were aloft as soon as the first beams of sunlight penetrated the forest, warming them and stimulating their flight to nearby nectar plants and pools of water to slake their collective thirst.

When I spoke with lepidopterist William Calvert, who had spent much of the winter attempting to census the monarch overwintering grounds of the Sierra Madre in Michoacán, he estimated that up to 30 million butterflies were concentrated in little more than 12 acres scattered over just five officially protected sanctuaries in the adjacent mountain ranges. Another of our traveling companions was Bob Pyle, founder of the Xerces Society for invertebrate conservation. He called it "the greatest aggregation of a living organism—and its deceased kin—anywhere in the world." An elderly Mexican peasant put it another way as he walked with me along the trail at El Rosario: "This is the monarch's sanctuary, but also their pantheon. . . ."

The old man was right. Tens of millions of monarchs survive the winter thanks to the peculiar microhabitat offered by Sierra El Campanario, but millions may die there as well. Over most winters, only 5 to 10 percent of the entire population is killed by the slicing beaks of orioles, the chomping bites of grosbeaks, the nibbles of some 40 other potential animal predators, or the chilling frosts of winter nights. But our local guide, Homero, told me tales of what happened to monarchs when blizzards hit the highest ranges in Michoacán. When one freezing snow arrived a few years ago, a third of all the monarchs were killed in one fell swoop.

Vulnerability to this kind of unpredictable event may seem at first glance to be par for the course—migratory butterflies must take their lumps with everyone else. The trouble is that migratory monarchs now aggregate and overwinter at so few sites because much of the formerly suitable habitat for them has been irrevocably altered. The monarchs are very finicky about forest canopy structure. In fact, they settle in only a few of the possible forest locations available to them in their wintering ranges in coastal California and the Transverse Neo-Volcanic Belt in central Mexico. Although these two regions have completely different sets of trees in their forests, both offer closed forest canopies providing a narrow range of direct solar radiation and reflected light, offering most monarchs a buffer from climatic extremes. The forest stands are also associated with winter nectar sources, enough standing water to serve thirsty monarchs, and cool moist air pools that keep them from further dehydration.

Intact forests of sacred oyamel fir now make up less than 2 percent of Mexico's total land area—and within that 2 percent, sizable forest patches with the necessary water, nectar-producing plants, and suitable temperatures cover an even smaller area. Many once favorable forest habitats have already been cleared for farming, or logged, or overgrazed by cattle, or otherwise degraded by bark beetles and dwarf mistletoe. (These last two are small-scale natural occurrences.) With only ten overwintering roosts totaling less than 100 acres suitable for monarch use in central Mexico—and only five of them legally protected—migrating butterflies have perhaps begun to aggregate in concentrations much higher than their historic sizes.

The same trend appears to be true for California's overwintering monarchs. More than 21 sites historically frequented by monarchs have been destroyed, and another 7 of the 15 remaining have been severely damaged by California land developers within the last few decades. Accordingly, Bob Pyle was successful in getting the World Conservation Union (IUCN) to decree that the monarch migrations are "threatened phenomena" even though the species itself, *Danaus plexippus*, is not globally endangered.

But there is another view of the large-scale monarch phenomenon Biologist Richard Vane-Wright has argued that the highly aggregated overwintering sites of monarchs in California and Mexico were in fact *caused* by the deforestation of 300 million acres of North America, a process that began in the centuries following Columbus. According to Vane-Wright, it was only *after* 1864, when deforestation was well advanced, that the monarchs began to expand their range and concentrate in just a few overwintering sites. The opening of forests into meadows and pastures, he says, actually increased the range and abundance of the milkweeds that monarch larvae and adults rely on for the alkaloids. (They sequester bitter-tasting poisonous milkweed alkaloids in their body tissues to protect them from predators—who get a dose of "one-trial learning" after eating one of these beauties and then vomiting it up.) Vane-Wright contends that "spectacular annual migration cycles, found only in the populations of North America monarchs, have evolved in their present form as a result of the cataclysmic ecological changes wrought by European colonists—[what he calls] the Columbus Hypothesis."

Whatever the case—whether the migratory phenomena of monarchs are on the increase or are "endangered"—an incredibly high proportion of all monarchs spend a third of the year gathered at just a few sites. Other migratory species face similar problems—most rely on a relatively small amount of habitat for a critical segment of their migratory cycle. If that habitat is vulnerable to development, or even to extreme fluctuations in weather, the migrants are like so many eggs in the proverbial basket.

But migrants such as monarchs are vulnerable in another sense as well because they have so far to go—so much territory to cross, so much time in which something may go wrong. When you consider the against-the-odds success of a long-distance migration lottery plus the need to pass through areas while their flowers are still providing nectar, it is amazing that migrating pollinators do not suffer more frequent catastrophic declines from natural disasters alone.

Buchmann, Stephen L., and Gary Paul Nabhan. 1996. *The Forgotten Pollinators.* Washington, D.C.: Island Press/Shearwater Books.

5

The Superorganism: Social Insects

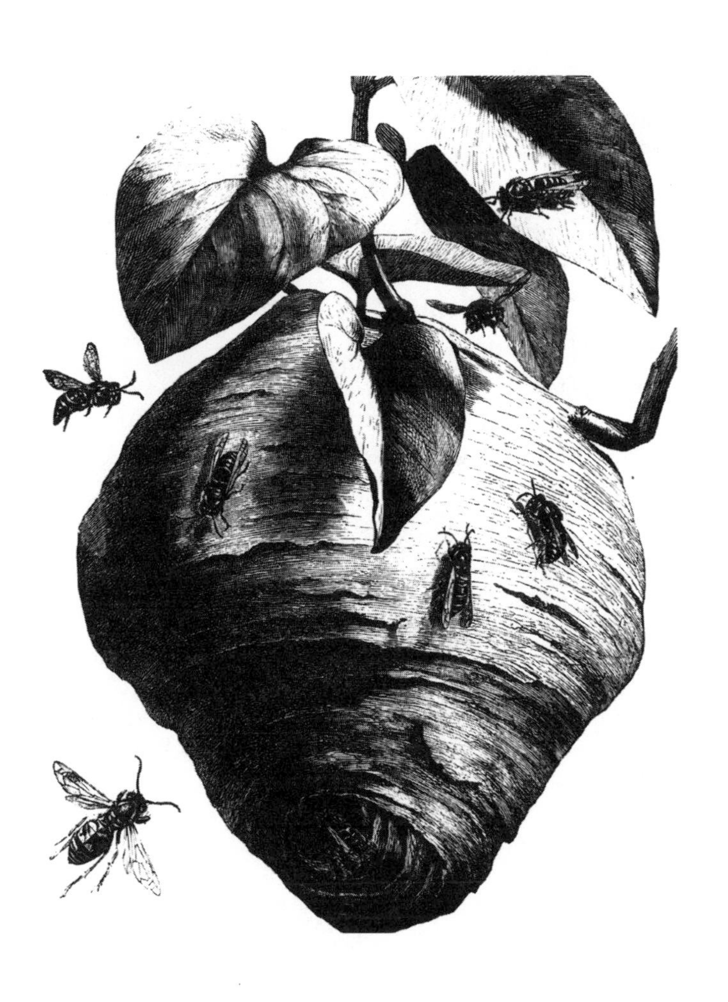

Since Biblical times, naturalists have marveled at the uncanny resemblance of the societies of termites, ants, bees, and wasps to those of humans. Organized around a queen (or, in the case of termites, a king and a queen), based on the principles of cooperation and self-sacrifice, capable of extraordinary architectural feats, and epitomizing efficiency and division of labor, social insect societies have been presumed to be the result of some sort of organizing "intelligence."

Darwin, worried about the meaning of a sterile (nonreproducing) worker caste for his theory of natural selection, was the first to suggest that the colony as a whole was analagous to a single organism in nonsocial animals. Starting in 1910, this concept was taken in a somewhat different direction by William Morton Wheeler, who considered the "superorganism" to be an "emergent property" of the complex interactions between members of a social insect colony. By the 1930s the "superorganism" (a.k.a. "supraorganism") had become common parlance (as an "analogy") in the new science of ecology, largely due to its adoption by the termite biologist and ecologist Alfred E. Emerson.

Today the concept is hardly used, and the colony as a whole is regarded as only one in a spectrum of units of selection. In fact, the social insect colony is no longer regarded as a harmonious, single entity but, rather, as a knotted tangle of competing interests, struggles for dominance, and uneasy compromises that only look like harmony when viewed from afar. This current view is by no means the final word. New information is coming in all the time. In addition to termites, ants, wasps, and bees, we now know about social beetles, aphids, and thrips, not to mention social mole rats and even social shrimp! As we learn more, our ideas about the evolutionary forces behind insect societies will themselves continue to evolve.

The Insect Societies

Edward O. Wilson

Who can doubt that this generation's driving force for the study of social insects is Harvard entomologist and sociobiologist E. O. Wilson (b. 1929)? And who can deny the seminal influence of his magisterial 1971 work, The Insect Societies? *Here, from that book, is Wilson's prescient ode to the social insects and where their study might lead us.*

Why do we study these insects? Because, together with man, hummingbirds, and the bristlecone pine, they are among the great achievements of organic evolution. Their social organization—far less than man's because of the feeble intellect and absence of culture, of course, but far greater in respect to cohesion, caste specialization, and individual altruism—is nonpareil. The biologist is invited to consider insect societies because they best exemplify the full sweep of ascending levels of organization, from molecule to society. Among the tens of thousands of species of wasps, ants, bees, and termites, we witness the employment of social design to solve ecological problems ordinarily dealt with by single organisms. The insect colony is often called a superorganism because it displays so many social phenomena that are analogous to the physiological properties of organs and tissues. Yet the holistic properties of the superorganism stem in a straightforward behavioral way from the relatively crude repertories of individual colony members, and they can be dissected and understood much more easily than the molecular basis of physiology.

A second reason for singling out social insects is their ecological dominance on the land. In most parts of the earth ants in particular are among the principal predators of other invertebrates. Their colonies, rooted and perennial like woody plants, send out foragers which comb the terrain day and night. Their biomass and energy consumption exceed those of vertebrates in most terrestrial habitats. Social insects are especially prominent in the tropics. In the seventeenth century Portuguese settlers called ants the "king of Brazil," and later travelers referred to them with such phrases as "the actual owners of the Amazon Valley" and "the real conquerors of

Brazil." Brazil, it was claimed, is "one great ants' nest." Similar impressions are invariably gained from other tropical countries. Ants in fact are so abundant that they replace earthworms as the chief earth movers in the tropics. Recent research has shown they are nearly as important as earthworms in cold temperate forests as well; in one locality in Massachusetts they bring 50 grams of soil to the surface per square yard each year and add one inch to the topsoil every 250 years. Termites are among the chief decomposers of dead wood and such cellulose detritus as leaf litter and humus in the tropics, and they, too, contribute significantly to the turning of the soil.

When considering ecology, it is useful to think of an insect colony as a diffuse organism, weighing anywhere from less than a gram to as much as a kilogram and possessing from about a hundred to a million or more tiny mouths. It is an animal that forages ameba-like over fixed territories a few square meters in extent. A colony of the common European ant *Tetramorium caespitum,* for example, contains an average of about 10,000 workers who weigh 6.5 g in the aggregate and control 40 m^2 of ground. The average colony of the American harvester ant *Pogonomyrmex badius* contains 5,000 workers who together weigh 40 g and patrol tens of square meters. The giant of all such "superorganisms" is a colony of the African driver ant *Anomma wilverthi,* which may contain as many as 22 million workers weighing a total of over 20 kg. During the statary phase of their cycle, columns of this species regularly patrol an area of between 40,000 and 50,000 m^2 in extent. When all of the resident ant populations are counted, the statistics are even more impressive. In Maryland, a single population of the mound-building ant *Formica exsectoides* comprised 73 nests covering an area of 10 acres and containing approximately 12 million workers. Since individual workers of this rather large species weigh 11.6 ± 0.13 mg, the total population weighed about 100 kg, and this was only one of many ant species inhabiting the same area, albeit the most abundant one. Termites have colonies of similar magnitudes, and in tropical habitats their populations approach densities comparable to those of ants. The savannas of Africa are dotted with great mounds of the fungus-growing macrotermitines, some 5 to 20 feet or more in height and containing 2 million workers. The mother of each colony is a grotesquely fattened queen weighing in excess of 10 g.

These superlatives can be made because of an adaptive radiation that took place for the most part between 50 and 100 million years ago in each of the major groups of social insects. In the social wasps, the ants, the social bees, and the termites, evolutionary convergence has resulted in the repeated appearance of the same basic design features: the systems of castes and labor roles changing according to age; the elaborate systems of chemical communication that typically include signals for alarm, recruitment, and recognition; the elaboration of nest structure to enhance temperature and humidity control; and others. One criterion of adaptive radiation that I use half seriously when thinking about evolution is: a group of species sharing common descent can be said to have truly radiated if one or more species is a specialized predator on the others. Ants have achieved this level with some distinction. Many of the army ants (subfamily Dorylinae) feed primarily on ants and other social insects, while all of the Cerapachyinae so far investigated feed entirely on ants. Among the bees, the meliponine *Lestrimelitta limao* specializes in robbing other meliponine species, while the large wasp *Vespa deusta* preys largely on colonies of the wasp genera *Ropalidia* and *Stenogaster.* Social parasitism, in which one species lives inside the nests of another and in some cases receives food and care unilaterally, occurs in all four of the major groups of social insects. It is tempting to speculate (and perhaps impossible to prove) that the social insects as a whole have employed all, or nearly all, of the social strategies permissible within the limits imposed by the arthropod brain and the peculiarities of their colonial system.

Wilson, Edward O. 1971. *The Insect Societies.* Cambridge, Mass.: The Belknap Press, Harvard University Press.

Section of an anthill (*Silver Wings and Golden Scales* by Anon., Cassell Petter & Galpin, London, 1889).

Morpho Eugenia

A. S. Byatt

Here, in a 1992 novella by British novelist A. S. Byatt entitled "Morpho Eugenia," from the book Angels & Insects, *we find the natural history of ants laid out as part of the Victorian fascination with collecting and social insects. Byatt (b. 1936) reads her entomology and has her favorite entomologists, and at least one of them, Chris O'Toole, chose her* Angels & Insects *as one of his ten favorite natural history classics in* BBC Wildlife *magazine.*

The penniless entomologist William Adamson, having lost his collections (and nearly his life) at sea, has landed at the estate of the Alabasters, where he agrees to sort out the old man's dilapidated natural history collections and give a few insect lessons to the children, while he writes his journals and gets ready to make his next collecting voyage. Of course, there is a beautiful woman involved too, Adamson's daughter Eugenia, which helps spur his interest. But he's seriously devoted to his science.

In this selection, Adamson sets about to build a formicary for the children. As he puts it together and the children come to see it, Byatt introduces a concept that she imagines the hero Adamson, as a far-sighted Victorian, might champion: the superorganism concept. Byatt, whose favorite insect passage is Maeterlinck's spirit of the hive (in The Life of the Bee*) for its unabashed anthropomorphism, weaves this idea into the narrative. The other interesting note is her anticipation of the cafeteria experiments that E. O. Wilson popularized in the 1950s. In these simple experiments, food is laid out for insects to determine what their favorite items are. Although Wilson is apparently the first to do this systematically and write about it, it is not at all far fetched to think that a clever and observant naturalist such as the fictional Adamson from the nineteenth century would have tried this.*

Matty Crompton reminded William of the promise she had extracted about the glass hive and the formicary. The glass hive was constructed under

William's direction, the width of the comb of honey, with an entrance hole for the bees cut in the nursery window, and black cloth curtains placed over its walls. The bees were procured from a tenant farmer and inserted, buzzing darkly, into their new home. For the ants, a large glass tank was carried from the nearest town, and set up on its own table on a green baize cloth. Matty Crompton said that she herself would accompany William in search of the ants themselves. She had observed trails of several sorts of ants in the elm coppice last Summer. They set out together with two buckets, various jars, boxes and test-tubes, a narrow trowel and several pairs of tweezers. She had a quick step, and was not given to conversation. She led William straight to what he immediately saw to be a very large Wood Ants' nest, the work of generation upon generation, backed up against an elm-stump, and thatched with a high dome of twigs, stalks and dry leaves. Little ragged chains of ants could be observed entering and leaving.

'I have attempted to keep these insects myself,' said Matty Crompton, 'but I have a deathly touch, it appears. No matter how beautiful a house I build, or how many flowers and fruits I offer, the creatures simply curl up and die.'

'You probably had not captured a Queen. Ants are social beings: they exist, it appears, only for the good of the whole nest, and the centre of the nest is the Queen ant whose laying and feeding the others all tend ceaselessly. They will kill her and drag her away, it is true, if she ceases to produce young—or abandon her, when she will rapidly starve, for she is unable to fend for herself. But they exist to lavish attention on her when she is in her prime, on her and her brood. If we are to make a mimic community, we must capture a Queen. The worker ants lose their will to live without the proximity of a Queen—they become immobile and listless, like young ladies in a decline, and then give up the ghost.'

'How shall we find a Queen? Must we break open the city? We shall do a great deal of damage . . .'

'I will look about and try to find a fairly recently established nest, a young community that can be transferred more or less entire.'

He paced up and down, turning over leaves with a stick, following small convoys of ants to their cracks and crannies in roots and earth. Matty Crompton stood watchfully by. She was wearing a brown stuff dress, severe and unornamental. Her dark hair was plaited around her head. She was good at keeping still. William felt a prick of pleasure at the return of his hunting, scanning self, which had been unexercised inside the walls of the Hall. Under his gaze the whole wood-floor became alive with movement, a centipede, various beetles, a sanguine shiny red worm, rabbit pellets, a tiny breast feather, a grass smeared with the eggs of some moth or butterfly, violets opening, conical entrance holes with fine dust inside, a swaying twig, a shifting pebble. He took out his magnifying lens and looked at a patch of moss, pebbles and sand, and saw a turmoil of previously invisible energies, striving, striving, white myriad-legged runners, invisible semi-transparent arthropods, button-tight spiderlings. His senses, and his mind attached to them, were like a magnetic field, pulled here and there. Here was a nest of Jet-black Ants, *Acanthomyops fuliginosus* who lived in small households inside the interconnected encampments of the Wood Ants. Here, on the edge of the coppice, was a trail of slave-making ants, *Formica sanguinea.* He had always wanted to study these in action. He said so to Matty Crompton, pointing out the difference between the Wood Ants, *Formica rufa,* with their muddy-brown heads and blackish-brown gasters, or hind parts, and the blood-red *sanguinea.*

'They invade the nests of the Wood Ants, and steal their cocoons, which they rear with their own, so that they become *sanguinea* workers. Terrible battles are fought by raiders and defenders.'

'They resemble human societies in that, as in many things.'

'The British slave-makers appear to be less dependent on their slaves than the Swiss *Formica rufescens* observed by Huber, who remarks that the workers of this species do no other work than capturing slaves, without whose labour their tribe would certainly become extinct, as all the child-rearing, and the food-gathering, are done by slaves. Mr Darwin observes that when these British Blood-red Ants migrate, they *carry* their slaves to the new home—but the more ferocious Swiss masters are so dependent, they require to be carried helplessly in the jaws of their slaves.'

'Maybe they are all perfectly content in their stations,' observed Matty Crompton. Her tone was neutral, so extraordinarily neutral that it would have been impossible to detect whether she spoke with irony or with conventional complacency, even if William had been giving her his complete attention, which he was not. He had found a meagre roof of thatch which he was ready to excavate. He took the trowel from her hands and removed several layers of earth, bristling with angry ant-warriors, littered with grubs and cocoons. A kind of seething attack accompanied his next moves, as he cut into the heart of the nest. Miss Crompton, on his instructions, gathered up the workers, grubs and cocoons in large clods of earth, interlayered with twigs and leaves.

'They bite,' she observed tersely, brushing her minute attackers from her wrists.

'They do. They make a hole with their mandibles and inject formic acid through their gaster, which they curve round, very elegantly. Do you wish to retreat?'

'No. I am a match for a few justifiably furious ants.'

So you could not say with the Fire Ants or the tucunderas in the forest, who made me suffer torments for weeks when I unwarily stirred them up. In Brazil the Fire Ant is King, they say, and rightly. It cannot be kept down, or diverted, or avoided—men leave their houses to escape its ravages.'

Matty Crompton, tightlipped, picked individual ants out of her cuffs and scattered them in the collecting boxes. William followed a tunnel, and came upon the brood-chamber of the ant Queen.

'Here she is. In her glory.'

Matty Crompton peered in.

'You would not suppose her to be of the same species as her rapid little servants—'

'No. Though she is less disproportionately gross than the termite Queens who are like huge inflated tubes, the size of haystacks compared to their docile little mates, who are in attendance in the same chamber, and the workers, who clamber all over them, cleaning and repairing and carrying away the endless succession of eggs as well as any debris.'

The Queen of the Wood Ants was only half as large again as her daughter-workers/servants. She was swollen and glossy, unlike the matt workers, and appeared to be striped red and white. The striping was in fact the result of the bloating of her body by the eggs inside it, which pushed apart her red-brown armour plating, showing more fragile, more elastic, whitish skin in the interstices. Her head appeared relatively small. William picked her up with his forceps—several workers came with her, clinging to her legs. He placed her on cottonwool in a collecting-box and directed Miss Crompton in the collection of various sizes of worker ants and grubs and cocoons from various parts of the nest.

'We should also take a sample of the earth and the vegetable matter, from which they have made their nest, and note what they appear to be eating—and the little girls may usefully experiment with their preference in foods, if they have patient natures, when they are in their new home.'

'Should we not search for male ants also?'

'There will be none, at this time of the year. They are only present in the nest in June, July and possibly August. They are born sometimes—it is thought—from eggs laid by unfertilised workers—a kind of parthenogenesis. They do not long survive the mating of the Queens in the Summer months. They are easy to recognise—they have wings and hugely developed eyes—and they do not *appear* to be in any way able to fend for themselves, or build, or forage. Natural Selection appears to have favoured in them the development of those skills which guarantee success in the nuptial dance, at the expense of the others—'

'I cannot help observing that this appears to be the *opposite* to human societies, when it is the woman whose success in that kind of performance determines their lives—'

'I have thought along those lines myself. There is a pleasing paradox in the bright balldresses, the *floating* of young girls in our world, and the dark erectness of the young men. In savage societies, as much as in birds and butterflies, it is the males who flaunt their beauty. But I do not know that the condition of the Queen here is much happier than that of the swarms of useless and disregarded suitors. I ask myself, are these little creatures, who run up and down, and carry, and feed each other lovingly, and bite enemies—are they truly individuals—or are they like the cells in our body, all parts of one whole, all directed by some mind—the Spirit of the Nest—which uses all, Queen, servants, slaves, dancing partners—for the good of the race itself, the species itself—'

'And do you go on, Mr Adamson, to ask *that* question about human societies?'

'It is tempting. I come from the North of England, where the scientific mill owners and the mine owners would like to make men into smoothly gliding parts of a giant machine. Dr Andrew Ure's *Philosophy of Manufacturers* wishes that workers could be trained to be co-operative—"to renounce their desultory habits of work, and to identify themselves with the unvarying regularity of the complex automaton." Robert Owen's experiments are the bright side of that way of thinking.'

'That is interesting, but it is not the same question,' said Miss Crompton. 'The will of the mill owners is not the Spirit of the Nest.'

William's brow furrowed as he thought this out. He said, 'It might be. If you were to suppose the mill owners in their machine-making to be equally in fact obeying the will of the Spirit of the Hive.'

'Ah,' said Matty Crompton with a kind of glee. 'I see where you are. A modern Calvinism by the back door, the nest door.'

'You think a great deal, Miss Crompton.'

'For a woman. You were about to add, "for a woman", and then refrained, which was courteous. It is my great amusement, thinking. I think as bees sun themselves, or ants stroke aphids. Do you not think we should provide an artificial ant-paradise with aphids, Mr Adamson?'

'Indeed we should. We should surround it with plants beloved of aphids if it can be contrived. If their presence can be tolerated in the schoolroom.'

The little girls gathered to observe the ants with mingled squeals of fascination and repulsion. The ants set about excavating and organising their new home with exemplary industry. Miss Mead, an elderly soft-faced person with thinning hair and sprouting hair pins, made little speeches to the little girls about the *kindness* of the ants, who laboured for the good of each other, who could be observed greeting passing sisters with little drinks of nectar from their stored supply, who caressed each other, and nursed their unborn sisters in the egg, or in their larval form, with loving care, moving them from dormitory to dormitory, cleaning and feeding with unselfish devotion. Margaret jabbed Edith in the side with a quick elbow and said, 'See, you are a little grub, you are just a little *grub.*'

'You are all three grubbier than you should be,' said Matty Crompton. 'You have spread the earth much further than necessary, well beyond your pinafores.'

Byatt, A. S. 1992. "Morpho Eugenia." *Angels & Insects.* London: Chatto & Windus.

The Spirit of the Hive

Maurice Maeterlinck

Maurice Maeterlinck (1862–1949) was a Nobel Prize–winning Belgian playwright and statesman who took a mystical interest in social insects. His first book on the subject, written in 1901, was the popular science bestseller of its day, The Life of the Bee, *from which this excerpt is taken. He later wrote books on termites (in 1926) and ants (1930).*

In 1920, Maeterlinck's bee story—in which he talks about the spirit of the hive, in effect, the superorganism—was passed around in synopsis form in Hollywood. When producer Sam Goldwyn read it, he is said to have remarked, "My God . . . the hero is a bee!" It was not just a bee, but a whole hive.

The queen . . . obeys, as meekly as the humblest of her subjects, the masked power, sovereignly wise, that for the present, and till we attempt to locate it, we will term the "spirit of the hive." But she is the unique organ of love; she is the mother of the city. She founded it amid uncertainty and poverty. She has peopled it with her own substance; and all who move within its walls—workers, males, larvae, nymphs, and the young princesses whose approaching birth will hasten her own departure, one of them being already designed as her successor by the "spirit of the hive"—all these have issued from her flanks.

What is this "spirit of the hive"—where does it reside? It is not like the special instinct that teaches the bird to construct its well planned nest, and then . . . seek other skies when the day for migration returns. Nor is it a kind of mechanical habit of the race, or blind craving for life, that will fling the bees upon any wild hazard the moment an unforeseen event shall derange the accustomed order of phenomena. On the contrary, be the event never so masterful, the "spirit of the hive" still will follow it, step by step, like an alert and quick-witted slave, who is able to derive advantage even from his master's most dangerous orders.

It disposes pitilessly of the wealth and the happiness, the liberty and life, of all this winged people; and yet with discretion, as though governed itself by some great duty. It regulates day by day the number of births, and contrives that these shall strictly accord with the number of flowers that brighten the country-side. It decrees the queen's deposition or warns her that she must depart; it compels her to bring her own rivals into the world, and rears them royally, protecting them from their mother's political hatred. So, too, in accordance with the generosity of the flowers, the age of the spring, and the probable dangers of the nuptial flight, will it permit or forbid the first-born of the virgin princesses to slay in their cradles her younger sisters, who are singing the song of the queens. At other times, when the season wanes, and flowery hours grow shorter, it will command the workers themselves to slaughter the whole imperial brood, that the era of revolutions may close, and work become the sole object of all. The "spirit of the hive" is prudent and thrifty, but by no means parsimonious. And thus, aware, it would seem, that nature's laws are somewhat wild and extravagant in all that pertains to love, it tolerates, during summer days of abundance, the embarrassing presence in the hive of three or four hundred males, from whose ranks the queen about to be born shall select her lover; three or four hundred foolish, clumsy, useless, noisy creatures, who are pretentious, gluttonous, dirty, coarse, totally and scandalously idle, insatiable, and enormous.

But after the queen's impregnation, when flowers begin to close sooner, and open later, the spirit one morning will coldly decree the simultaneous and general massacre of every male. It regulates the workers' labours, with due regard to their age; it allots their task to the nurses who tend the nymphs and the larvae, the ladies of honour who wait on the queen and never allow her out of their sight; the house-bees who air, refresh, or heat the hive by fanning their wings, and hasten the evaporation of the honey that may be too highly charged with water; the architects, masons, wax-workers, and sculptors who form the chain and construct the combs; the foragers who sally forth to the flowers in search of the nectar that turns into honey, of the pollen that feeds the nymphs and the larvae, the propolis that welds and strengthens the buildings of the city, or the water and salt required by the youth of the nation. Its orders have gone to the chemists who ensure the preservation of the honey by letting a drop of formic acid fall in from the end of their sting; to the capsule-makers who seal down the cells when the treasure is ripe, to the sweepers who maintain public places and streets most irreproachably clean, to the bearers whose duty it is to remove the corpses; and to the amazons of the guard who keep watch on the threshold by night and by day, question comers and goers, recognise the novices who return from their very first flight, scare away vagabonds, marauders and loiterers, expel all intruders, attack redoubtable foes in a body, and, if need be, barricade the entrance.

Finally, it is the spirit of the hive that fixes the hour of the great annual sacrifice to the genius of the race: the hour, that is, of the swarm; when we find a whole people, who have attained the topmost pinnacle of prosperity and power, suddenly abandoning to the generation to come their wealth and their palaces, their homes and the fruits of their labour; themselves content to encounter the hardships and perils of a new and distant country. This act, be it conscious or not, undoubtedly passes the limits of human morality. Its result will sometimes be ruin, but poverty always; and the thrice-happy city is scattered abroad in obedience to a law superior to its own happiness. Where has this law been decreed, which, as we soon shall find, is by no means as blind and inevitable as one might believe? Where, in what assembly, what council, what intellectual and moral sphere, does this spirit reside to whom all must submit, itself being vassal to an heroic duty, to an intelligence whose eyes are persistently fixed on the future?

It comes to pass with the bees as with most of the things in this world; we remark some few of their habits; we say they do this, they work in such and such fashion, their queens are born thus, their workers are virgin, they swarm at a certain time. And then we imagine we know them, and ask nothing more. We watch them hasten from flower to flower, we see the constant agitation within the hive; their life seems very simple to us, and bounded, like every life, by the instinctive cares of reproduction and nourishment. But let the eye draw near, and endeavour to see; and at once the least phenomenon of all becomes overpoweringly complex; we are confronted by the enigma of intellect, of destiny, will, aim, means, causes; the incomprehensible organisation of the most insignificant act of life.

Maeterlinck, Maurice. 1901. *The Life of the Bee.* New York: Dodd, Mead and Co. (Translation by Alfred Sutro.)

The Termite Queen in Her Egg Chamber

Gary Larson

The termite queen in her egg chamber.

Hive Mind

Kevin Kelly

Kevin Kelly (b. 1952) is executive editor of Wired *magazine, former publisher and editor of* Whole Earth Review, *and a longtime beekeeper. He is also a synthesizer of ideas. His book* Out of Control *plumbs biology, systems theory, chaos theory, information science, economics, cyberphilosophy, and religion in search of an all-encompassing theory for reinventing human civilization. Here he expresses a late-twentieth-century version of Maeterlinck's "spirit of the hive," connecting social insect colonies, artificial intelligence, and a computer-age version of "emergent properties."*

The beehive beneath my office window quietly exhales legions of busybodies and then inhales them. On summer afternoons, when the sun seeps under the trees to backlight the hive, the approaching sunlit bees zoom into their tiny dark opening like curving tracer bullets. I watch them now as they haul in the last gleanings of nectar from the final manzanita blooms of the year. Soon the rains will come and the bees will hide. I will still gaze out the window as I write; they will still toil, but now in their dark home. Only on the balmiest day will I be blessed by the sight of their thousands in the sun.

Over years of beekeeping, I've tried my hand at relocating bee colonies out of buildings and trees as a quick and cheap way of starting new hives at home. One fall I gutted a bee tree that a neighbor felled. I took a chain saw and ripped into this toppled old tupelo. The poor tree was cancerous with bee comb. The further I cut into the belly of the tree, the more bees I found. The insects filled a cavity as large as I was. It was a gray, cool autumn day and all the bees were home, now agitated by the surgery. I finally plunged my hand into the mess of comb. Hot! Ninety-five degrees at least. Overcrowded with 100,000 cold-blooded bees, the hive had become a warm-blooded organism. The heated honey ran like thin, warm blood. My gut felt like I had reached my hand into a dying animal.

The idea of the collective hive as an animal was an idea late in coming. The Greeks and Romans were famous beekeepers who harvested respectable yields of honey from homemade hives, yet these ancients got almost every fact about bees wrong. Blame it on the lightless conspiracy of bee life, a secret guarded by ten thousand fanatically loyal, armed soldiers. Democritus thought bees spawned from the same source as maggots. Xenophon figured out the queen bee but erroneously assigned her supervisory responsibilities she doesn't have. Aristotle gets good marks for getting a lot right, including the semiaccurate observation that "ruler bees" put larva in the honeycomb cells. (They actually start out as eggs, but at least he corrects Democritus's misguided direction of maggot origins.) Not until the Renaissance was the female gender of the queen bee proved, or beeswax shown to be secreted from the undersides of bees. No one had a clue until modern genetics that a hive is a radical matriarchy and sisterhood: all bees, except the few good-for-nothing drones, are female and sisters. The hive was a mystery as unfathomable as an eclipse.

I've seen eclipses and I've seen bee swarms. Eclipses are spectacles I watch halfheartedly, mostly out of duty, I think, to their rarity and tradition, much as I might attend a Fourth of July parade. Bee swarms, on the other hand, evoke another sort of awe. I've seen more than a few hives throwing off a swarm, and never has one failed to transfix me utterly, or to dumbfound everyone else within sight of it.

A hive about to swarm is a hive possessed. It becomes visibly agitated around the mouth of its entrance. The colony whines in a centerless loud drone that vibrates the neighborhood. It begins to spit out masses of bees, as if it were emptying not only its guts but its soul. A poltergeist-like storm of tiny wills materializes over the hive box. It grows to be a small dark cloud of purpose, opaque with life. Boosted by a tremendous buzzing racket, the ghost slowly rises into the sky, leaving behind the empty box and quiet bafflement. The German theosophist Rudolf Steiner writes lucidly in his otherwise kooky *Nine Lectures on Bees:* "Just as the human soul takes leave of the body . . . one can truly see in the flying swarm an image of the departing human soul."

For many years Mark Thompson, a beekeeper local to my area, had the bizarre urge to build a Live-In Hive—an active bee home you could visit by inserting your head into it. He was working in a yard once when a beehive spewed a swarm of bees "like a flow of black lava, dissolving, then taking wing." The black cloud coalesced into a 20-foot-round black halo of 30,000 bees that hovered, UFO-like, six feet off the ground, exactly at eye level. The flickering insect halo began to drift slowly away, keeping a constant six feet above the earth. It was a Live-In Hive dream come true.

Mark didn't waver. Dropping his tools he slipped into the swarm, his bare head now in the eye of a bee hurricane. He trotted in sync across the yard as the swarm eased away. Wearing a bee halo, Mark hopped over one fence, then another. He was now running to keep up with the thundering animal in whose belly his head floated. They all crossed the road and hurried down an open field, and then he jumped another fence. He was tiring. The bees weren't; they picked up speed. The swarm-bearing man glided down a hill into a marsh. The two of them now resembled a superstitious swamp devil, humming, hovering, and plowing through the miasma. Mark churned wildly through the muck trying to keep up. Then, on some signal, the bees accelerated. They unhaloed Mark and left him standing there wet, "in panting, joyful amazement." Maintaining an eye-level altitude, the swarm floated across the landscape until it vanished, like a spirit unleashed, into a somber pine woods across the highway.

"Where is 'this spirit of the hive' . . . where does it reside?" asks the author Maurice Maeterlinck as early as 1901. "What is it that governs here, that issues orders, foresees the future . . . ?" We are certain now it is not the queen bee. When a swarm pours itself out through the front slot of the hive, the queen bee can only follow. The queen's daughters manage the election of where and when the swarm should settle. A half-dozen anonymous workers scout ahead to check possible hive locations in hollow trees or wall cavities. They report back to the resting swarm by dancing on its contracting surface. During the report, the more theatrically a scout dances, the better the site she is championing. Deputy bees then check out the competing sites according to the intensity of the dances, and will concur with the scout by joining in the scout's twirling. That induces more followers to check out the lead prospects and join the ruckus when they return by leaping into the performance of their choice.

It's a rare bee, except for the scouts, who has inspected more than one site. The bees see a message, "Go there, it's a nice place." They go and return to dance/say, "Yeah, it's *really* nice." By compounding emphasis, the favorite sites get more visitors, thus increasing further visitors. As per the law of increasing returns, them that has get more votes, the have-nots get less. Gradually, one large, snowballing finale will dominate the dance-off. The biggest crowd wins.

It's an election hall of idiots, for idiots, and by idiots, and it works marvelously. This is the true nature of democracy and of all distributed governance. At the close of the curtain, by the choice of the citizens, the swarm takes the queen and thunders off in the direction indicated by mob vote. The queen who follows, does so humbly. If she could think, she would remember that she is but a mere peasant girl, blood sister of the very nurse bee instructed (by whom?) to select her larva, an ordinary larva, and raise it on a diet of royal jelly, transforming Cinderella into the queen. By what karma is the larva for a princess chosen? And who chooses the chooser?

"The hive chooses," is the disarming answer of William Morton Wheeler, a natural philosopher and entomologist of the old school, who founded the field of social insects. Writing in a bombshell of an essay in 1911, Wheeler claimed that an insect colony was not merely the analog of an organism, it is indeed an organism, in every important and scientific sense of the word. He wrote: "Like a cell or the person, it behaves as a unitary whole, maintaining its identity in space, resisting dissolution . . . neither a thing nor a concept, but a continual flux or process."

It was a mob of 20,000 united into oneness.

In a darkened Las Vegas conference room, a cheering audience waves cardboard wands in the air. Each wand is red on one side, green on the other. Far in back of the huge auditorium, a camera scans the frantic attendees. The video camera links the color spots of the wands to a nest of computers set up by graphics wizard Loren Carpenter. Carpenter's custom software locates each red and each green wand in the auditorium. Tonight there are just shy of 5,000 wandwavers. The computer displays the precise location of each wand (and its color) onto an immense, detailed video map of the auditorium hung on the front stage, which all can see. More importantly, the computer counts the total red or green wands and uses that value to control software. As the audience wave the wands, the display screen shows a sea of lights dancing crazily in the dark, like a candlelight parade gone punk. The viewers see themselves on the map; they are either a red or green pixel. By flipping their own wands, they can change the color of their projected pixels instantly.

Loren Carpenter boots up the ancient video game of Pong onto the immense screen. Pong was the first commercial video game to reach pop consciousness. It's a minimalist arrangement: a white dot bounces inside a square; two movable rectangles on each side act as virtual paddles. In short, electronic ping-pong. In this version, displaying the red side of your wand moves the paddle up. Green moves it down. More precisely, the Pong paddle moves as the average number of red wands in the auditorium increases or decreases. Your wand is just one vote.

Carpenter doesn't need to explain very much. Every attendee at this 1991 conference of computer graphic experts was probably once hooked on Pong. His amplified voice booms in the hall, "Okay guys. Folks on the left side of the auditorium control the left paddle. Folks on the right side control the right paddle. If you think you are on the left, then you really are. Okay? Go!"

The audience roars in delight. Without a moment's hesitation, 5,000 people are playing a reasonably good game of Pong. Each move of the paddle is the average of several thousand players' intentions. The sensation is unnerving. The paddle usually does what you intend, but not always. When it doesn't, you find yourself spending as much attention trying to anticipate the paddle as the incoming ball. One is definitely aware of another intelligence online: it's this hollering mob.

The group mind plays Pong so well that Carpenter decides to up the ante. Without warning the ball bounces faster. The participants squeal in unison. In a second or two, the mob has adjusted to the quicker pace and is playing better than before. Carpenter speeds up the game further; the mob learns instantly.

"Let's try something else," Carpenter suggests. A map of seats in the auditorium appears on the screen. He draws a wide circle in white around the center. "Can you make a green '5' in the circle?" he asks the audience. The audience stares at the rows of red pixels. The game is similar to that of holding a placard up in a stadium to make a picture, but now there are no preset orders, just a virtual mirror. Almost immediately wiggles of green pixels appear and grow haphazardly, as those who think their seat is in the path of the "5" flip their wands to green. A vague figure is materializing. The audience collectively begins to discern a "5" in the noise. Once discerned, the "5" quickly precipitates out into stark clarity. The wand-wavers on the fuzzy edge of the figure decide what side they "should" be on, and the emerging "5" sharpens up. The number assembles itself.

"Now make a four!" the voice booms. Within moments a "4" emerges. "Three." And in a blink a "3" appears. Then in rapid succession, "Two . . . One . . . Zero." The emergent thing is on a roll.

Loren Carpenter launches an airplane flight simulator on the screen. His instructions are terse: "You guys on the left are controlling roll; you on the right, pitch. If you point the plane at anything interesting, I'll fire a rocket at it." The plane is airborne. The pilot is . . . 5,000 novices. For once the auditorium is completely silent. Everyone studies the navigation instruments as the scene outside the windshield sinks in. The plane is headed for a landing in a pink valley among pink hills. The runway looks very tiny.

There is something both delicious and ludicrous about the notion of having the passengers of a plane collectively fly it. The brute democratic sense of it all is very appealing. As a passenger you get to vote for everything; not only where the group is headed, but when to trim the flaps.

But group mind seems to be a liability in the decisive moments of touchdown, where there is no room for averages. As the 5,000 conference participants begin to take down their plane for landing, the hush in the hall is ended by abrupt shouts and urgent commands. The auditorium becomes a gigantic cockpit in crisis. "Green, green, green!" one faction shouts. "More red!" a moment later from the crowd. "Red, red! REEEEED!" The plane is pitching to the left in a sickening way. It is obvious that it will miss the landing strip and arrive wing first. Unlike Pong, the flight simulator entails long delays in feedback from lever to effect, from the moment you tap the aileron to the moment it banks. The latent signals confuse the group mind. It is caught in oscillations of overcompensation. The plane is lurching wildly. Yet the mob somehow aborts the landing and pulls the plane up sensibly. They turn the plane around to try again.

How did they turn around? Nobody decided whether to turn left, or right, or even to turn at all. Nobody was in charge. But as if of one mind, the plane banks and turns wide. It tries landing again. Again it approaches cockeyed. The mob decides in unison, without lateral communication, like a flock of birds taking off, to pull up once more. On the way up the plane rolls a bit. And then rolls a bit more. At some magical moment, the same strong thought simultaneously infects five thousand minds: "I wonder if we can do a 360?"

Without speaking a word, the collective keeps tilting the plane. There's no undoing it. As the horizon spins dizzily, 5,000 amateur pilots roll a jet on their first solo flight. It was actually quite graceful. They give themselves a standing ovation.

The conferees did what birds do: they flocked. But they flocked self-consciously. They responded to an overview of themselves as they co-formed a "5" or steered the jet. A bird on the fly, however, has no overarching concept of the shape of its flock. "Flockness" emerges from creatures completely oblivious of their collective shape, size, or alignment. A flocking bird is blind to the grace and cohesiveness of a flock in flight.

At dawn, on a weedy Michigan lake, ten thousand mallards fidget. In the soft pink glow of morning, the ducks jabber, shake out their wings, and dunk for breakfast. Ducks are spread everywhere. Suddenly, cued by some imperceptible signal, a thousand birds rise as one thing. They lift themselves into the air in a great thunder. As they take off they pull up a thousand more birds from the surface of the lake with them, as if they were all but part of a reclining giant now rising. The monstrous beast hovers in the air, swerves to the east sun, and then, in a blink, reverses direction, turning itself inside out. A second later, the entire swarm veers west and away, as if steered by a single mind. In the 17th century, an anonymous poet wrote: ". . . and the thousands of fishes moved as a huge beast, piercing the water. They appeared united, inexorably bound to a common fate. How comes this unity?"

A flock is not a big bird. Writes the science reporter James Gleick, "Nothing in the motion of an individual bird or fish, no matter how fluid, can prepare us for the sight of a skyful of starlings pivoting over a cornfield, or a million minnows snapping into a tight, polarized array. . . . High-speed film [of flocks turning to avoid predators] reveals that the turning motion travels through the flock as a wave, passing from bird to bird in the space of about one-seventieth of a second. That is far less than the bird's reaction time." The flock is more than the sum of the birds.

In the film *Batman Returns* a horde of large black bats swarmed through flooded tunnels into downtown Gotham. The bats were computer generated. A single bat was created and given leeway to automatically flap its wings. The one bat was copied by the dozens until the animators had a mob. Then each bat was instructed to move about on its own on the screen following only a few simple rules encoded into an algorithm: don't bump into another bat, keep up with your neighbors, and don't stray too far away. When the algorithmic bats were run, they flocked like real bats.

The flocking rules were discovered by Craig Reynolds. a computer scientist working at Symbolics, a graphics hardware manufacturer. By tuning the various forces in his simple equation—a little more cohesion, a little less lag time—Reynolds could shape the flock to behave like living bats, sparrows, or fish. Even the marching mob of penguins in *Batman Returns* were flocked by Reynolds's algorithms. Like the bats, the computer-modeled 3-D penguins were cloned en masse and then set loose into the scene aimed in a certain direction. Their crowdlike jostling as they marched down the snowy street simply emerged, out of anyone's control.

So realistic is the flocking of Reynolds's simple algorithms that biologists have gone back to their hi-speed films and concluded that the flocking behavior of real birds and fish must emerge from a similar set of simple rules. A flock was once thought to be a decisive sign of life, some noble formation only life could achieve. Via Reynolds's algorithm it is now seen as an adaptive trick suitable for any distributed vivisystem, organic or made.

Wheeler, the ant pioneer, started calling the bustling cooperation of an insect colony a "superorganism" to clearly distinguish it from the metaphorical use of "organism." He was influenced by a philosophical strain at the turn of the century that saw holistic patterns overlaying the individual behavior of smaller parts. The enterprise of science was on its first steps of a headlong rush into the minute details of physics, biology, and all natural sciences. This pell-mell to reduce wholes to their constituents, seen as the most pragmatic path to understanding the wholes, would continue for the rest of the century and is still the dominant mode of scientific inquiry. Wheeler and colleagues were an essential part of this reductionist perspective, as the 50 Wheeler monographs on specific esoteric ant behaviors testify. But at the same time, Wheeler saw "emergent properties" within the superorganism superseding the resident properties of the collective ants.

Wheeler said the superorganism of the hive "emerges" from the mass of ordinary insect organisms. And he meant emergence as science—a technical, rational explanation—not mysticism or alchemy.

Wheeler held that this view of emergence was a way to reconcile the reduce-it-to-its-parts approach with the see-it-as-a-whole approach. The duality of body/mind or whole/part simply evaporated when holistic behavior lawfully emerged from the limited behaviors of the parts. The specifics of how superstuff emerged from baser parts was very vague in everyone's mind. And still is.

What was clear to Wheeler's group was that emergence was a common natural phenomena. It was related to the ordinary kind of causation in everyday life, the kind where *A* causes *B* which causes *C*, or 2 + 2 = 4. Ordinary causality was invoked by chemists to cover the observation that sulfur atoms plus iron atoms equal iron sulfide molecules. According to fellow philosopher C. Lloyd Morgan, the concept of emergence signaled a different variety of causation. Here 2 + 2 does not equal 4; it does not even surprise with 5. In the logic of emergence, 2 + 2 = apples. "The emergent step, though it may seem more or less saltatory [a leap], is best regarded as a qualitative change of direction, or critical turning-point, in the course of events," writes Morgan in *Emergent Evolution*, a bold book in 1923. Morgan goes on to quote a verse of Browning poetry which confirms how music emerges from chords:

> And I know not if, save in this, such gift be
> allowed to man
> That out of three sounds he frame, not a
> fourth sound, but a star.

We would argue now that it is the complexity of our brains that extracts music from notes, since we presume oak trees can't hear Bach. Yet "Bachness"—all that invades us when we hear Bach—is an appropriately poetic image of how a meaningful pattern emerges from musical notes and generic information.

The organization of a tiny honeybee yields a pattern for its tinier one-tenth of a gram of wing cells, tissue, and chitin. The organism of a hive yields integration for its community of worker bees, drones, pollen and brood. The whole 50-pound hive organ emerges with its own identity from the tiny bee parts. The hive possesses much that none of its parts possesses. One speck of a honeybee brain operates with a memory of six days; the hive as a whole operates with a memory of three months, twice as long as the average bee lives.

Ants, too, have hive mind. A colony of ants on the move from one nest site to another exhibits the Kafkaesque underside of emergent control. As hordes of ants break camp and head west, hauling eggs, larva, pupae—the crown jewels—in their beaks, other ants of the same colony, patriotic workers, are hauling the trove east again just as fast, while still other workers, perhaps acknowledging conflicting messages, are running one direction and back again completely empty-handed. A typical day at the office. Yet, the ant colony moves. Without any visible decision making at a higher level, it chooses a new nest site, signals workers to begin building, and governs itself.

The marvel of "hive mind" is that no one is in control, and yet an invisible hand governs, a hand that emerges from very dumb members. The marvel is that more is different. To generate a colony organism from a bug organism requires only that the bugs be multiplied so that there are many, many more of them, and that they communicate with each other. At some stage the level of complexity reaches a point where new categories like "colony" can emerge from simple categories of "bug." Colony is inherent in bugness, implies this marvel. Thus, there is nothing to be found in a beehive that is not submerged in a bee. And yet you can search a bee forever with cyclotron and fluoroscope, and you will never find the hive.

This is a universal law of vivisystems: higher-level complexities cannot be inferred by lower-level existences. Nothing—no computer or mind, no means of mathematics, physics, or philosophy—can unravel the emergent pattern dissolved in the parts without actually playing it out. Only playing out a hive will tell you if a colony is immixed in a bee. The theorists put it this way: running a system is the quickest, shortest, and only sure method to discern emergent structures latent in it. There are no shortcuts to actually "expressing" a convoluted, nonlinear equation to discover what it does. Too much of its behavior is packed away.

That leads us to wonder what else is packed into the bee that we haven't seen yet? Or what else is packed into the hive that has not yet appeared because there haven't been enough honeybee hives in a row all at once? And for that matter, what is contained in a human that will not emerge until we are all interconnected by wires and politics? The most unexpected things will brew in this bionic hivelike supermind.

Kelly, Kevin. 1994. *Out of Control: The Rise of Neo-biological Civilization.* Reading, Mass.: Addison-Wesley.

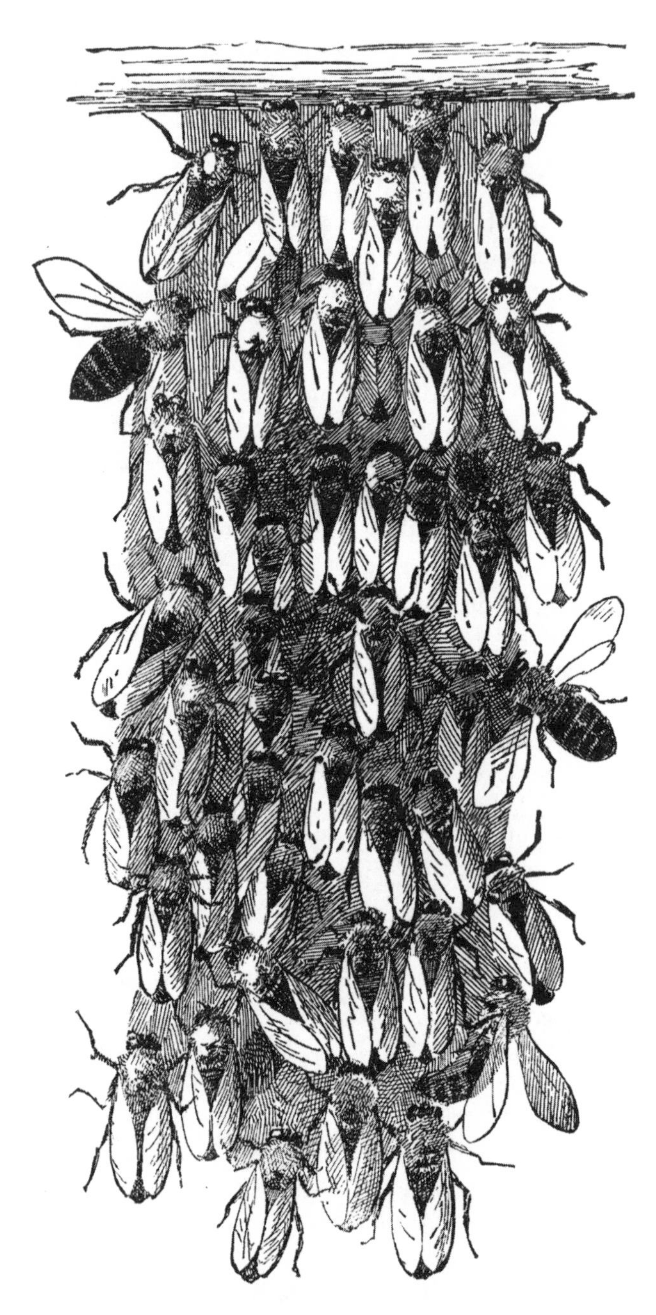

The honey bee, *Apis mellifera:* a curtain of wax-workers (*Tenants of an Old Farm* by Henry C. McCook, George W. Jacobs & Co., Philadelphia, 1884).

6

Insect Architecture

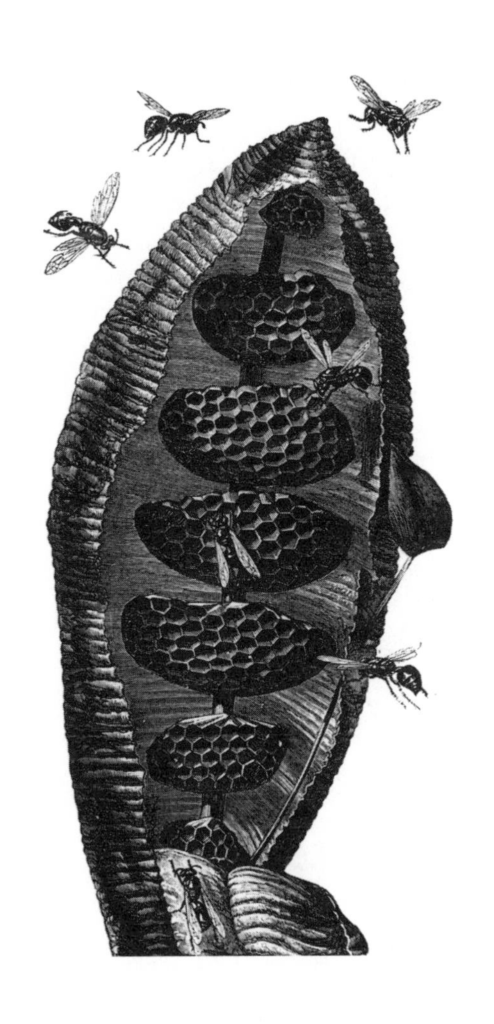

Insects are builders. The extent and variety of their handiwork is astonishing. Some insects construct and inhabit their own "shells" from sand or pieces of gravel, which in some cases resemble the shells of snails right down to the whorls. Others harness a huge workforce to build edifices on a human scale, such as the towering termite mounds of Africa. Tent caterpillars live together with their brothers and sisters in woven structures in trees and bushes. Solitary wasps and bees construct elaborate nests for their young, sometimes tunneled out of the soil, other times carefully crafted from home-made cement. Paper wasps literally make paper from wood fibers laboriously scraped from trees and fence posts and mixed with saliva; they use the paper to construct fabulously ornate nests. Some ants build earthen homes complete with vast networks of underground tunnels and simple yet effective dehumidification and air-conditioning systems; others inhabit whole trees, dispersed throughout hollowed-out thorns, stems, and branches. Still others live in cacao pods, or build homes woven from the silk of their own larvae. Army ants build temporary bivouacs of their own bodies, comprised of worker ants with their legs hooked together. And other insects construct traps, snares, pits, and hiding places in order to catch their prey.

Insects have been building, evolving, perfecting their structures for millions of years—long before birds and mammals began to build. Although humans have taken architecture to new heights and depths, we still have much to learn from the simple elegance of insect architecture.

Some Accounts of the Termites

Henry Smeathman

Henry Smeathman wrote a lengthy account to botanist Sir Joseph Banks, then the president of the Royal Society, to tell the world about the termites, or so-called white ants, he had witnessed in his travels through tropical west Africa. The letter was read to the Royal Society and published in 1781. In this excerpt, we see how Smeathman's eye for detail and careful description informed the scientific world about the incredible creations of the termites. He was also the first to announce that in termite society there was not only a queen, as in ant societies, but a king who lived side by side with the queen. This fact was doubted for nearly a hundred years after Smeathman's report until Fritz Müller confirmed it in a letter to Nature *in 1874.*

Clement's Inn,
Sir, Jan. 23, 1781.

Of a great many curious parts of the creation I met with on my travels in that almost unknown district of Africa called Guinea, the Termites, which by most travellers have been called White Ants, seemed to me on many accounts most worthy of that exact and minute attention which I have bestowed upon them. . . .

The size and figure of their buildings have attracted the notice of many travellers, and yet the world has not hitherto been furnished with a tolerable description of them, though their contrivance and execution scarce fall short of human ingenuity and prudence; but when we come to consider the wonderful economy of these insects, with the good order of their subterraneous cities, they will appear foremost on the list of the wonders of the creation, as most closely imitating mankind in provident industry and regular government. . . .

The Termites resemble the Ants also in their provident and diligent labour, but surpass them as well as the Bees, Wasps, Beavers, and all other animals which I have ever heard of, in the arts of building. . . .

The *Termes bellicosus* being the largest species is most remarkable and best known on the Coast of Africa. It erects immense buildings of well-tempered clay or earth, which are contrived and finished with such art and ingenuity, that we are at a loss to say, whether they are most to be admired on that account, or for their enormous magnitude and solidity. . . .

The nests of this species are so numerous all over the island of Bananas, and the adjacent continent of Africa, that it is scarce possible to stand upon any open place, such as a rice plantation, or other clear spot, where one of these buildings is not be seen within fifty paces, and frequently two or three are to be seen almost close to each other. In some parts near Senegal, as mentioned by Mons. Adanson, their number, magnitude, and closeness of situation, make them appear like the villages of the natives*: and you have yourself seen them perhaps still more numerous, though not so large, in New Holland.

These buildings are usually termed hills, by natives as well as strangers, from their outward appearance, which is that of little hills more or less conical, generally pretty much in the form of sugar loaves, and about ten or twelve feet in perpendicular height above the common surface of the ground†.

These hills continue quite bare until they are six or eight feet high; but in time the dead barren clay, of which they are composed, becomes fertil-

---

* "But of all the extraordinary things I observed, nothing struck me more than certain eminences, which, by their height and regularity, made me take them at a distance for an assemblage of negroes huts or a considerable village, and yet they were only the nests of certain insects. They are round pyramids from eight to ten feet high, upon nearly the same base, with a smooth surface of rich clay, excessively hard and well built."—Adanson's *Voyage to Senegal.*

† The labourers are not quite a quarter of an inch in length; however, for the sake of avoiding fractions, and of comparing them and their buildings with those of mankind more easily, I estimate their length or height so much, and the human standard of length or height, also to avoid fractions, at six feet, which is likewise above the height of men. If then one labourer is = to one-fourth of an inch = to six feet, four labourers are = to one inch in height = 24 feet, which multiplied by 12 inches, gives the comparative height of a foot of their building = 288 feet of the building of men, which multiplied by 10 feet, the supposed average height of one of their nests is = 2880 of our feet, which is 240 feet more than half a mile, or near five times the height of the great pyramid; and, as it is proportionably wide at the base, a great many times its solid contents. If to this comparison we join that of the time in which the different buildings are erected, and consider the Termites as raising theirs in the course of three or four years, the immensity of their works sets the boasted magnitude of the ancient wonders of the world in a most diminutive point of view, and gives a specimen of industry and enterprize as much beyond the pride and ambition of men as St. Paul's Cathedral exceeds an Indian hut.

ized by the genial power of the elements in these prolific climates, and the addition of vegetable salts and other matters brought by the wind; and in the second or third year, the hillock, if not over-shaded by trees, becomes, like the rest of the earth, almost covered with grass and other plants; and in the dry season, when the herbage is burnt up by the rays of the sun, it is not much unlike a very large hay-cock.

Every one of these buildings consists of two distinct parts, the exterior and the interior.

The exterior is one large shell in the manner of a dome, large and strong enough to inclose and shelter the interior from the vicissitudes of the weather, and the inhabitants from the attacks of natural or accidental enemies. It is always, therefore, much stronger than the interior building, which is the habitable part divided with a wonderful kind of regularity and contrivance into an amazing number of apartments for the residence of the *king* and *queen,* and the nursing of their numerous progeny; or for magazines, which are always found well filled with stores and provisions.

I shall forbear at this time entering into a very minute account of the inside of these wonderful buildings, as the bare recital might appear tedious; though I flatter myself, that when I have an opportunity of communicating it to the publick at large, the readers will follow me through an exact description of them with pleasure.

These hills make their first appearance above ground by a little turret or two in the shape of sugar loaves, which are run a foot high or more. Soon after, at some distance, while the former are increasing in height and size, they raise others, and so go on increasing the number and widening them at the base, till their works below are covered with these turrets, which they always raise the highest and largest in the middle, and by filling up the intervals, between each turret, collect them as it were into one dome.

They are not very curious or exact about these turrets, except in making them very solid and strong, and when by the junction of them the dome is compleated, for which purpose the turrets answer as scaffolds, they take away the middle ones entirely, except the tops (which joined together make the crown of the cupola) and apply the clay to the building of the works within, or to erecting fresh turrets for the purpose of raising the hillock still higher; so that no doubt some part of the clay is used several times, like the boards and posts of a mason's scaffold.

When these hills are at about little more than half their height, it is always the practice of the wild bulls to stand as centinels upon them, while the rest of the herd is ruminating below. They are sufficiently strong for that purpose, and at their full height answer excellently as places to look out. I have been with four men on the top of one of these hillocks.

Termite mounds (from Smeathman 1781).

Whenever word was brought us of a vessel in sight, we immediately ran to some Bugga Bug hill, as they are called, and clambered up to get a good view, for upon the common surface it was seldom possible to see over the grass or plants, which, in spite of monthly brushings, generally prevented all horizontal views at any distance.

The outward shell or dome is not only of use to protect and support the interior buildings from external violence and the heavy rains; but to collect and preserve a regular degree of genial warmth and moisture which seems very necessary for hatching the eggs and cherishing the young ones.

The *royal chamber,* which I call so on account of its being adapted for, and occupied by, the *king* and *queen,* appears to be in the opinion of this little people of the most consequence, being always situated as near the center of the interior building as possible, and generally about the height of the common surface of the ground, at a pace or two from the hillock. It is always nearly in the shape of half an egg or an obtuse oval within, and may be supposed to represent a long oven.

In the infant state of the colony, it is not above an inch or thereabout in length; but in time will be increased to six or eight inches or more in the clear, being always in proportion to the size of the *queen,* who, increasing in bulk as in age, at length requires a chamber of such dimensions.

This singular part would bear a long description, which I shall not trouble you with at present, and only observe, that its floor is perfectly horizontal; and in large hillocks, sometimes an inch thick and upward of solid clay. The roof also, which is one solid and well-turned oval arch, is generally of about the same solidity, but in some places it is not a quarter of an inch thick, this is on the sides where it joins the floor, and where the doors or entrances are made level therewith at pretty equal distances from each other.

These entrances will not admit any animal larger than the soldiers or labourers, so that the *king*, and the *queen* (who is, at full size, a thousand times the weight of a *king*) can never possibly go out.

The royal chamber, if in a large hillock, is surrounded by an innumerable quantity of others of different sizes, shapes, and dimensions; but all of them arched in one way or another, sometimes circular, and sometimes elliptical or oval.

These either open into each other or communicate by passages as wide, and being always empty are evidently made for the soldiers and attendants, of whom it will soon appear great numbers are necessary, and of course always in waiting.

These apartments are joined by the magazines and nurseries. The former are chambers of clay, and are always well filled with provisions, which to the naked eye seem to consist of the raspings of wood and plants which the Termites destroy, but are found in the microscope to be principally the gums or inspissated juices of plants. These are thrown together in little masses, some of which are finer than others, and resemble the sugar about preserved fruits, others are like tears of gum, one quite transparent, another like amber, a third brown, and a fourth quite opaque, as we see often in parcels of ordinary gums.

These magazines are intermixed with the nurseries, which are buildings totally different from the rest of the apartments; for these are composed entirely of wooden materials, seemingly joined together with gums. I call them the nurseries because they are invariably occupied by the eggs, and young ones, which appear at first in the shape of labourers, but white as snow. These buildings are exceedingly compact, and divided into many very small irregular-shaped chambers, not one of which is to be found of half an inch in width. They are placed all round the royal apartments, and as near as possible to them.

When the nest is in the infant state, the nurseries are close to the royal chamber; but as in process of time the queen enlarges, it is necessary to enlarge the chamber for her accommodation; and as she then lays a greater number of eggs, and requires a greater number of attendants, so it is necessary to enlarge and encrease the number of the adjacent apartments; for which purpose the small nurseries which are first built are taken to pieces, rebuilt a little farther off a size bigger, and the number of them encreased at the same time.

Thus they continually enlarge their apartments, pull down, repair, or rebuild, according to their wants, with a degree of sagacity, regularity, and foresight, not even imitated by any other kind of animals or insects that I have yet heard of.

There is one remarkable circumstance attending the nurseries, which I must not at this time omit. They are always found slightly overgrown with *mould,* and plentifully sprinkled with small white globules about the size of a small pin's head. These at first I took to be the eggs; but, on bringing them to the microscope, they evidently appeared to be a species of mushroom, in shape like our eatable mushroom in the young state in which it is pickled. They appear, when whole, white like snow a little thawed and then frozen again, and when bruised seem composed of an infinite number of pellucid particles, approaching to oval forms and difficult to separate; the mouldiness seems likewise to be the same kind of substance.*

The nurseries are inclosed in chambers of clay, like those which contain the provisions, but much larger. In the early state of the nest they are not bigger than an hazel-nut, but in great hills are often as large as a child's head of a year old.

The disposition of the interior parts of these hills is pretty much alike, except when some insurmountable obstacle prevents; for instance, when the *king* and *queen* have been first lodged near the foot of a rock or of a tree, they are certainly built out of the usual form, otherwise pretty nearly according to the following plan.

The royal chamber is situated at about a level with the surface of the ground, at an equal distance from all the sides of the building, and directly under the apex of the hill.

It is on all sides, both above and below, surrounded by what I should call the *royal apartments,* which have only labourers and soldiers in them, and can be intended for no other purpose than for these to wait in, either to guard or serve their common father and mother, on whose safety depends the happiness, and, according to the negroes, even the existence of the whole community.

* Mr Konig, who has examined these kind of nests in the East Indies, in an Essay upon the Termites, read before the Society of Naturalists of Berlin, conjectures, that these mushrooms are the food of the young insects. This supposition implies, that the old ones have a method of providing for and promoting their growth; a circumstance which, however strange to those unacquainted with the sagacity of these Insects, I will venture to say, from many other extraordinary feats I have seen of them, is not very improbable.

These apartments compose an intricate labyrinth, which extends a foot or more in diameter from the *royal chamber* on every side. Here the nurseries and magazines of provisions begin, and, being separated by small empty chambers and galleries, which go round them or communicate from one to the other, are continued on all sides to the outward shell, and reach up within it two-thirds or three-fourths of its height, leaving an open area in the middle under the dome, which very much resembles the nave of an old cathedral: this is surrounded by three or four very large Gothic-shaped arches, which are sometimes two or three feet high next the front of the area, but diminish very rapidly as they recede from thence like the arches of aisles in perspectives, and are soon lost among the innumerable chambers and nurseries behind them.

All these chambers, and the passages leading to and from them, being arched, they help to support one another; and while the interior large arches prevent them falling into the center, and keep the area open, the exterior building supports them on the outside.

There are, comparatively speaking, few openings into the great area, and they for the most part seem intended only to admit that genial warmth into the nurseries which the dome collects.

The interior building or assemblage of nurseries, chambers, etc. has a flattish top or roof without any perforation, which would keep the apartments below dry, in case through accident the dome should receive any injury and let in water; and it is never exactly flat and uniform; because they are always adding to it by building more chambers and nurseries; so that the divisions or columns between the future arched apartments resemble the pinnacles upon the fronts of some old buildings, and demand particular notice as affording one proof that for the most part the insects project their arches, and do not make them, as I imagined for a long time, by excavation.

Smeathman, Henry. 1781. "Some Accounts of the Termites Which Are Found in Africa and Other Hot Climates. In a letter from Mr. Henry Smeathman, of Clement's Inn, to Sir Joseph Banks, Bart. P.R.S." *Philosophical Transactions of the Royal Society of London,* LXXI(I):139–92.

The Hometown of the Army Ants

William Beebe

New York Zoological Society scientist and author William Beebe (1877-1962) devoted much of his working life to the tropics. In Edge of the Jungle, *Beebe watches that extraordinary tropical spectacle, the army ant. He marvels at the architecture of the army ants' home—themselves (their own bodies), called the "bivouac." In this sometimes amusing selection, the army ants have adopted an unusual site for their bivouac.*

I sat at my laboratory table at Kartabo, and looked down river to the pink roof of Kalacoon, and my mind went back to the shambles of Pit Number Five. I was wondering whether I should ever see the army ants in any guise other than that of scouting, battling searchers for living prey, when a voice of the jungle seemed to hear my unexpressed wish. The sharp, high notes of white-fronted antbirds—those white-crested watchers of the ants—came to my ears, and I left my table and followed up the sound. Physically, I merely walked around the bungalow and approached the edge of the jungle at a point where we had erected a small outhouse a day or two before. But this two hundred feet might just as well have been a single step through quicksilver, hand in hand with Alice, for it took me from a world of hyoids and syrinxes, of vials and lenses and clean-smelling xylol, to the home of the army ants.

The antbirds were chirping and hopping about on the very edge of the jungle, but I did not have to go that far. As I passed the doorless entrance of the outhouse I looked up, and there was an immense mass of some strange material suspended in the upper corner. It looked like stringy, chocolate-colored tow, studded with hundreds of tiny ivory buttons. I came closer and looked carefully at this mushroom growth which had appeared in a single night, and it was then that my eyes began to perceive and my mind to record, things that my reason besought me to reject. Such phenomena were all right in a dream, or one might imagine them and tell them to children on one's knee, with wind in the eaves—wild tales to be laughed at and forgotten. But this was daylight and I was a scientist; my eyes were in excellent order, and my mind rested after a dreamless sleep; so I had to record what I saw in that little outhouse.

This chocolate-colored mass with its myriad ivory dots was the home, the nest, the hearth, the nursery, the bridal suite, the kitchen, the bed and board of the army ants. It was the focus of all the lines and files which ravaged the jungle for food, of the battalions which attacked every living creature in their path, of the unnumbered rank and file which made them known to every Indian, to every inhabitant of these vast jungles.

Louis Quatorze once said, *"L'Etat, c'est moi!"* but this figure of speech becomes an empty, meaningless phrase beside what an army ant could boast,—*"La maison, c'est moi!"* Every rafter, beam, stringer, window-frame and door-frame, hall-way, room, ceiling, wall and floor, foundation, superstructure and roof, all were ants—living ants, distorted by stress, crowded into the dense walls, spread out to widest stretch across tie-spaces. I had thought it marvelous when I saw them arrange themselves as bridges, walks, hand-rails, buttresses, and sign-boards along the columns; but this new absorption of environment, this usurpation of wood and stone, this insinuation of themselves into the province of the inorganic world, was almost too astounding to credit.

All along the upper rim the sustaining structure was more distinctly visible than elsewhere. Here was a maze of taut brown threads stretching in places across a span of six inches, with here and there a tiny knot. These were actually tie-strings of living ants, their legs stretched almost to the breaking-point, their bodies the inconspicuous knots or nodes. Even at rest and at home, the army ants are always prepared, for every quiescent individual in the swarm was standing as erect as possible, with jaws widespread and ready, whether the great curved mahogany scimitars of the soldiers, or the little black daggers of the smaller workers. And with no eyelids to close, and eyes which were themselves a mockery, the nerve shriveling and never reaching the brain, what could sleep mean to them? Wrapped ever in an impenetrable cloak of darkness and silence, life was yet one great activity, directed, ordered, commanded by scent and odor alone. Hour after hour, as I sat close to the nest, I was aware of this odor, sometimes subtle, again wafted in strong successive waves. It was musty, like something sweet which had begun to mold; not unpleasant, but very difficult to describe; and in vain I strove to realize the importance of this faint essence—taking the place of sound, of language, of color, of motion, of form.

I recovered quickly from my first rapt realization, for a dozen ants had lost no time in ascending my shoes, and, as if at a preconcerted signal, all simultaneously sank their jaws into my person. Thus strongly recalled to the realities of life, I realized the opportunity that was offered and planned for my observation. No living thing could long remain motionless within the sphere of influence of these six-legged Boches, and yet I intended to spend days in close proximity. There was no place to hang a hammock, no over-hanging tree from which I might suspend myself spider-wise. So I sent Sam for an ordinary chair, four tin cans, and a bottle of disinfectant. I filled the tins with the tarry fluid, and in four carefully timed rushes I placed the tins in a chair-leg square. The fifth time I put the chair in place beneath the nest, but I had misjudged my distances and had to retreat with only two tins in place. Another effort, with Spartan-like disregard of the fiery bites, and my haven was ready. I hung a bag of vials, notebook, and lens on the chairback, and, with a final rush, climbed on the seat and curled up as comfortably as possible.

All around the tins, swarming to the very edge of the liquid, were the angry hosts. Close to my face were the lines ascending and descending, while just above me were hundreds of thousands, a bushel-basket of army ants, with only the strength of their thread-like legs as suspension cables. It took some time to get used to my environment, and from first to last I was never wholly relaxed, or quite unconscious of what would happen if a chair-leg broke, or a bamboo fell across the outhouse.

Beebe, William. 1926. *Edge of the Jungle.* Garden City, N.Y.: Garden City Publishing Co.

The New Zealand Glow-worm

F. W. Edwards

Imagine floating in a boat on a large, underground river, passing through a pitch-black passage that opens into a large subterranean cavern. And there, illuminating the darkness of the great arched ceiling, you observe tens of thousands of glowing points of light, like stars in the night sky. Meet these luminous fly larvae, glow-worms of a sort, that, like luminescent deep-sea organisms, have evolved light-emitting organs to attract their prey into their sticky webs.

One of the most remarkable insects in the New Zealand fauna is the luminous larva which has long been known as the 'New Zealand Glow-worm'. Though widely distributed through both islands of New Zealand in suitable localities—damp ravines, crevices, tunnels, etc.—this larva has attracted most attention in certain localities where it lives in great numbers in caves. Of these caves the most famous are those at Waitomo, in the centre of the North Island, which are visited by many tourists. In the damper parts of the Waitomo caves, through which an underground river passes, the 'glow-worms' occur in immense numbers on the walls, roof, and stalactites, the myriad lights producing a scene of impressive beauty.

The 'New Zealand Glow-worm' is in no way related to the glow-worms of Europe, but is a Dipterous larva. The adult insect was first reared many years ago by Mr. G.V. Hudson of Wellington, who has recently re-investigated the life-history of the insect and published an account of it in which he supplements his own early account as well as that of Norris. Further observations (as yet unpublished) were made by Mr. A.L. Tonnoir (now of Canberra) when at Canterbury a few years ago.

From the accounts of these observers the following summary is compiled:

The larva forms a web of fine threads, to which it adds globules of hygroscopic glutinous matter. It glides to and fro within a strong tube-like central mucilaginous thread, from which it can easily pass to any part of the web. From the web numerous loose threads hang down, and are provided with sticky globules at distances of 2–3 mm., thus appearing like strings of beads. Regarding these threads Tonnoir states (in litt.), 'I have counted as many as fifty threads hanging from one glow-worm web, and in the caves they are about 5 inches to 1 foot long, and some can reach up to several feet. In other places where exposed to wind they never reach but a few centimeters'.

The luminous organ is situated in the terminal abdominal segment of the larva, and the production of light is under the control of the insect. According to Hudson the larvae shine most brilliantly on dark damp nights, and the light is brightest just before daybreak. They cease to shine on cold nights, and are very susceptible to vibrations of the air; it is said that in the caves even the vibrations due to talking will cause the larvae to put out their lights. The structure of the photogenic organ has been investigated by Wheeler and Williams, who found it to be formed of the swollen ends of the four Malpighian tubules, which lie close together just ventral to the rectum, and are embedded in a 'reflector' made up of tracheal epithelium (not derived from the fat-body).

The food of the larvae has been found to consist largely if not entirely of small insects caught in the webs, and it seems clear that the function of the hanging threads is to act as 'fishing-lines', and that of the light to attract the prey. When a small fly is caught on one of the hanging threads, the thread is 'withdrawn slowly by a twisting movement of the anterior part of the body as if it were winding the thread round its thorax'. The prey is actually eaten (not merely sucked dry) by the larva. The prey taken appears to be mainly small Diptera, in the caves almost entirely Chironomid midges which hatch out from the water below. (A photograph reproduced by the New Zealand Government Tourist Department, showing a Geometrid moth on one of the threads, is a fake.)

The prey is rendered powerless on coming into contact with the viscous globules on the threads, but whether this is due to the physical or chemical properties of the liquid has not yet been ascertained. In view of the recent work of Mansbridge on the biology of the British Ceroplatinae and Macrocerinae, it seems very probable that the fluid of the globules will be found to contain oxalic acid in sufficient concentration to be lethal to the prey of the 'glow-worm'.

The larva does not form any cocoon, the pupa being merely slung in the web by a few strong threads, which, though merely part of the larval web, may, on drying, appear to be part of the pupa itself (as was thought by Hudson). Before pupation all the hanging threads have disappeared from the web (so that the pupa, as well as the adult when it emerges, is not liable to contact with the viscous and probably poisonous droplets). The pupa, like the larva, is luminous at the tip of the abdomen; 'they can put their lights on or off at will; sometimes the light is just as strong as that of the worms' (Tonnoir). The adult female is also luminous for at least two days after emergence, but the male is not so. Hudson thought that the light produced by the female might serve as an attraction to the male, but experiments he made to ascertain if this were so gave entirely negative results. It would not appear that the light produced by the pupa and imago is of any definite importance to the species.

Edwards, F. W. 1934. "The New Zealand Glow-worm." *Proceedings of The Linnean Society of London* 146:3–9.

Guatemalan Web-Spinning Cave Flies

O. F. Cook

It is well known that spiders catch flies in webs. What is not so well known is that some flies spin webs to catch their own prey. The reader will note that the larvae of these Guatemalan flies earn their livings through a less spectacular version of that employed by their New Zealand cousins, described in the previous excerpt. It is unlikely that this combination of resemblance and close relationship is accidental, and so these glowing, web-spinning Guatemalan cave larvae provide us with a plausible glimpse of an earlier stage in the evolution of their kin across the Pacific.

The limestone mountains of the Department of Alta Verapaz, in eastern Guatemala, abound in caves, most of them as yet quite unexplored. Ancient remains show that some of the caves were used for burial places in prehistoric times, which may account for the aversion of the present Indian population to entering this underground world. Two caves on the Trece Aguas coffee estate near Senahú were visited by the writer on March 30, 1906, to see whether they contained millipeds or other cave-dwelling arthropods.

In one of the caves, which was very dry, a few human teeth were found with small circular mounds of earth where ancient pottery vessels had crumbled, though in some cases the rims remained. The other cave, which was entered by crawling through a low narrow passage, partly filled with water, had also been used for burial purposes and one of the chambers showed a few rude designs traced in black, something after the manner of Mayan hieroglyphics. There were several large chambers, some of them with lofty roofs and extensive deposits of stalactites and stalagmites. The air was very damp owing to wet walls and dripping water. It was in one of the inner chambers of this cave, probably at least 100 yards from the entrance, that curious fringelike webs were noticed hanging from the roof. A sloping floor brought us up close to the webs, and the light of an acetylene lamp rendered the glistening threads very conspicuous against a background of complete darkness.

The general plan of these webs is entirely unlike that of any spider or other web-building arthropod of the upper world, and could be used only in caves or in very sheltered recesses of forests. The only familiar objects to which the webs can be compared are the rope signals that are hung near bridges and railroad tunnels to avoid accidents to train crews. The construction is simple but rather extensive, the webs being usually over a foot long and sometimes nearly 2 feet. Usually the same general direction is kept, along the roof of the cave, but sometimes there is a simple curve and return.

The whole structure is supported from the roof of the cave by a few perpendicular strands, rather irregularly spaced, usually about 2 inches long, and often 2 or 3 inches apart. The ends of these supports are connected by a horizontal cable. Where the roof of the cave is uneven the lengths of the supports are varied, so as to maintain the horizontal direction of the cable. The ends of the cable are drawn up and attached to the roof, and there is only a little sagging between the supports. The remainder of the web consists of a fringe of perpendicular threads attached to the cable above and with the lower ends hanging free. The threads of this fringe are 2 or 3 inches long, and from about 1 mm. to 3 mm. apart. A diagram, kindly prepared by Mr. W. E. Chambers of the Bureau of Plant Industry, illustrate[s] the plan and appearance of the web. The drawing shows a small section of the fringe with a part of the horizontal cable and one of the vertical supports.

The cable and its supports were very slender and had the appearance of ordinary spider-webs, but the threads that formed the pendant fringe were much thicker, perhaps 0.5 mm. in diameter, and appeared as though filled or heavily coated with water. The thickening of the threads did not reach the junction with the cable, but began about 5 mm. below, with great regularity.

The construction of such a web implies, of course, the possession of a highly specialized spinning instinct. Indeed, without observing the operation it is not easy to understand how the webs are built unless we suppose that at least the supporting framework of the structure is first laid out on the ceiling of the cave, to be dropped into the pendent position afterward, perhaps when the heavy fringe is added. But even on this assumption the provision for keeping the cable horizontal by varying the lengths of the supports would involve a high order of instinctive skill. The stretching of the cable by carrying a thread along the wall would not seem so difficult, but more talent would be required to carry the supporting threads up to the ceiling from the cable or to let them down from above to meet the cable.

When the pendent threads were gathered upon the finger they formed a mass of slime, which shows that the material is very unlike the silk of

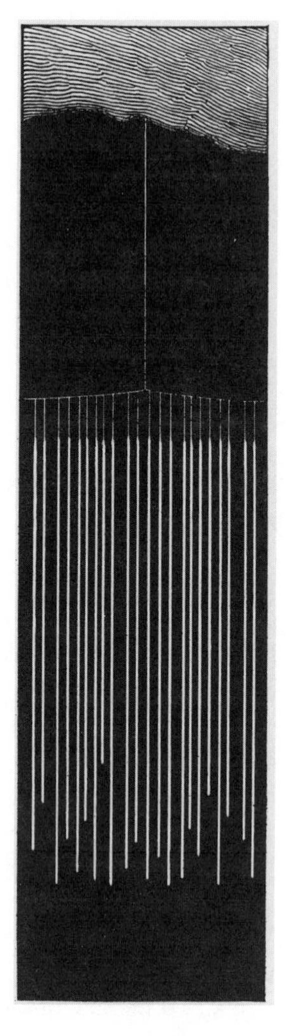

Diagram of web of cave-dwelling fly larva (from Cook 1913).

spiders. Yet the webs evidently serve the same purpose of trapping insects. Several small insect were found embedded in the slime, from which they could be squeezed out by slight pressure. Mosquitoes and other soft-bodied forms, which have the habit of seeking dark roosting-places, probably furnish most of the victims, but one of the webs had caught a small beetle. If an investigation of the insect life of the caves were to be undertaken, these webs might afford considerable assistance in trapping the small insects that flit along the roofs of the caverns.

The larvae which were evidently the builders of these curious structures, were slender, transparent, vermiform creatures about 20 mm. long. They were found in all cases lying along the main cable of the nest, on which they seemed to slide back and forth, with considerable speed.

The attention of Mr. H.S. Barber, of the Bureau of Entomology who visited Guatemala in the same season, was called to these webs and he saw some of them in another cave near Trece Aguas. At first he was inclined to believe that the spinning larvae might belong to the family Tipulidae, but he now considers it more probable that they are Mycetophilidae, as several other members of this family are known to spin webs or to live in web-like tubes of slime. The larval characters of this group of flies are so little known that a definite identification of the animals found in the webs is at present out of the question. But as no webs of similar construction seem to have been described, Mr. Barber has urged me to publish my notes on the subject. . . .

Nothing in the way of a specialized subterranean fauna was found in the caves, unless it be the larvae that spin these webs, and even these may not be confined to the eaves. Other webs that may have been made by the same kind of larvae were seen afterward in open recesses in the side of cliffs along the road between Senahú and Sepacuité, though not in condition to compare with the much more perfect structures seen in the caves.

Cook, O. F. 1913. "Web-Spinning Fly Larvae in Guatemalan Caves." *Journal of the Washington Academy of Sciences* III(7):190–93.

Caddisfly Houses and Net Traps

Bernd Heinrich

Caddisfly larvae build secure underwater houses that they carry with them and, in some species, even use to catch fish. Bernd Heinrich (b. 1940), professor of zoology at the University of Vermont in Burlington, likes to watch the caddisfly larvae moving on the river bottom near his cabin in Maine.

Each year I teach a course titled Winter Ecology to a group of a dozen students that gathers for three weeks at my cabin in the Maine woods. The snow is often deep then, and temperatures dip to minus 30 degrees Celsius. Yet we always find life, especially when we go to the trouble of chopping through the thick ice of one of the many local streams or brooks.

Lying face down, we peer through our "window" to the sandy and debris-strewn stream bottom. After looking long and carefully, we almost always see a small movement here, another there. Caddisfly larvae, their legs protruding from camouflaged protective cases or houses are, even in mid-winter, walking about and feeding much as they do the rest of the year.

The caddisflies (order Trichoptera) are an ancient group that has changed little in 250 million years. The adults superficially resemble drab moths, and the Lepidoptera (moths and butterflies) are thought to derive from them. In fact, the gills most aquatic insect larvae use for breathing are thought to have been the structures from which insect wings later evolved. Caddisfly larvae are grub-like creatures that are protected from fish and other predators by the cocoon-like cases they construct around themselves. Like the cocoons spun by moth larvae, the cases that caddisfly larvae make are spun by glands from the mouthparts. Unlike moth cocoons, which serve to hide and to physically protect the pupae, caddisfly cases are made to shelter and conceal the insects throughout their active larval life, and they are never sealed off as cocoons are. They are analogous to the shell that a snail carries.

Peering onto the bottom of Alder Stream, which runs below my cabin, we usually see no insects at first. Then gradually hundreds of caddisfly larvae in their artfully constructed cases come into focus. They are busily consuming leaves and other plant debris that have fallen into the stream during the fall.

The most common type of "house" here is that of a species that glues twigs, spruce and fir needles, and bits of wood lengthwise into a tube. The larva's head, thorax, and six legs protrude from one end; the caddisfly walks along the bottom, dragging its house along. On slight provocation it pulls its house up over itself, so that only a small, neat bundle of stream debris remains to be seen.

We found several species of larvae in Alder Stream, having very different cases. One type had its tubular silk domicile covered with sand grains and attached beneath a sheet of sand pebbles glued together to look just like the loose, pebbly gravel on which they were found. We found another species living among vegetation; its cases were made of small bits of organic debris that were attached crossways rather than lengthwise. Among the many species that carry their houses on their backs, there is great variation in building material and construction style. In some, the cases are strikingly beautiful, symmetrical, horn-like tubes. In one species that uses sand grains, the tube coils and ends up resembling a snail shell. The animals never outgrow their houses. As caddisflies grow, they continue to add new material at the mouth-end, and most cases are gradually flaring tubes.

The shape and composition of the cases are at least functionally related. One species of predatory larvae bites off plant stems to a specific length and then binds them with silk about itself in a spiral ribbon. The result is a sleek and cryptic case which is easily maneuvered through thick vegetation. In contrast, some stream-dwelling, herbivorous larvae attach sticks at right angles to the long axis of the case. Like anchor hooks, these laterally protruding sticks probably aid in attaching the case to the weedy substrate of their feeding grounds in a current. In swift, sandy bottom streams, most grazers and detritivores weigh their cases down with pebbles and attach them to the bottom with silk anchor lines.

Not all caddisfly larvae carry their houses with them. Some have evolved structures from their cases that serve as a dwelling place and as a trap for prey. Members of the families Hydropsychidae, Philoptamidae, and Psychomiidae build bag-like nets that they occupy at one end and that act as a seine. The larvae wait at the bottom, feeding upon food particles swept into the trap by the currents.

These nets may have originated by greatly enlarging the case entrance with silk while deleting the addition of sticks or other substrate. The result was a "leaky case" that came to be used as a seine. In the genus *Neuroclipsis* (Psychomiidae), the seine net is shaped like a cornucopia with the wide

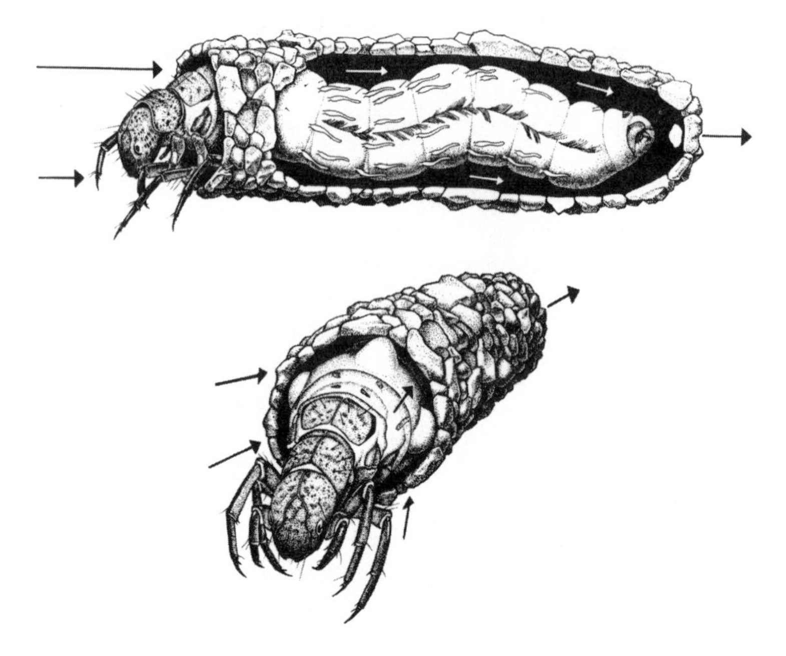

The circulation of water through the case of a typical case-making caddis larva. Illustration by Anker Odum (*Larvae of the North American Caddisfly Genera [Trichoptera]*, 2nd edition, Glenn B. Wiggins, University of Toronto Press, Toronto, 1996).

mouth where the food comes in and the narrow end the larva's place of residence. Both the entrance and the tip of the cone are attached to vegetation; as the water flows through, the seine bows downstream past the anchored tip where the larva waits.

Some net-spinning caddisflies lay down as many as seventy threads at a time, in one sweep of the mouthparts. The threads are microscopic and they trap microscopic prey, primarily diatoms and bacteria.

Whenever my students and I seine with our own nets over the stream bottom under the thick ice of Alder Stream we bring up caddisfly larvae in strange and wonderful cases, but we haven't yet netted a netter. There is still much to look forward to in our exploration of other streams where there will be other species, maybe even one that seines with its own artfully constructed net. The insect treasures are often hard to find, which makes them all the more enticing.

Heinrich, Bernd. 1996. "Insect Architects: Caddisfly Larvae." *Wings* 19, (Fall) no. 2:11–13.

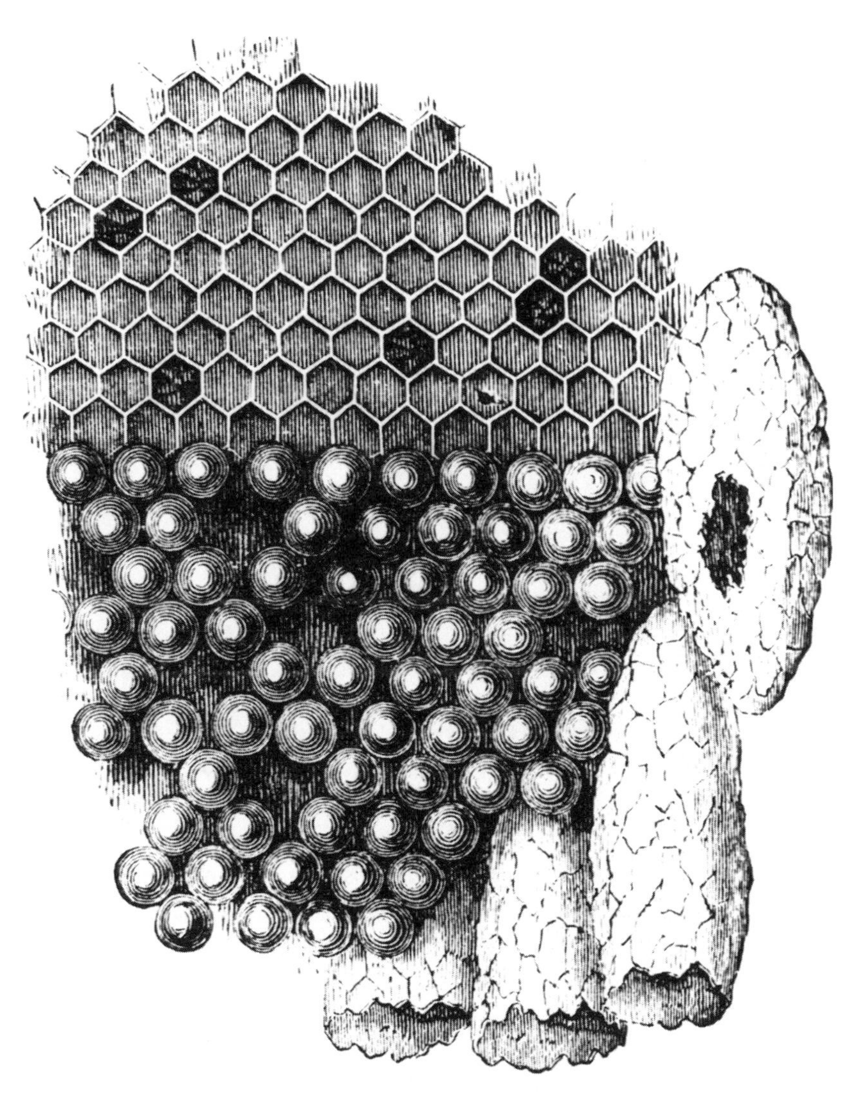

Honeycomb with eggs occupying the cells (*Silver Wings and Golden Scales* by Anon., Cassell Petter & Galpin, London, 1889).

Bee Cells

Karl von Frisch

Why do bees build perfect hexagonal prisms for their wax comb houses? How do they manage to do it? What are their tools and what do they use to measure with? Here we enter into the fine points of bee joinery in a modern scientific account, courtesy of 1973 Nobel Prize winner, the bee master himself Karl von Frisch (1886–1982). This is from Animal Architecture, *prepared with the collaboration of his son, Otto von Frisch.*

In contrast to the paper combs of wasps, which normally are horizontal and have cells only on their undersides, the combs of bees are made of wax, hang vertically, and have cells on both sides separated by a wall in the middle. They serve for the rearing of the brood and for the storage of honey and pollen. The same kind of cell is used for all these purposes. Within the building precinct of a hive, the queen bee lays her eggs in the central areas of the combs; only the more peripheral parts are used as containers for pollen and honey. The larger cells for the rearing of drones are usually placed in the lower sections of the combs. Even larger cells, shaped somewhat like spruce cones, which are designed for the raising of queens, are usually added a little to one side or at the lower edge of the main cell matrix.

The walls of the main body of connected cells form regular hexagonal prisms. This is a noteworthy fact. After all, the bees might build their cells with rounded walls as the bumblebees do or as they themselves build for the cradles of their queens. Or they could base their architectural style on some other geometrical configuration. However, if the cells were round or, say, octagonal or pentagonal, there would be empty spaces between them. This would not only mean a poor utilization of space; it would also compel the bees to build separate walls for all or part of each cell, and entail a great waste of material. These difficulties are avoided by the use of triangles, squares, and hexagons. The three-, four-, and six-sided shapes [can be configured] to enclose the same area. Provided their depth was the same, such cells would therefore hold the same volume. But of the three geometrical figures equal in area, the hexagon has the smallest circumference. This means, of course, that the amount of building material required for cells of the same capacity is the least in the hexagonal construction, and hence that such a pattern is the most economical design for warehouses.

In most combs, one edge of the hexagonal prism points to the top, and another to the ground. Exceptionally, combs are found in which there are horizontal walls at the top and bottom of the cells. The question whether the customary arrangement has a static advantage that increases the load-bearing capacity of the cells has never been investigated, as far as I know. Dr. Georg Kirchner of Frankfurt am Main very kindly agreed to look into it for me as an engineering problem. His computations gave the same result for both arrangements and do not help us to explain why the bees prefer one to the other. The geometry of the shape and depth of the cell bottoms and the manner in which they dovetail into each other contributes, no doubt, a great deal to the stability of the comb. Anyone lifting a full honey-comb for the first time will find it amazingly heavy. A comb measuring 37 by 22.5 centimeters can hold more than four pounds of honey. Yet in the manufacture of such a comb, the bees use only about forty grams of wax. The relationship between the construction of a comb and its strength would seem to be a worthwhile subject for study.

Building can proceed quite rapidly if necessary. When bees start building, they first attach themselves to each other in chains. Soon they form themselves into a dense ball, the building cluster within which they maintain a temperature of 35°C. (95°F.)—the temperature needed for the secretion of wax. In honeybees, wax is formed only in glands on the underside of four abdominal segments. Where these segments meet, wax appears in the shape of two small flakes between the ventral scales. To pick up a flake, the bee uses the greatly enlarged first tarsal joint of her hind leg, which carries on its inner side a small brush for the collection of pollen. She impales the little flake on the last row of bristles, pulls it out of its skin pocket, and passes it forward to be taken over by her front legs and mandibles. Holding it in front of her mouth, the bee kneads and mixes it thoroughly with a secretion of her salivary gland. This treatment, combined with the right temperature, gives it the necessary homogeneous consistency and that degree of plasticity at which it can best be molded.

Building starts at the ceiling, which in a normal hive means at the top of a frame. Bees usually begin at two or three different places at once and construct sections that taper downward. They do not build one complete cell after another. While the lateral walls of the first cells are gradually being added to, new adjoining cells are being started lower down. As these triangular sections are enlarged laterally, they gradually coalesce from the top down.

The joins are so skillfully made that no traces of the separate beginnings remain visible. This is even more remarkable when one considers that many bees are employed in the building of each individual cell and that they often relieve each other at intervals of no more than half a minute or so. Apparently, each bee immediately comprehends what stage the construction has reached at the place where she starts to work and continues accordingly. When the pace is fast, the individual sections grow rapidly, both vertically and laterally, so that a contiguous building front is reached in a short time.

Modern beekeepers insert artificial walls with a stamped-on hexagonal pattern into the frames they want their bees to fill with combs. This speeds up the building program, partly, of course, because the bees need less wax. However, the cells built without such aids are no less regular. It has often been maintained that this regularity was not a particularly remarkable achievement since, under the influence of lateral pressure, the cells would of necessity acquire the shape with the smallest surface area. This is not so. From the very beginning, the cells are built in this, the most practical, shape. Usually a rhomb-shaped section of the base is made first, followed by the first parts of two adjoining cell walls. Next, another rhomb is added and two more cell walls are started; thereafter, the third rhombic section is erected, which completes the hexagon, and the two last walls. Right from the start, the cell walls meet at the correct angle of 120°. Admittedly, the regular hexagonal shape is hidden at the top by a ring of wax which contains material for further use, but as soon as this bulky wax ring is carefully removed, it shows at once. This pattern must be imprinted into the minds of the bees by the forces of heredity. The bees even apply it as useless ornamentation to the outer surfaces of queen cells in the course of their modeling work. The same regular hexagonal shape characterizes the combs of the paper wasps and those of the tropical wasp species that are modeled in clay.

It is not only the exact shape of the cells that depends upon the skill of their builders; skill is just as much needed to vary the size of the cells for worker bees and drones, to manufacture such extraordinarily thin walls, and to orient them accurately in space. None of these things just "happen," they are the result of work directed to a purpose.

The cell walls are built with a gradient of about 13° from base to opening. This is sufficient to prevent the thick honey from running out. The distance from the cell wall to that opposite is 5.2 millimeters in a worker cell, and 6.2 millimeters in a drone cell. The thickness of the cell walls is 0.073 millimeter, with a tolerance of no more than 0.002 millimeter. What truly astounding precision! Economy in the use of building material is thus taken to the utmost limit. Human craftsmen could not do work of this nature without the use of carpenters' squares and sliding gauges.

Measuring instruments of the bees. The bee's own head serves as a plummet to determine the line of gravity. It rests on two pivots forming part of the outer skeleton of the thorax, and its center of gravity lies below this articulated connection. Hence, if a bee sits with her head pointing upward, its heavier, lower part will be pulled toward the thorax by the force of gravity. In a downward position, the head is automatically rotated in the opposite direction. These gravity pulls are accurately registered by a tactile organ consisting of a set of highly sensitive bristles on the tips of these pivots. Any position at an angle to the vertical is registered by a characteristic distribution of pressure on the set of sensory hairs. This is the way bees control both their own position in space and the position of the comb, which is always built vertically downward.

It has been possible to prove experimentally the importance of these sensory organs in the bees' necks for their building activities and for the correct orientation of the cell walls. These organs can be put out of action by coating the bristles with a warm mixture of wax and rosin which hardens on cooling and by which they are completely immobilized. Setting up this experiment was a time-consuming operation. It was necessary to guard against the possibility of results being affected by bees outside the usual age groups taking part in building. Therefore, all bees in the experimental colony—some thousand individuals—had to be treated. The effort proved worth while. The bees were in a building mood and immediately gathered in a building cluster. But nothing happened. They behaved like workmen whose tools have been confiscated: they did not work. The gumming-up of their neck bristles had clearly not impaired their general well-being, because the flying-out and food-gathering activities of the hive were normal. Nor was there a reduction in their production of wax, but the flakes were allowed to drop to the ground unused. In one experiment in which a building cluster was observed for a whole month, no more than three pitiful little cells with irregular walls were produced during the first two weeks. Then came a heat wave. The coating substance started to melt, and the tips of the sensory hairs reappeared. The result was that the bees, within the next four days, after producing some irregular transitional shapes, managed once more to build more or less normal cells. The experiment indicates that for the correct positioning of their cell complex in space, the bees need the organs in their necks.

The search for the instrument used by the bees to determine with such impressive accuracy the size of the cells and the angles and dimensions of the cell walls has not led to conclusive results so far. However, experiments on queens have produced some interesting evidence. A curious fact that has long been known about bees is the way in which sex is determined. A queen can either lay unfertilized eggs, which produce males (drones), or fertilized eggs, which produce females (queens or workers). She possesses a sperm pouch where the male sperm, acquired by her on her nuptial flight, is stored for the rest of her life. From this she can release sperm cells the moment an egg slides past its exit duct. As the larvae of the drones grow bigger than those of worker bees, the queen lays her unfertilized eggs in the larger drone cells. If the tips of her front legs are amputated, the queen continues her egg-laying unabated, but she can no longer distinguish between the two types of cells and produces a terrible muddle of fertilized and nonfertilized eggs. From this experiment, one may infer that the queen measures the size of the cells in which she is about to lay her eggs with the tips of her front legs; hence it is probable that the worker bees do likewise.

The problem of discovering how the building bees gauge the thickness of the cell walls seemed, if anything, even more difficult to solve. Obviously they must be able to measure, for how else could they keep to the exact thickness of 0.073 millimeters (0.094 millimeters for drone cells)? Yet this particular puzzle has been solved.

When the bees start building up a cell wall, they first make a roughly kneaded ring of wax at its upper end and gradually roll it out into a thin lamella by planing motions of the mandibles. Throughout this operation, they constantly measure the thickness of the wall and shave off surplus wax. From the way they do this, one might think that the workers had received a thorough training in physical mechanics. Only the basic principle involved can be explained here very briefly. The bee presses her mandibles against the cell wall and thereby produces an indentation. When she withdraws her mandibles, the wall returns to its original shape. During this process, she continually examines the relevant section of the wall with the tips of her antennae. These tips are endowed with special sense organs consisting of three rings of tactile cells equipped with curved hairs, each surrounding a cell sensitive to tactile and chemical stimuli and carrying a spike. This spike is pushed into the wall while the bristles of the rings are placed upon it. In this manner, the organ registers the course of deflection and recovery. Under the conditions prevailing—a given consistency of the wax, a constant temperature of 35°C. (95°F.), and a predetermined shape of the cell—the speed of this movement depends entirely on the thickness of the wall and reflects it accurately. When the tips of the antennae where these sense organs are located were experimentally removed, this abruptly put an end to the bee's precision work. The bee still managed the basic architecture of the cells, but the walls got either too thin, or, more often, too thick [unlike the] regular edges of normal newly built cells.

Storage cells for pollen are left open. Honey cells are closed with a wax lid when they are full. The wax necessary for this is kept ready for use as a thick ring around the lip of the cell, and can be quickly rolled over the opening from all sides when it is required. The breeding cells, too, are covered with a domed lid of wax for the twelve days of pupation. Below the wax lid, the larvae themselves spin a dense cap of silk threads. Under this twofold cover, the metamorphosis of larva into winged bee can proceed undisturbed.

Orientation of combs by the earth's magnetic field. By putting wooden frames into his hives, the beekeeper determines the direction in which the combs will be built. But the interesting question is, what happens under natural conditions, in a hollow tree, for example? How is it possible that the thousands of bees, starting work without delay after taking possession, and often completing large parts of their new combs overnight, end up, not with a chaotic muddle, but with a well-laid-out, regular structure?

In this context, the German ethologist Professor Martin Lindauer and his co-worker Dr. Martin Oehmke have made a remarkable discovery. They placed a swarm from a conventional hive into a cardboard cylinder that had no frames and whose flight hole was in the center of the floor. Neither the cylinder nor the position of the flight hole gave any pointer to the orientation of the combs in the old hive. Yet, within a few hours, the bees had produced the beginnings of neat, parallel combs whose orientation corresponded to that of the combs in the original colony, or deviated by no more than a few degrees. How was this possible? Recently it has been found that the orientation of bees is influenced by the magnetic field of the earth. Though the experimenters had carefully removed all directional landmarks from the bees in the round hives, they had not been able to take away the natural compass of the magnetic field. Their assumption that this was the bees' means of orientation was confirmed by further experiments. A swarm was moved first to one cylindrical hive and transferred from there into another, similar one. The combs were placed in exactly the same compass direction in both hives. Next, the natural magnetic field was artificially disturbed and deflected by 40° between the first and second transfer with the aid of a magnet placed outside the second hive; now the direction of the new combs differed from the previous direction by the same angle of 40°.

Before this discovery, it had occurred to neither beekeepers nor scientists that bees laid out their combs with a compass. But now that they have demonstrated their ability to do so, it is easy to see what a useful accomplishment this is. For when a swarm has taken possession of a dark hollow in a tree, thousands of workers start building at the same time at different corners of the new abode, and there is no foreman or site architect to tell each worker what to do. However, if they all have the urge and the capacity to orient their combs in space exactly as in their old home, a well-ordered structure will be the result. The manner in which the bees perceive the magnetic field of the earth is still a mystery.

Bees' glue (propolis). Though wax is the chief building material of bees, it is not the only one. They also use resin for filling in gaps and holes. This same material also has an entirely different use. When a mouse or other small animal invades the hive in search of honey and is killed by the valiant defenders, its dead body is covered with a layer of resin which effectively shuts out the air and so preserves and mummifies it. Wax is produced by the bees themselves, while resin, called propolis, has to be collected. The bees gnaw it with their mandibles from the sticky buds of certain trees, transfer it to the "baskets" on their hind legs, and carry it home like pollen. However, this material, which is viscous and sticky, cannot be transferred to the leg baskets in flight in the manner of pollen. The bees have to alight somewhere to do this. It is surprising that they can handle it at all, but the same secretion of their mandibular glands that helps to improve the working properties of wax is of use here too.

The behavior of a bee arriving home with her baskets full of propolis is quite different from that of a bee carrying in pollen. The pollen collector seeks an empty cell to unload her harvest. But a bee that carries propolis makes her way to the building site where her product is needed and very unobtrusively offers it to the building workers. Sometimes she may sit or slowly walk about for hours. When the building workers need propolis, they come along and gnaw off the required small amounts from the tough lump in her basket. At times propolis is mixed with wax to make it go further.

The chief time for collecting propolis comes in late summer or in autumn when the nights get cool and draughty places in the hive become menacingly noticeable. During that period the bees are hard at work filling in cracks and crevices with glue to preserve the precious heat of the hive. Sometimes the bees have been so busy that when spring returns the beekeeper can hardly separate the frames or prize them from the walls of the hive.

In southern countries, heat rather than cold is the danger since the wax structures will melt at high temperatures. Here again, the bees know how to cope with the problem. In the hot volcanic areas of Salerno in southern Italy, bees were observed mixing propolis into the wax they used for building in order to raise its melting point.

Manifold are the benefits and pleasures we owe to the bees. A less well-known reason for our gratitude is the fact that in all probability propolis was one of the secret ingredients that the Italian makers of violins added to the lacquer to improve the sound of their instruments.

von Frisch, Karl (with the collaboration of Otto von Frisch). 1974. *Animal Architecture.* New York: Harcourt Brace Jovanovich. (Translation by Lisbeth Gombrich.)

7

Go Forth and Multiply: Mating and Reproduction

Amid all the strange behaviors and fanciful morphologies (body shapes) devised by living things, the strangest and most fanciful are those associated with the process whereby, ultimately, male sperm meets female ovum. Sex. Think, for instance, of orchid flowers, peacock feathers, or the chorus of frogs on a summer night. Darwin had an explanation for this, summarized by the term "sexual selection," which posits that sexual structures and behaviors in males and females of a species are in constant evolutionary motion, one changing and the other evolving to keep up, resulting in a rapid, never-ending dance down infinite roads of unexpected filigreed complexity. Take this process and multiply it by the teeming diversity of insect species, and you have a recipe for a universe of wondrous form and behavior.

The Hostile Madness of Love

Maurice Maeterlinck

This is the dramatic tale of the mating of honeybees, as only Maurice Maeterlinck (1862–1949)—Belgian poet, dramatist, and statesman—could write it. This excerpt is from The Life of the Bee, *published before Maeterlinck won the Nobel Prize. It was the most successful popular science book of its day.*

Very few, I imagine, have profaned the secret of the queen-bee's wedding, which comes to pass in the infinite, radiant circles of a beautiful sky. But we are able to witness the hesitating departure of the bride-elect and the murderous return of the bride.

However great her impatience, she will yet choose her day and her hour, and linger in the shadow of the portal till a marvelous morning fling open wide the nuptial spaces in the depths of the great azure vault. She loves the moment when drops of dew still moisten the leaves and the flowers, when the last fragrance of dying dawn still wrestles with burning day, like a maiden caught in the arms of a heavy warrior; when through the silence of approaching noon is heard, once and again, a transparent cry that has lingered from sunrise.

Then she appears on the threshold—in the midst of indifferent foragers, if she have left sisters in the hive; or surrounded by a delirious throng of workers, should it be impossible to fill her place.

She starts her flight backwards; returns twice or thrice to the alighting-board; and then, having definitely fixed in her mind the exact situation and aspect of the kingdom she has never yet seen from without, she departs like an arrow to the zenith of the blue. She soars to a height, a luminous zone, that other bees attain at no period of their life. Far away, caressing their idleness in the midst of the flowers, the males have beheld the apparition, have breathed the magnetic perfume that spreads from group to group till every apiary near is instinct with it. Immediately crowds collect, and follow her into the sea of gladness, whose limpid boundaries ever recede. She, drunk with her wings, obeying the magnificent law of the race that chooses her lover, and enacts that the strongest alone shall attain her in the solitude of the ether, she rises still; and, for the first time in her life, the blue morning air rushes into her stigmata, singing its song, like the blood of heaven, in the myriad tubes of the tracheal sacs, nourished on space, that fill the centre of her body. She rises still. A region must be found unhaunted by birds, that else might profane the mystery. She rises still; and already the ill-assorted troop below are dwindling and falling asunder. The feeble, infirm, the aged, unwelcome, ill-fed, who have flown from inactive or impoverished cities, these renounce the pursuit and disappear in the void. Only a small, indefatigable cluster remain, suspended in infinite opal. She summons her wings for one final effort; and now the chosen of incomprehensible forces has reached her, has seized her, and bounding aloft with united impetus, the ascending spiral of their intertwined flight whirls for one second in the hostile madness of love. . . .

Here it is evidently nature's wish, in the interests of crossed fertilisation, that the union of the drone and the queen-bee should be possible only in the open sky. But her desires blend network-fashion, and her most valued laws have to pass through the meshes of other laws, which, in their turn, the moment after, are compelled to pass through the first.

In the sky she has planted so many dangers—cold winds, storm-currents, birds, insects, drops of water, all of which also obey invincible laws—that she must of necessity arrange for this union to be as brief as possible. It is so, thanks to the startlingly sudden death of the male. One embrace suffices; the rest all enacts itself in the very flanks of the bride.

She descends from the azure heights and returns to the hive, trailing behind her, like an oriflamme, the unfolded entrails of her lover. Some writers pretend that the bees manifest great joy at this return so big with promise—Büchner, among others, giving a detailed account of it. I have many a time lain in wait for the queen-bee's return, and I confess that I have never noticed any unusual emotion except in the case of a young queen who had gone forth at the head of a swarm, and represented the unique hope of a newly founded and still empty city. In that instance the workers were all wildly excited, and rushed to meet her. But as a rule they appear to forget her, even though the future of their city will often be no less imperilled. They act with consistent prudence in all things, till the moment when they authorise the massacre of the rival queens. That point reached, their instinct halts; and there is, as it were, a gap in their foresight.—They appear to be wholly indifferent. They raise their heads; recognise, probably, the murderous tokens of impregnation; but, still mistrustful, manifest none of the gladness our expectation had pictured. Being positive in their ways, and slow at illusion, they probably need further proofs before permitting themselves to rejoice. Why endeavour to render too logical, or too human, the feelings of little creatures so different from ourselves? Neither among the bees nor among any other animals that have a ray of our intellect, do things happen with the precision our books record. Too many circumstances remain unknown to us. Why try to depict the bees as more perfect than they are, by saying that which is not? Those who would deem them more interesting did they resemble ourselves, have not yet truly realised what it is that should awaken the interest of a sincere mind. The aim of the observer is not to surprise, but to comprehend; and to point out the gaps existing in an intellect, and the signs of a cerebral organisation different from our own, is more curious by far than the relating of mere marvels concerning it.

But this indifference is not shared by all; and when the breathless queen has reached the alighting-board, some groups will form and accompany her into the hive; where the sun, hero of every festivity in which the bees take part, is entering with little timid steps, and bathing in azure and shadow the waxen walls and curtains of honey. Nor does the new bride, indeed, show more concern than her people, there being not room for many emotions in her narrow, barbarous, practical brain. She has but one thought, which is to rid herself as quickly as possible of the embarrassing souvenirs her consort has left her, whereby her movements are hampered. She seats herself on the threshold, and carefully strips off the useless organs, that are borne far away by the workers; for the male has given her all he possessed, and much more than she requires. She retains only, in her spermatheca, the seminal liquid where millions of germs are floating, which, until her last day, will issue one by one, as the eggs pass by, and in the obscurity of her body accomplish the mysterious union of the male and female element, whence the worker-bees are born. Through a curious inversion, it is she who furnishes the male principle, and the drone who provides the female. Two days after the union she lays her first eggs, and her people immediately surround her with the most particular care. From that moment, possessed of a dual sex, having within her an inexhaustible male, she begins her veritable life; she will never again leave the hive, unless to accompany a swarm; and her fecundity will cease only at the approach of death.

Prodigious nuptials these, the most fairylike that can be conceived, azure and tragic, raised high above life by the impetus of desire; imperishable and terrible, unique and bewildering, solitary and infinite. An admirable ecstasy, wherein death supervening in all that our sphere has of most limpid and loveliest, in virginal, limitless space, stamps the instant of happiness in the sublime transparence of the great sky; purifying in that immaculate light the something of wretchedness that always hovers around love, rendering the kiss one that can never be forgotten; and, content this time with moderate tithe, proceeding herself, with hands that are almost maternal, to introduce and unite, in one body, for a long and inseparable future, two little fragile lives.

Profound truth has not this poetry, but possesses another that we are less apt to grasp, which, however, we should end, perhaps, by understanding and loving. Nature has not gone out of her way to provide these two "abbreviated atoms," as Pascal would call them, with a resplendent marriage, or an ideal moment of love. Her concern, as we have said, was merely to improve the race by means of crossed fertilisation. To ensure this she has contrived the organ of the male in such a fashion that he can make use of it only in space. A prolonged flight must first expand his two great tracheal sacs; these enormous receptacles being gorged on air will throw back the lower part of the abdomen, and permit the exsertion of the organ. There we have the whole physiological secret—which will seem ordinary enough to some, and almost vulgar to others—of this dazzling pursuit and these magnificent nuptials.

Maeterlinck, Maurice. 1901. *The Life of the Bee.* New York: Dodd, Mead and Co. (Translation by Alfred Sutro.)

The Synchronous Flashing of Fireflies

John Buck and Elisabeth Buck

The flashing of fireflies is, to all who witness it and no matter their age, a magical occurrence. How much more so the perfectly synchronized flashing—on, off, on, off—of thousands of fireflies on a single tree! Imagine taking a boat down a lazy river on a hot night in Malaysia, rounding a bend, and seeing before you just such a tree, lighting up the dark like a flashing beacon. Although it is now known that it is clearly for the purpose of attracting females, no one knows for sure why a male firefly would participate in such an aggregation and help his rivals to get mates instead of going off and being a loner. Still, this firefly phenomenon has parallels in other male insect aggregations that give off coordinated auditory or olfactory signals. The current speculation, by no means proven, is that while synchronized males may be cooperatively improving the overall signal (and their ability to attract females from far away), they may still be competing for the potential mates that come in. But, for now, let's just enjoy the mystery as we read one of the more graphic accounts.

Reports of synchronous rhythmic flashing by fireflies in South-East Asia have been appearing for more than two hundred years. Smith's description of displays along the Chao Phraya (Meinam) River south of Bangkok will serve to introduce the phenomenon: "Imagine a tree thirty-five to forty feet high thickly covered with small ovate leaves, apparently with a firefly on every leaf and all the fireflies flashing in perfect unison at the rate of about three times in two seconds, the tree being in complete darkness between the flashes . . . Imagine a tenth of a mile of river front with an unbroken line of *Sonneratia* trees with fireflies on every leaf flashing in unison, the insects on the trees at the ends of the line acting in perfect unison with those between. Then, if one's imagination is sufficiently vivid, he may form some conception of this amazing spectacle." The flashing, Smith continues, is confined to male fireflies and ". . . occurs hour after hour, night after night, for weeks or even months . . . ".

This behaviour is strikingly different from that of most fireflies in other parts of the world: in fact, such large-scale, long-lasting, concerted rhythmic activity is seemingly unique in the animal kingdom. For this reason the phenomenon has been a perennial source of interest tinged with scepticism. Yet the many accounts have not led to a consensus about certain important observational details of the synchrony, or to satisfying explanations of either the physiological mechanism of the mass flashing or of its biological significance.

On the basis of a recent visit to Thailand and Borneo we have reported elsewhere some photographic and photometric measurements on a Thai species of firefly indicating that the synchrony of flashing varies less than 16 msec in a cycle of 560 msec, and have described observations on the build-up of synchrony under laboratory conditions which shed some light on the physiology of this firefly's "sense of rhythm". In this article we shall consider some more speculative aspects of Oriental concerted flashing related to its distribution and possible function. Two general premises are involved. First, the whole behaviour pattern, since it represents the product of countless millennia of evolutionary selection, must have a definite and important function in the life of the participating firefly species. Secondly, Oriental mass synchronism is a complex of behaviours (congregation *per se*, congregation in trees, choice of trees near water, flashing, synchrony) which may or may not be functionally related and which probably have quite different relative importances.

Riverbank Firefly Congregations

Morrison and Smith claimed that the male Thai fireflies fly out from the adjoining jungle each evening ". . . for the purpose of engaging in this nightly display". This statement implies that the trees are populated anew each evening, an idea which we believe to be incorrect. Not only were riverbank trees near Bangkok full of fireflies in early evening, but we also found many of the insects present in mid-morning in a tree which we had studied the previous evening. Fireflies have also been seen by day in trees harbouring synchronizing swarms in New Britain and in non-synchronous trees in Jamaica. In Sarawak we found the tree inhabitants still flashing at midnight and at 6 p.m., as did Haneda in New Britain. Resident fireflies thus account for at least part of the successive evening congregations. Fireflies were indeed seen flying to communal trees in both Thailand and Sarawak, but at the same time some were also seen flying away. The possible significance of these flights will be discussed later.

Practically nothing is known about the longevity of fireflies in Nature, but even at carefully controlled humidities and temperatures captive adults of most species die in a few days. This suggests that congregations that maintain themselves for weeks or months must be replenished constantly. Because a West African firefly species has been found to breed throughout the year, and the adults of several species in Jamaica and in New Britain are active in every month, tropical firefly trees could presumably be maintained for long periods, perhaps indefinitely, by recruits from a population in a more or less steady state of reproduction.

So far as it is known, all adult fireflies come from pupae in the soil—the pupae come, in turn, from larvae that live on or in the soil, or in water. Haneda found mating pairs and larvae on the ground in the Botanical Garden at Rabaul near silk-cotton trees in which fireflies were flashing synchronously. However, the mangroves bordering the swamp rivers of Thailand and Borneo stand in mud that is scoured by tidal currents so swift that it seems impossible that larvae, even if aquatic, could survive on the riverbank. Thus the swarms of fireflies in the trees must migrate from the interior swamp.

Good firefly trees were not common along the Chao Phraya River, but even in Sarawak, where the terrain is much closer to virgin swamp and firefly trees were more abundant, there were many mangroves of the same kind as those in which firefly swarms occurred, and similarly situated, which were completely free of the insects. The observed distribution of the firefly population therefore requires more than random flight from hinterland to the edge of a river. When watching male Thai fireflies in a darkroom we discovered that there is a period, before the flying specimens alight and synchronism begins to build up, when they are definitely attracted to each others' light. This positive phototaxis suggests that male fireflies arriving at the fringe of riverbank trees would be attracted by flashing congregations there, or, if no previous swarm existed, might themselves form a nucleus by mutual photic attraction. Such assemblies might be expected to build up competitively, leading eventually to one or a few large swarms that had outdrawn nearby smaller centres because of higher mean light emission. Many potential display areas may therefore exist for each that develops to a spectacular level. The corollary that a good display tree would take some time to establish agrees with other indications that the assemblages are not renewed nightly.

The fact that flashing behaviour changes (from rapid twinkling to a steady tempo of two per second) after the male fireflies alight is not a peculiarity of the particular Thai species or of the perching habit. Flash patterns in resting fireflies are usually quite different from those in flight. What is unusual in tree fireflies is the maintenance of steady spontaneous flashing while perched—almost all roving fireflies flash irregularly, if at all, while at rest.

In sum, riparian firefly trees may be viewed as quasi-permanent assemblies, formed by way of photic attraction and maintained by recruitment. . . .

Aggregation and Mating

Since it is firmly established that male and female of most, if not all, roving-type fireflies are brought together for mating by individual-to-individual photic signalling, it is natural to enquire whether the flashing congregations in trees have any sexual significance. Both Morrison and Smith specifically denied this possibility but their stand was based on the misapprehension that female fireflies are always wingless and thus would be unable to reach the trees. The fact is that females of the riverbank species we studied in Thailand and Borneo are present in the trees, and in large numbers. The same situation was found in firefly trees in New Britain by Haneda. The females do not participate in synchronous flashing with the males, but this perhaps re-inforces the suspicion that they are in the trees for another purpose.

In the Jamaican trees large numbers of coupling pairs were found and the suggestion was made that the aggregating behaviour serves to promote mating. Haneda similarly felt that the synchronous flashing of the New Britain male fireflies calls the females. In Thailand we unfortunately concentrated so hard on observing and recording the various kinds of flashes in the trees that it was not until we had returned to the United States that we realized the significance of possible non-luminous activity (most fireflies cease flashing while mating). We did find a coupling pair on the gunwhale of our canoe after we had beaten the foliage with our insect nets in making a mass collection, and we observed that such beatings always flushed out many more animals than we would have expected from the density of flashing. Thus mating presumably does occur in the firefly trees of Thailand as in Jamaica and New Britain.

One further tenuous indication of mating habits comes from eye size. It is fairly well established that in firefly genera in which a roving male seeks a sessile female the eyes of the male are much larger than those of the female; but if the female takes a mobile part in the courtship the eyes are more nearly equal in size. In all the species of riverbank fireflies we have examined the eyes of the males and females are about equal. This finding is compatible with the idea that females as well as males are attracted to the congregations. Nets placed around a small firefly tree might give evidence of the postulated two-way (but not necessarily luminous) traffic and might make it possible to ascertain whether incoming females are primarily virgin and departing females mated.

Possible Rationale of Mass Mating

If fireflies do convene at the water's edge for mating, the utility of trees as fixed and dry sites for assembly is obvious, particularly since the insects do not mate on the wing. The assembly trees would presumably be selected by chance, and mangroves, being common, would be used often. However, it is not obvious why the riverbank should be chosen for assembly in the first place or why the fireflies in question should have evolved indiscriminate centralized mating instead of the system of individual courtships, which is not only the more usual plan among fireflies but would appear to be more efficient in terms of dispersal of the species.

In most species of roving firefly, males in an already widely distributed population search on the wing for stationary females. Recognition and homing depend on the pair being able to see each other nearly continuously in order to distinguish the signal code of their own species from those of other species patrolling the same terrain. Our experience with American, Caribbean and upland Malayan forms indicates, as might be expected, that roving fireflies are found primarily on open ground, such as meadows, forest glades, road clearings, savannah and jungle trails.

There could scarcely be imagined a terrain less favourable for line-of-sight recognition than the flat, tangled, mangrove-nypa swamps of South-East Asia in their virgin state. A firefly tree on a watercourse, however, might provide a sufficiently bright and large beacon, perhaps enhanced by reflexions, to attract fireflies that wander out into the clear over the water, and might provide enough opportunities for mating to compensate for the (assumed) long flights required of the mated females for egg dispersal. We propose therefore that, in tropical swampland, mass mating has evolved instead of pair courtship because sustained photic communication between individuals is impossible.

Buck, John, and Elisabeth Buck. 1966. "Biology of Synchronous Flashing of Fireflies." *Nature* 211 (Aug. 6), no. 5049:562–64.

Sex on the Brain

James E. Lloyd

Insect courtship and sex can be exceedingly strange and radically different from species to species. It is one of the tasks of evolutionary biology to explain these widely different particulars in terms of broad, unifying evolutionary principles, asking the question: What do these strange mating systems have in common? Utilizing a string of wonderful examples from nature, entomologist James E. Lloyd (b. 1933) here first explains why biologists are obsessed with sex and then discusses how most mating behaviors can be understood as the net effect of the selfish interests of individual genes.

Laymen get the impression that biologists have an inordinate preoccupation with sex. We are immoderate, and it is excusable: sexual behavior is the key to understanding biological species. Only when he turns his attention to mating behavior does the biologist begin to use more than inferential evidence for species "boundaries." Most basic and applied investigations ultimately depend upon a knowledge of species, whether for acquiring species-pure samples, or for identifying and manipulating vulnerable points in the ecology of some competing organism. Questions that are addressed to the way species found their origin, got to be the way they are, and exist today, begin and end with a discussion of biological species.

Gene flow is the phenomenon at the center of understanding and definition of biological species. In the final analysis, when reduced to its smallest moment of flux, to its irreducible whit of displacement, genes flow down an aedeagus and into another individual, and the genes of 2 parents flow together. (Sometimes the actual mechanics are not exactly like this—genes are handed over in sacks, left on posts, or squirted into the surrounding medium.) Mating behavior arranges for and accomplishes gene flow. It comprises the activities and events that take place as the animals seek, identify, win over and appraise, and finally accept partners in reproduction. Thus sexual behavior, in all its intimate and diverse detail in the animal kingdom, becomes a necessitous obsession with biologists. Further, because the biologies of most organisms are constructed around sexual success, there is more at stake in understanding mating behavior than "merely" straightening out species. This knowledge is fundamental to understanding biology at all, and its most important principles.

Natural selection is the choreographer, composer, and lyricist of the entire sexual performance. It brings about change in gene (= allele) frequencies, and this, in a reasonable working definition, is evolution. Natural selection occurs when certain genetic sorts of individuals in a population leave a greater number of progeny than do others. As simple and old-hat as it sounds, this elementary fact can be used with considerable reward and success when one addresses mating behavior studies. Surprisingly, many published studies, speculations, and conclusions, indicate that not all biologists understand, use, or profit from this simple, old but fresh biological verity. One practical application or approach-plan is the one-gene-analysis-model (OGAM) that Richard Dawkins presented at a popular level in his book *The Selfish Gene,* as did David Barash in his highly readable *Sociobiology and Behavior.* This technique has been used by some biologists for more than a decade. The OGAM pits 2 opposing phenotypes against each other in the reproductive game. For purposes of simplification, the competing phenotypes are based on 2 allelic forms of a gene (or as more commonly expressed, upon the 2 "genes" that are competing for a locus). Simple genetic logic is followed, or rather is pushed through to its seeming endpoint. The conclusion, as to what should or should not be, is not final or binding on nature: it merely provides a guide and prevents certain kinds of errors, raises suspicions of certain explanations or observations, suggests lines of research to be followed, and provides a sound criterion for recognizing significant observations on natural phenomena. The OGAM places in proper probability perspective some erroneous explanations that otherwise seem credible or plausible (such as the heart-warming story of the mutualistic yucca-moth, that appears in ecology texts).

To begin with a simple example, consider a species of beetle in which the female emits a pheromone that males smell at a distance and approach. Males with long antennae (gene L+) are more successful in getting to females than males with shorter ones (gene L-)—the former detect lower levels of the pheromone and are able to track it better. It is obvious that unless the disadvantages of the longer antennae outweigh the advantages, as measured in reproductive success, L+ will gradually replace L-. (We really don't need the OGAM to come to this plain, straightforward conclusion.)

In a more complicated case, males of S.E. Asian *Pteroptyx* fireflies congregate in great numbers in trees and flash in synchrony. Males and females are attracted to the pulsing trees, and it has been speculated that by synchronizing their flashes these males are providing a huge beacon-tree *to help* (proposed context of selection) other members of their species get to a gathering place quickly and thereby avoid predation by bats. In other words, the synchrony is said to be a group adaptation (biotic as opposed to organic), evolved and maintained in the context of group benefit. Let S+ result in synchronous flashing, and its competitor for that locus on the chromosome (S-) not produce such behavior. By devoting more of their activity (energy, attention) to their own reproductive success, males with S- will find and inseminate more females, and leave more S- progeny than their rivals leave S+, in each generation. In fact, we would predict that the synchronizing behavior should be lost; indeed, it should never have evolved. But we observe that males do flash in synchrony, and, therefore, conclude that selection producing this behavior must be acting in some context other than that proposed. Synchrony must be doing something for the competing, S+-bearing and -perpetuating male. The assisted individuals, those using the beacon-*effect* to reach the tree, are but cueing in on a conspicuous and highly relevant marker for locating available eggs to fertilize. As a working hypothesis the benevolent-beacon theory is certainly worse than none at all—it flies in the face of simple genetics—and we are guided by the OGAM to seek other explanations before setting out across the Pacific to study beacon-trees.

Bedbugs and kin are reproductively eccentric. Males inject sperm through the female body wall and into the hemocoel, where evolutionarily new structures within (paragenital system) conduct as well as store sperm prior to fertilization. Males of the genus *Afrocimex* have external paragenital structures like, in fact in some respects, *more* developed than those of conspecific females. Males are found with copulation scars where other males have jabbed them, and with spermatozoa within. Males of *Xylocoris* mount mounted males and inject them with sperm, some of which finds its way into the sperm ducts of the prime-positioned males and hence into the females with their ejaculate. These developments were not recognized as belonging to male competition—the latter circumstance might correctly be called autocuckoldry—but instead it was even imagined that male bedbugs would evolve helper ducts *to assist* the alien sperm (uplifting larceny from pilferage to grand theft!). Let D+ build the helper duct and D- not do so. D+ cannot gain. Although it may sometimes help other D+, it will on occasion, to one extent or another, contribute to the success of D-, thereby reducing its own proliferation. Because D-, on the other hand, will selfishly keep all possible fertilizations, it will always deny passage to D+. It is ironic that such a theory could be proferred when it appears that male competition probably was a major selective force resulting in the evolution of internal fertilization and of the extra-genital, traumatic insemination of these bugs.

In these examples the underlying assumption that permitted and led to the originally speculated "adaptations" was that behavior occurs for the benefit of the population or species. One of the greatest values of the OGAM is that it will often uncover this unlikely, if not completely erroneous, assumption. It forces one to focus on the selfish gene, its fate, and its consequences. Selfish genes instruct their gene machines, to use Dawkin's metaphor, to be competitive in every aspect of their mating behavior, from the beginning search to final fertilization—to yield nothing without a net reproductive gain. Consider these insectan examples: sperm put into a dung fly female by a male is largely pushed aside by sperm from the next male. A male heliconiid butterfly puts a chemical on the female that deters other males. After copulating, males of many insects, including flies and Lepidoptera, put plugs behind their sperm which prevent the entrance of sperm from subsequent males. Male walkingsticks remain with and ride their females, as living chastity belts, for days or weeks, and a *Parnassius* butterfly glues his genitalia to those of a female. In some Diptera, males inject accessory secretions (matrone) that inhibit the mating behavior of the female by affecting components in the central nervous system. Males of a firefly inject flashes into the coded patterns of rivals, possibly making them appear to be those of a different species. The external "superfemale" genitalia on *Afrocimex* males may also perpetrate sexual dirty tricks, directing the copulatory thrusts of rivals to the wrong spot, more or less masturbating them as they compete to fertilize the eggs of the contested females. Males of a New Guinea longhorned grasshopper insert raspberries (i.e. Bronx cheers) into the rhythmic, "nasal", bleating songs of nearby males. Male army-worm moths emit a pheromone that inhibits the responses of rivals to available females. Males of the horseshoe crab pry and shove each other as they gather around and press up against the female of a tandem-pair at the oviposition site, perhaps stealing fertilizations from the consort by releasing sperm when the eggs are laid. Finally, in more straightforward competition, males of other species, such as scarabs and phengodid beetles, bite, horn, pinch, shove, kick, and maim each other in the presence of females.

If, in the analysis of any behavior, one finds that success in the competition doesn't seem to correlate with reproductive success, (i.e. that certain males seem to be fighting while others are doing the mating), understanding OGAM, he will not conclude that they were simply "naturally aggressive" (a meaningless expression), but pause in coming to a conclusion, and give the behavior closer scrutiny. When a particular mating strategy is observed in males, a rival counter-strategy will be sought, as will ecological imperatives that make certain other strategies potent. For example, a male firefly that makes the pattern of a rival appear to be that of a sympatric species has his strategy safely anchored in reproductive isolation. When it was found that the male heliconiid put a deterrent chemical on the female, Gilbert addressed himself to the question of why males continued to be deterred by the chemical and did not evolve out of the trap. The ethologist that is trying to construct an ethogram, or the sociobiologist making a sociogram, should also notice that unless competitive situations simulating possible natural events and triggers are arranged, a critical part of the catalogue is being omitted. Similarly, if mating analyses are always made under unnatural, crowded conditions in the laboratory, one cannot expect to see the entire repertory of behavior.

The mating behavior discussed so far has been rather simple. To round out this overview, I should mention insects with different ecologies and phylogenetic backgrounds. In insects that require resources that are spatially limited such as carrion, bracket fungi, dung, stream oviposition sites, or perhaps even certain insects themselves, male success may depend upon the ability to fight, to take over and defend these locations or territories on them or nearby. When resources are not limiting, and populations dense, males may easily reach females, presenting each female with many suitors and the opportunity to exercise choice. If such ecological circumstances exist for some time, females may force males into leks where they must compete by singing, flashing, strutting, or butting heads for hours. Under other circumstances, no better understood, males may bring resources or tokens to females, such as seeds, dead flies, and empty balloons and the nature of the associated male rivalry is different, as is its theoretical interest and application. Other phylogenetic and ecological circumstances have resulted in females dispensing with males and sex, seasonally or completely, as in certain aphids, weevils, and flies, the Surinam roach in Florida, and an Australian grasshopper, to mention a few. There is even a coccid that is a self-fertilizing hermaphrodite. Sexual parasitism, in which females of a parthenogenic species copulate with males of another, such as the obligate, sexually parasitic spider beetle *Ptinus mobilis,* may be based on the purloining of nutritional ejaculate. Other explanations, that the egg requires a sperm trigger for development (easily corrected through evolutionary time), or that the female is preventing the mating of a potential mother of the ecological competitors of her progeny, are by themselves inadequate and unsatisfying.

To the naturalist and the theorist, and each should have some of the other in him, among the more interesting behaviors are those in which he sees an evolutionary tracking and one-ups-man-ship, or evidence that this has taken place. In male-male contests it is obvious that for every new strategy or ploy that evolves, a counter move may be expected sooner or later. As observers we never know when we have tuned in, in evolutionary time: is it sooner or later? Going from insect to insect is like being in a time machine in which one remains stationary as coevolutionary phylogeny moves past. Theorists and observers alike, but not the animals themselves though they may reach the point, are pursuing the so-called evolutionarily stable strategy (ESS). This is the space-time-genome coordinate at which no new mutant strategy can displace the existing strategies.

By adding yet another dimension . . . I can indicate some behavior that I believe must exist and that may be very common, and some counter-behavior that may have evolved to subvert it. This will return to our point of departure, the species problem, and will link the biological species with all of their subtle sexual competition and complication, to those reminders of a simpler day, the morphological cabinet species of the museum taxonomist. In many animals, probably most insects, the energetic contributions made by the 2 parents to each individual offspring are quantitatively different, with females giving more. This fact has stimulated a great flurry of theoretical activity and animal inspection over what the consequences for mating behavior ought to be. Presently it is pretty generally concluded that this fundamental and ancient asymmetry should result in hot-to-trot males which have been selected to drop their sperm and dash to the next female, and choosy (coy) females that are very particular and selective with respect to their mates. The generalization seems legitimate, and to go a long way in explaining, for example, male coyness observed in species in which there are sex role reversals and the males have taken on rearing chores generally found in females. With the occurrence of cannibalism in mantids and some few others, these males may sometimes be coy also.

Selection for haste in males, and coyness in females, results in what amounts to competition *between* the sexes. Males may be selected to bypass any choice that the females attempt to exercise, and then females selected to maintain their options, to not be misled or to have their choices subverted. If males subdue and seduce females with true aphrodisiacs, females may be expected to escape sooner or later in evolutionary time. And after sperm has been placed in a female, she should manipulate it: store, transfer (from chamber to chamber), use, eat, or dissolve it, as she makes additional observations on males. Females may accept and store sperm from a male for insurance that they will get a mate, and then become choosy. And if the male ejaculate contains nutritional elements which he contributes to *his* zygotes, females should evolve to get this from him for free (recall sexual

(Top) Wheel bugs *(Arilus cristatus)* stacked in a sexual encounter with two males (top) competing to fertilize the eggs of the female. According to J. E. Lloyd, similar interactions may have been significant in the evolutionary development of traumatic insemination in bedbugs. Males are touching antennae, and the upper one has his wings spread (to maintain his precarious perch?). The female may be evaluating them, or simply awaiting the victor of whatever contest is being waged.
(Bottom) Enigmatic copulation wheel in the damselfly *Ischnura ramburii.* The male is on top. In odonates (dragonflies and damselflies), the male grasps the female behind the head with terminal appendages and the female then places her terminal gonopore over the sperm, which has been placed on "accessory" genitalia beneath the basal segments of the male's abdomen. How could such an awkward position evolve? What could be the intermediate stages of the evolution of this wheel? Sexual competition or selection, says J. E. Lloyd, may be a crucial part of the story. Assuming an ancestral gonopore-gonopore connection (or even a spermatophore placed on the ground), the male could have pushed this connection against his underside to prevent rivals from prying him loose or slipping in between, to keep the female from escaping, or to force sperm into her. Seizing the female behind the head would have evolved later, perhaps from restraining moves by males. Photographs from J. E. Lloyd in *The Florida Entomologist* (1979).

parasitism). Then males should be selected to prevent it, and to make sure that their genes are used with their expensive, nutritional contributions. It is possible for sperm to be manipulated in the female (e.g. sex determination in Hymenoptera). Female reproductive morphology often includes sacs, valves, and tubes that could have evolved in this context. In fact, it is possible that some reported examples of sperm competition are actually cases of sperm manipulation.

Given that females, to one extent or another, subvert male interests by the internal manipulation of ejaculate, it is not inconceivable that males will have evolved little openers, snippers, levers and syringes that put sperm in the places females have evolved ("intended") for sperm with priority usage—collectively, a veritable Swiss Army Knife of gadgetry! Remember copulation in the bedbug and the male blade? Males of some scutellerid Hemiptera have large, bizarre genitalia, half the size of the female abdomen. In *Hotea* they are spiky and heavily sclerotized, and apparently tear their way through the vagina and body cavity to reach the spermatheca. Also recall the diverse shapes of male genitalia that taxonomists have exploited for decades. These variations and elaborations may in many instances have evolved to bypass female resistance and sperm manipulation, and represent present or past sexual success strategies. Carrying this line of reasoning a bit farther, it may be possible to make certain inferences and predictions about sexual selection and courtship, and even ecology, on the basis of the diversity (adaptive radiation) of male and female genitalia within a group. This suggests that intraspecific variation, perhaps polymorphism, is to be expected and in many instances another source of taxonomic confusion.

Recognition that sexual conflict-coevolution occurs, and the consequent revolutionary perspective of mating and courtship behavior, followed from an appreciation of natural selection: the genteel view that assumed the individual got mated to serve the good of the species and to prevent its extinction was a conceptual desert, and a researcher's dead end.

Lloyd, J. E. 1979. "Mating Behavior and Natural Selection." *The Florida Entomologist* 62, no. 1:17–34.

The Courtship Gifts of Balloon Flies

Edward L. Kessel

According to some human traditions, when a male suitor proposes to his prospective bride, he presents her with a gift, typically an engagement ring. When entomologists talk about "nuptial gifts," however, they are referring to a curious parallel that has cropped up repeatedly in diverse insect groups: Prior to mating, male insects present females with a gift. In most species the gift is something nutritious, like a freshly killed prey item, but in some species it can be (as far as we humans can tell) entirely symbolic. What's going on?

One of our best opportunities for answering this question comes from the balloon flies, extraordinary creatures that, when the entire range of species is surveyed, seem to present the full spectrum of "nuptial gift" behavior. The delightful paper excerpted here represents one of the first attempts to arrange the differing versions of nuptial gift-giving in different species of balloon flies into a reasonably plausible evolutionary sequence.

In 1875, while visiting in the Swiss Alps, Baron Osten-Sacken made the first recorded observations on what we now call balloon flies. On sunny days, between the hours of about 9 and 10 o'clock, he noticed brilliant, silvery flashes among the sunbeams which penetrated the shadows of the fir forest. When he concentrated his attention on the flashes he concluded that they must be produced by insects, so he determined to catch some and added an extension to his net handle. The white flashing objects seemed to dodge with agility and it was only after several swings that he was certain he had netted one of them. However, when he looked into the net, he was at first astonished to see only an inconspicuous dull-colored fly, much smaller than he had anticipated. He assumed that this one had been captured by accident, along with the one he had tried for. So he examined his net carefully and it was only then that he noticed the white, filmlike packet of sparkling material on the gauze. He tried to pick it up, but it was so light that his breath carried it away. He pursued it and finally eased it into a vial. This was the first of the so-called fly-balloons to be taken.

To make certain that the balloons actually were associated with the flies, Osten-Sacken caught one specimen after another, always with the same results. Each time there was the balloon in the net along with the fly. And examining the flies with his hand lens, he observed that invariably they were males.

It was Becker (1888) who described this fly that Osten-Sacken had discovered, and he named it *Hilara sartor. Hilara* is a genus of the family Empididae which is made up of the dance flies, so called because of the habit indulged in by many species of dancing on the wing in assemblages which sometimes reach huge proportions. All of the balloon-making flies belong to the Empididae.

Becker was interested in natural history as well as taxonomy, so he joined Osten-Sacken in speculation as to what the significance of the balloon might be. In fact he went to observe the flies in their home environment, but they flew too high for him to get close enough to make accurate observations. As a result he reached the erroneous conclusion that the males carry the white structures on their backs. He also supposed that these structures serve as decorations which attract the attention of the females, and of course in this he was not far wrong.

That same year Mik (1888) also went to see these flies in action, and apparently he had either better eyesight or a taller stepladder than Becker because he was correct in insisting that the flies carry the balloons below their bodies, where they are suspended by the legs. He observed, moreover, that these structures are considerably flattened, this flattening being much more pronounced than in any of the later-discovered species. This observation led him to propose the fantastic theory that the male hilaras use the flattened objects as aeronautical surfboards on which to cavort among the sunbeams.

The next entomologist to speculate on the significance of the balloons was Verhoff (1894). He concluded that they must serve as warning signals to birds and predaceous insects. There is no evidence that these flies are distasteful to birds; on the contrary I have seen birds attack swarms of related species of empidids. In regard to the second part of Verhoff's contention, namely, that their scintillating balloons serve to protect them from predaceous insects, this is nearer the truth, although not in the sense that Verhoff had in mind. The predaceous insects in this case are not other forms as one might suppose, but rather the females of the male's own species. Empidid flies are generally predaceous, this being true for the female as well as the male. And the potential empidid bride might be more hungry than amorous, at least in the early stages of courtship. We now know that the balloons serve three functions: (1) to provide a "come hither" invitation to the female instead of a warning; (2) to distract the female from her predaceous inclinations once they have embraced; (3) to serve as a stimulus to mating. What are some of the details of this fascinating story, and how were they discovered?

Evolutionary Stages

The problem presented to the early workers by the case of *H. sartor* was so perplexing because it represents the end stage in a complex evolutionary sequence. But as more and more species of these balloon-making flies were discovered, the significance of the light-reflecting structures which they carry in their aerial dances became apparent. Hamm (1928) suggested three more or less distinct stages, and Melander (1940) added a fourth. While I find it necessary to disagree with these authors as to the origin of the material from which the balloon is made, I nevertheless accept the four general stages which they have suggested. However, during the eight years that I have been making observations on balloon flies, I have attempted to analyze still further the details of this evolutionary sequence and have come to recognize eight stages in all. These are as follows:

Stage 1: The first stage is simply that of a predaceous species in which the male does not bear a wedding gift for his bride. There are numerous species of empidids, e.g., those of the genus *Tachydromia* which represent this basic stage. The prey is captured and devoured by the sexes independently. This stage differs in no way from the cases of ordinary predaceous species, such as those which occur in other insect families. Sometimes the prey may belong to the same species as the captor, and the female may, on occasion, include the male among her victims.

Stage 2: In the second stage the male avoids any cannibalistic attention on the part of the female by carrying with him, ready for presentation at the moment they embrace, a wedding present in the form of a juicy insect. The male first catches the prey, and then he flies about with it in search of a mate. *Empimorpha comata* and certain species of *Rhamphomyia* and *Empis* represent this stage. Often the prey carried by the bride-hunting male is as large as he is, and on one occasion I captured a male carrying two bibionid flies, each almost as large as himself. This was surprising inasmuch as it would be supposed that the instinctive urge to accomplish the prey-catching step in this mating pattern would be satisfied by the capture of one fly. In this second stage the prey is simply eaten by the female, the gift serving merely to occupy her attention as she passively accepts the male during the time she is thus distracted. The mating lasts as long as the food does.

Stage 3: While the third stage is similar to the second, it differs in two important respects. As in the second, the male first catches the prey, but in this stage he does not go in search of a female. Instead, he joins other males, bearing gifts, in an aerial dance of eligible bachelors. When a female approaches, the males do not take after her, but instead continue to dance in the swarm. The female, on approaching such a swarm, must take the initiative, and it is worth while noting at this point that this is a feature common to all of the remaining stages in the series of mating patterns to be described. In this third stage, the prey has become the stimulus for mating, and it no longer serves merely as a distraction to save the male's life. The female enters the swarm and selects one of the dancing males, they embrace and she accepts the gift, and the two continue to fly in the swarm. In some species the female feeds on the prey while copulation is under way and the pair are still flying. In other forms the female indicates that her primary interest is in mating, for she abstains from feeding until such time as the two flies leave the swarm and settle on some nearby vegetation. Here she has her meal and copulation again continues as long as the food lasts. Certain species of *Hilara* and *Rhamphomyia* represent this stage.

Stage 4: In the fourth stage the male bears a wedding gift which, as before, consists of a prey, but now something has been added. Pasted haphazardly here and there against the prey are shiny viscid globules or, as in certain cases, the prey is more or less entangled in silken threads. Here we are dealing with a number of species of *Hilara.* In this stage it is still the prey which serves as the mating stimulus for the female, the simple wrappings being unimportant in this regard. These entangling structures help to quiet the newly captured prey and doubtless that is their sole function.

Stage 5: In the fifth stage the simple wrappings of the fourth stage have been elaborated to form the complex structure which we call a balloon, and this structure shares with the relatively large prey the function of stimulating mating. This stage is illustrated by two balloon-making forms of the genus *Empis* which occur in western United States. These are *Empis aerobatica* Melander and *Empis bullifera* Kessel and Kessel. (My wife and I have published on the latter.) . . .

It was in the spring of 1949 that we discovered the first representatives of the species that was to be known as *E. bullifera.* We named it "bullifera" because of its bubble-bearing activities. To date these flies seem not to have been seen anywhere except on our home property (five acres of woods at Novato, in Marin County, California, about thirty miles north of San Francisco). They have recurred there every year since they were discovered.

These flies have been observed mostly as mating pairs, drifting in lazy flight, first in one direction and then in another, back and forth within the clearing under the trees, their glistening white balloons producing flashes whenever they penetrate a sunbeam. Sometimes the flight becomes so slow that the balloon appears to hang almost motionless in one spot. Then, without warning, the flies will make one to several quick turns, sometimes to leave the clearing altogether, but more often to resume their casual flight nearby. Always they can be followed easily by the conspicuous balloon which they carry.

If one looks at the balloon with the naked eye, it appears to be made up of fine, shiny, white fibers which are wound spirally into a thin-walled, hollow sphere. Covering the front of the balloon (front in the sense that the fly normally carries the balloon with this end forward) and somewhat entangled in its threads and globules is the prey. We have found the prey of this species to consist of midges, psocids, or tiny spiders. In this fifth stage of the evolution of the balloon-making habit the prey is still large in comparison with the size of the balloon and it is still fed on by the female. . . .

Stage 6: In the sixth stage the prey again serves as the stimulus for the male to construct the complex balloon. Now, however, the prey no longer serves as food for the female. It is somewhat smaller than in the preceding stage and it is probable that the male consumes what fluid it contains when he captures it. At any rate it is no longer edible when the sexes embrace and the male passes the balloon to the female. In this stage, however, the prey, still being conspicuous enough to be easily noticed, doubtless continues to share with the balloon the function of stimulating copulation. . . .

Stage 7: In stage 7 the prey is yet smaller; in fact it is always minute and so delicate that it would seem to be useless as food even for the male when first captured. The abdomen is always collapsed and is often fragmented into tiny pieces which are plastered into the front surface of the balloon. One may conclude that in this stage the prey still serves as the stimulus for the male to construct the complex balloon inasmuch as he always begins by catching the prey. But certainly the prey no longer serves as food for the female; moreover, because of its minute size and the fact that the female pays no attention to it while in the process of manipulating the balloon, it would seem that it has entirely lost its significance in courtship. It may be assumed, therefore, that in this stage the balloon remains as the sole stimulus for copulation.

Stage 7 is well illustrated by *Empimorpha geneatis* Melander, a species which in 1946 I discovered was a balloon-maker. This was the species which initiated my study of balloon flies, and these early observations were published a year later. One morning, while looking out of our kitchen window,

I observed a number of conspicuous white objects scintillating in the morning sunshine as they zigzagged back and forth close by a Monterey pine. They were flying at an altitude of some fifteen or twenty feet, and it was only by tying my net handle to a sturdy surf-casting rod and then mounting to the top of a six-foot stepladder that I was able to bring any of the specimens to net. They proved to be all males and the white object which each carried was a delicate, balloon-like structure which invariably had a minute dipteran or hymenopteran plastered into its anterior surface. When swept roughly into the net the flies invariably drop their balloons, but when care is exercised and the stroke of the net is gentle and true, they sometimes retain possession of these structures and can be transferred to a container while still grasping them tightly. As a rule these balloons are somewhat flattened, suggestive of the flattened structures already mentioned in connection with the introductory comments about *Hilara sartor*. In *E. geneatis*, however, the prey is always present although crushed and dry and so invariably unfit for food. These minute captured insects are always oriented so that the head, and often the thorax as well, projects free from the balloon, being directed forward and upward. And this projecting part of the prey's body is used as a handle which the male always seems to have hold of with one or more of his feet. In addition, these insects sometimes have the tarsi of two of their legs hooked together under the balloon. I was able to ascertain that the eligible bachelors of this species do not discard their balloons when they retire, unsuccessful from the standpoint of marriage, from the morning dance. Such a male keeps his balloon for another try at mating the following day, although he is able to make another, smaller balloon, in case the first is lost.

During the first years of my study I was able to observe the mating activities of this species. On several occasions I saw a female join the dancing group of males and select a mate. As the two embraced they lost altitude for a moment, and then floated off together to settle on nearby vegetation. Examination of the paired flies revealed that the balloon had been transferred to the female. In the act of mating the male hangs from the vegetation by his front feet, his middle legs supporting the female's thorax, and his hind legs holding her abdomen. The female holds onto the balloon with all of her feet; never does she grasp the prey as a handle for the balloon in the manner of the male. Instead, she keeps turning the balloon from one position to another during the mating period, apparently entirely unconcerned with the prey and certainly making no attempt to feed on it. This tiny insect is absolutely unfit for food; its abdomen is always ruptured and plastered into the balloon's surface. In fact I am of the opinion that *E. geneatis* has ceased to capture prey for food in the manner of its ancestors. Never did I find flies of this species in possession of prey other than the tiny ones crushed into the balloons, whereas again and again I have found them feeding on manzanita flowers. Each time they were observed, the fly was obtaining nectar through a puncture which it had made through the corolla with its proboscis. Here, as on all other occasions when the males were observed alone, they were carrying their balloons with them.

Stage 8: In this last stage of the evolutionary sequence the male no longer needs to capture a prey in order to be stimulated to construct a balloon. So the gift package includes no prey whatever, not even a dried-up, inedible one such as is present in stages 6 and 7. I have not yet had the opportunity to study any of the species representing this last stage, but *Hilara sartor,* which was discussed at the beginning of this paper, is an example. One cannot blame Osten-Sacken and his colleagues for some of their wild speculations as to the significance of the balloons of this species, for these forms were entirely without prey and the earlier stages in the evolutionary sequence of these breeding-behavior patterns were not yet recognized, let alone understood. Another fly which represents this eighth stage is *Hilara granditarsus* Curran which Melander (1940) observed carrying balloons in Alberta. As in the case of *H. sartor* there was of course no prey included in the gift package.

Kessel, Edward L. 1955. "The Mating Activities of Balloon Flies." *Systematic Zoology* 4, no. 3:97–103.

How to Win Mates and Influence Enemies

John Alcock

John Alcock (b. 1942) is Regents' Professor of Biology at Arizona State University, a behavioral ecologist, and the author of a highly respected textbook, Animal Behavior. *He is also a talented natural history writer whose books include* Sonoran Desert Spring, Sonoran Desert Summer, *and* The Masked Bobwhite Rides Again. *His enthusiasm and aesthetic appreciation for his subject are obvious in the following accounts of battling beetles.*

Despite the transient dip in temperature late in the month, June was even hotter than average. The mean maximum for the month was a little over 106 degrees, and July has taken up the challenge to match or surpass that mark. On red-hot Usery Mountain this morning many saguaros are adorned with fruits that have burst open to reveal deep scarlet flesh. Near the mountaintop, a large insect buzzes noisily past at eye level. The creature cruises on unswervingly for a bit, then turns and ascends until it is in position to drop clumsily onto a fruit atop the trunk. As it lands, another individual scrambles from one fruit to another, its huge orange-knobbed antennae held up in a victory sign. The two beetles, for this is what they are, come face-to-face and grapple with their pincer jaws. The smaller competitor barely makes brief contact with the larger beetle before turning from the unequal contest to tumble away, falling into a crevice between two fruits and then scrambling out of sight of its pursuer.

The battling aeronauts are nearly two inches long, and their size, sweeping antennae, and magnificent color pattern more than compensate for their apparent lack of aerial agility. The orange elytral wing covers are attractively banded in black and covered with a thick but translucent wax that gives the insects the appearance of having been lacquered. Perhaps their most impressive features are their jaws, which protrude from their heads like the robust pincers of a pair of pliers designed for some arcane purpose. It requires no great insight to realize why this insect has the Latin name

Male wood-boring beetle stands guard, by Marilyn Hoff Stewart (from Alcock 1990).

Dendrobias mandibularis. As is characteristic for long-horned, wood-boring beetles of the family Cerambycidae, the big-mandibled beetle also possesses antennae that are substantially longer than its body. Had the Egyptians known about this beetle, they surely would have honored it along with the sacred scarabs, which inspired so many of their artful rings and seals.

Another fight begins on the saguaro top, and one beetle is thrown or falls into space. As it drops it lifts its colorful elytra high in the air and unfolds its pliant transparent wings, which had been stored neatly under the elytral covers. With a mighty whirring the beetle struggles to gain altitude; having succeeded, it turns to sail away on a single-minded, single-direction journey. Its great antennae stream behind like windblown hairs.

A graduate student of mine, Steven Goldsmith, spent several scorching Junes and Julys watching the big-jawed beetles at a study site just off the Beeline Highway. Just off the highway means almost close enough to touch the many cars and trucks roaring along the Beeline. Tractor-trailers sent gusts of superheated wind and exhaust blasting into the desert shrubs on the shoulder of the roadway. Steve somehow became habituated to the unnerving traffic and the heat and calmly carried on, rather like the starlings that casually search for scraps of food on freeways. Necessity was the mother of habituation, in this case, because Steve had to locate a site with a large beetle population and the Beeline was the only place where he could find the numbers he needed.

The beetles transform themselves from wood-boring grubs to magnificent adults in midsummer. After emerging from the trunks of palo verdes, they search for food, which they often find in desert broom. This shrub typically grows sparsely along desert washes, but it does well at roadsides, thanks to extra water from pavement runoff. The Beeline Highway had several shoulder strips of unusually large desert brooms, each shrub growing luxuriantly to heights of two meters or more.

The big-jawed beetles and a host of other insects flock to roadside desert brooms. The plants provide rich supplies of insect food, especially of a sugary sap that oozes from little wounds in the stems of the shrub. The sap spills out into a small area that often attracts a beetle, which laps up the nutritious fluid.

Steve took advantage of the abundance of beetles drawn to broom food at the Beeline site to answer some questions about the natural history of *Dendrobias.* He was particularly interested in why only the larger males in this highly variable species possess the megajaws that give the species its Latin name. The females and the smaller males look very similar in all respects, including their modest mandibles. By measuring the lengths of the jaws of males of different sizes, Steve established that males could be placed in one of two distinct categories: the "majors" are medium to large beetles with jaws that are large in relation to their body, and the "minors" are small to medium beetles with proportionally small jaws. The two types evidently use different formulae for the development of their jaws; majors invest relatively more in pincers than do minors, as is seen when the data are controlled for the standard effects of increasing body size on jaw development. The minors appear to use the same schedule for jaw production employed by females. Indeed, minor males must be carefully inspected to distinguish them from the other sex.

When we find an animal whose males have some distinctive structure that might be used in fighting, the rule is that it is in fact employed to this end, and *Dendrobias* is no exception to the rule. Major males are formidable fighters, thanks in part to their large jaws and in part to their large body size. They snap at, butt, and grapple with other majors on split saguaro fruits and at the ooze sites on desert broom, both of which are limited in number and valuable to males because they attract hungry females, potential mates. Large majors defeat smaller majors, forcing them to abandon an ooze; minor males do not even try to gain access to a feeding site if it is occupied by a major. As a result of the fighting advantage enjoyed by larger males, majors both occupy ooze sites far more often than do minors and account for a disproportionately large percentage of all matings.

Thus we have a species whose receptive females are concentrated at specific feeding locations, little oozing wounds on desert broom small enough to be defended by a single male. Earlier I presented the argument that male animals generally try to secure as many mates as possible because the number of females inseminated usually is closely correlated with the number of offspring a male produces. Therefore, we expect that mate-seeking males will go where they can find the greatest possible number of receptive females. If *emerging* females are receptive and males can identify likely emergence sites, males will try to defend these sites, as do empress butterflies and digger bees. If *feeding* females are sometimes receptive and males can identify especially good feeding sites, males will try to defend these sites, as do *Dendrobias* males.

Knowing the importance of mating success for males and the importance of ooze sites as a food resource for females helps explain the remarkable differences in body design between major males and females of *Dendrobias mandibularis.* There has been competition among males for control of these scarce, valuable, and defensible spots because winners have enjoyed greater than average copulatory success. The competition for mating sites has favored large males with effective fighting equipment, with the evolutionary result that some males now have the capacity to develop large bodies and larger jaws. The males that achieve the potential to be big-bodied and big-mandibled leave more descendants than do smaller individuals with more modest mandibles, thanks to the nature of the feeding ecology of females of their species.

If large males produce more surviving offspring than do small ones, why are there any minor males at all? This puzzle would be resolved if it were true that all *Dendrobias* males have the *potential* to be behemoths with massive mandibles. In order for males to achieve large size, however, they probably have to be fortunate enough to have access to plenty of high-quality food when they are larval grubs. The immature stages of *Dendrobias* feed on palo verde trunks and branches while boring through the host plant. Just as is true for digger bees, the amount of food they consume determines the size they achieve as grubs, and this in turn determines the size of the adult beetle because insects cannot grow any larger after they become adults.

Minor males probably have had bad luck as larvae. Stunted by a shortage of nutritious tissues in the palo verdes that were their homes before metamorphosis, these males fail to become big-bodied adults. Their developmental systems somehow "sense" that they are destined to be small, and so these mechanisms do not allocate the energy to produce large adult jaws, which would be ineffective weapons when wielded by a small individual. If, however, a male grub can secure relatively large amounts of nutritious wood in its palo verde, the beetle's developmental program not only yields a large adult after metamorphosis but also adopts the pathway that ultimately produces massive pincers, which are useful in combat if employed by a large-bodied male.

This explanation for the two types of male cerambycids is speculative, although much evidence from other insects supports the general proposition that adult body size is dependent on food consumed during the larval phase. Furthermore, many examples are known from the insects, such as digger bees, and other animals in which body size or physiological condition correlates with the competitive tactics employed by a male.

Minor males may have experienced misfortune as grubs, and they may defer to their larger rivals at ooze sites on desert broom, but they do not throw in the towel when it comes to mating. Although they assiduously avoid majors, who might well amputate their limbs in a fight, minors wander through desert broom and are quick to mount a female should they encounter one. Female *Dendrobias* will sometimes copulate with males away from a feeding site, and so minors enjoy some reproductive success. In other words, small males seem to do the best they can with the hand dealt them, as we would expect if their behavior is the product of selection for individuals that try to mate as much as possible.

The proposed relationship between localized food resources used by receptive females and the evolution of male fighting weaponry used in defending food resources can be tested. One prediction from this hypothesis is that other male cerambycids will behave like *Dendrobias* if, and only if, the females of these species become concentrated in particular places by their resource requirements. Conversely, other long-horned, wood-boring beetles whose females are not concentrated in space by the need to exploit a patchily distributed resource should have a very different mating system.

Both predictions are supported by at least some examples, although there are surprisingly few studies of cerambycids. One resource-defending species, *Monochamus scutellatus*, is a beetle whose egg-laying females prefer to oviposit in the thickest portions of fallen white pine logs. Aggressive males focus their attention on these parts of the logs, and large individuals win control of territories at these spots. Territorial males reap a reproductive reward because egg-laden females will mate with males that monopolize superior oviposition sites.

Steve Goldsmith studied another Sonoran Desert species, *Perarthrus linsleyi*, whose receptive females are widely distributed through their environment. These entirely unspectacular little cerambycids feed on the stems of brittlebush as larvae. When adults appear, they switch over to creosote bush, from whose flowers males and females extract nectar.

Happily for Steve, creosote bush is one of the commonest and most widely distributed desert plants in Arizona and is not restricted to polluted roadside verges. Therefore, he was able to conduct his study of this species on a quiet rocky hillside in west Phoenix, far from thundering trucks. On Shaw Butte, as elsewhere in the Phoenix area, creosotes are abundant, producing a plethora of small yellow flowers in late March and April. Because many creosote bushes come into flower more or less at the same time, hungry female cerambycids have a great number of plants to choose among. Therefore, there are no small feeding patches with large concentrations of receptive females, and the males of this long-horned, wood-boring beetle are not territorial; instead they fly from bush to bush, hunting for and trying to contact as many females as possible during their allotted time. A male's mating success, and thus the number of descendants he produces, appears to depend primarily on his searching ability. Males fly much more often than females as they race other males to find mates scattered through the desert.

Steve's work provides additional support for the principle that the factors that determine the distribution of females have great influence on the mating tactics of males. The lives of insects may appear chaotic and indecipherable at first glance, but Steve's considerable patience and his willingness to cope with summer on the Beeline enabled him to bring a little more order to the world of long-horned, wood-boring beetles. Not that we should rest upon Steve's laurels. Well over a thousand species of cerambycids inhabit North America and thousands more live elsewhere. Each species represents a chance to test ideas about the evolution of behavior, and it will take the work of many more dedicated graduate students before we can say that we have made good use of the diversity within even this single group of animals.

A year after Steve completed his study, road crews began to widen the Beeline Highway in some stretches and to "clean up" the shoulders in others, all as part of a general plan to accommodate the increasing numbers of Phoenix residents, many of whom sensibly wish to get out of town on summer weekends and up into the mountains that can be reached by going north on the Beeline. Steve's beetle site has been graded into oblivion and the desert broom swept away in a few minutes of roadwork, erasing a miniature ecosystem populated for a few weeks each summer by one of the most magnificent creatures in the Sonoran Desert.

Alcock, John. 1990. *Sonoran Desert Summer.* Tucson: University of Arizona Press.

Fatal Attractions

May Berenbaum

A chapter on sex in insects would not be complete without the thoughts of entomologist May Berenbaum. Here's one of her "Buzzwords" columns on the subject from American Entomologist *magazine.*

If you went to public high school in the early seventies, as I did, then you probably are a survivor of an educational experiment called "health class." I took health class in ninth grade, as did every ninth grader in Pennsylvania, because the powers that be in the state decreed that all ninth graders had to take health class in order to receive a high school diploma. The curriculum was designed to acquaint us with the hazards of sex, drugs, and inadequate personal hygiene. As I recall, we saw a lot of movies and filmstrips that I think were intended to frighten us. They couldn't have been exceedingly effective because I don't remember much about them. What I remember most clearly about health class was the textbook, and I remember that because the nameless student who had used that particular book the year before it was assigned to me had been thoughtful enough to pencil in all of the obscene terms for the various parts of both male and female reproductive tracts, very few of which I knew before, right next to all of the technical terms. Thus, overall, I have to say I found the class dull, but definitely educational.

Driver education class, on the other hand, was terrifying. We saw movies in that class, too, but these films were so frightening that I ended up literally not getting behind the wheel of a car for eleven years after passing the course and getting my license. For the most part, these films depicted unremittingly horrific scenes of highway carnage. Time has dimmed these memories somewhat, and I do drive on occasion around town, but, thanks to a film called *Signal 30,* produced, I think, by the Ohio Traffic Safety Bureau, I don't think I'll ever drive in the state of Ohio.

It seems to me that secondary school educators in the state of Pennsylvania missed a golden opportunity to instill morality through terror merely by virtue of their choice of films. I don't know if they still teach health class to ninth graders in Pennsylvania, but if they do, I recommend that they show a few nature documentaries instead of the movies they showed us. A brief glimpse into the reproductive habits of insects would be enough to put anyone off sex for a long time. Take the courtship ritual of *Calopteran discrepans,* for example. Sivinski describes a presumably typical encounter:

"Males mount dorsally, between the females' slightly spread wings. Examination of coupled pairs showed the sickle-like male mandibles bite through the humeral angle (shoulder) of the female's right elytron. Up to 3 males were found upon a female's back. When such masses were picked up they clung together and were separated only with some effort, leaving bleeding wounds in the female's elytra."

In what can be described only as masterful understatement, Sivinski observes that such "love bites . . . illuminate the different reproductive interests of the sexes."

This sort of mangling actually appears fairly routine among insects. In a study of a dozen species of Nearctic Gomphidae, for example, 88 to 100% of the females examined "had 2–6 holes in their heads resulting from the grip of male abdominal appendages." The gomphids are apparently far rougher than the aeschnids, the male of which merely "gouges the dorsal surface of the female's compound eyes." *Hagenius brevistylus,* North America's largest gomphid, earns distinction of a sort by exhibiting "the most severe head damage due to mating attempts so far discovered in any dragonfly." In this species, "the laterodistal spines of the male epiproct gouged the edge of the female's compound eyes, and punctured the exoskeleton in . . . 32 percent of the females in which the male cerci also punctured the head. A proximodorsal ridge on each side of the male epiproct often . . . cracks the lateral corners of the female occiput. Finally, a distal spine and a mediolateral spine on each male cercus punctures the rear of the female head (postgenae). The pressure of the male grip splits the exoskeleton between the holes made by the cercal spines, resulting in a vertical split in each postgena. Thus a maximally damaged female would have 6 holes of varying sizes punched in her head."

I expect a young, impressionable, female high school student who grows up associating words like "gouge," "puncture," "split," and "punch" with the act of copulation might never yield to temptation, even after twenty years of marriage.

Actually, the girls have it relatively easy in the insect world—though they may be disfigured for life, at least they survive these encounters. There are innumerable accounts of sexual encounters among insects that leave males dead. I don't just mean those stories about praying mantids, which may be somewhat exaggerated. Male *Tribolium* beetles die a particularly horrible kind of death when they're maintained in all-male groups. Male *T. castaneum* beetles kept with females live on average 50 weeks; those in all-male groups die after only 15 weeks. These males die with a "hard whitish deposit at the tip of the abdomen," the apparent result of solidification of seminal fluids upon contact with air. When food particles adhere to the fluids, what results is a solid mass that interferes with the various and sundry functions of the nether end of the abdomen. Then there's the sad fate of *Julodimorpha bakewelli*, an Australian buprestid. These rugose shiny beetles mistake the shiny surfaces of a 370-ml beer bottle (called a "stubbie") with a female buprestid and attempt to mate with it, invariably with less than satisfying results. Entomologists Gwynne and Rentz conducted a short experiment by placing four bottles on the ground: within thirty minutes, six beetles had arrived to hit on the bottles. The hazards of this behavior are that the beetles don't give up; one male apparently died as the result of attack by ants, "biting at the soft portions of his everted genitalia."

It might be argued that knowledge of these fatal attractions would be of little relevance to people—that small, crawling animals have little to do with human sexual practices. Remarkably enough, that's not always the case. There are the occasional pheromonally-mediated interactions between mate-seeking male moths and humans that are doomed to failure (e.g., Cameron, E. A., 1995, *J. Chem. Ecol.* 21: 385–386), but of possibly more widespread occurrence is an unusual convergence of sex, invertebrates, and humans in a form of fetishist known as the "crush-freak."

To define "crush-freak," I refer to the definitive source on the subject, the *American Journal of the Crush-Freak* (1993):

"This is a very unique sexual fantasy, which is part of the foot-fetish. In the 'Crush-Freak's' mind he wishes himself tiny—insectlike—and wants to be stepped on and squashed by the foot of a woman. There are a number of variations on this fetish fantasy. Some of us want only to be stepped on barefoot, some only want to be crushed under the pump of the shiny, high-heeled shoe. Others want to create scenarios in which the female imagines that one male is a bug, and gets her boyfriend to stomp him. Many fetishists must see a female step on a tiny living thing; an insect makes a fine surrogate for the 'Crush-Freak.' "

The *American Journal of the Crush-Freak* is edited by Jeff Vilencia, an aspiring filmmaker and self-avowed crush-freak who, in the biographical information appearing in his journal, admits to fantasizing about "being a bug." I learned of Jeff Vilencia and of his unusual entomological interests when he called me a few months ago after reading about our departmental Insect Fear Film Festival in an article that appeared in *Modern Maturity* magazine (official publication of the American Association of Retired Persons) at his mother's house. Jeff was kind enough to send me a copy of his award-winning short film, "Smush," approximately eight minutes of actress Erika Elizondo crushing earthworms first with bare feet and then with her mother's black, stiletto-heeled pumps. This film was recognized at the Toronto International Film Festival in 1993, the Helsinki Film Festival of 1994, and, somewhat less surprisingly, at the Sick and Twisted Film Festival of 1995, and was written up in the *New York Post* (21 March 1995) and the *Washington Post* (11 December 1993).

Technically speaking, this sort of sexual encounter really has adverse consequences only for the small invertebrate, so I suppose "Smush" really wouldn't be suitable for showing to adolescents to demonstrate to them the hazards of unprotected sex. To be honest, I'm not exactly sure just what the appropriate audience would be for this film. I feel a little bad that I have trouble appreciating the aesthetics of the film because Jeff Vilencia certainly appreciates entomologists. In fact, in his journal he even reviews and rates entomological publications. Of course, he doesn't use the same criteria that, say, I might use in reviewing such a text; his "criteria for inclusion" are that the books must be written by a woman and "that there must be one or more good 'Crush' references."

The issue of the *American Journal of the Crush-Freak* in my possession contains two such book reviews. *Of Bug Busters,* by Bernice Lifton (1991, Avery, Garden City Park, NY). Vilencia excerpts six references to insect crushing, with annotations. Such annotations are often quite succinct—e.g., "p. #212, Ch. 13 ANTS, SPIDERS, AND WASPS . . . *QUOTE:* 'Try to kill the biting spider without squashing it beyond recognition . . . ' *OKAY.*" This is not to say he's entirely uncritical in his praise of entomological texts, though. In his review he states his disappointment that the author uses the term "squash" instead of "squish"; evidently he finds "squish" a more evocative term. He concludes the review with the note, "we can only hope that Ms. Lifton is a young sexy babe with a size 9 or 10 shoe, and loves to step on bugs!"

Rhonda Wassingham Hart also received accolades from Jeff Vilencia for her book *Bugs, Slugs & Other Thugs* (Storey, Pownal, VT). Vilencia's review ends with what must be the ultimate praise for an entomological text—"I can only say that I would love to be a bug in her garden so she could step on me."

I guess the point of all of this is that what is erotic is largely in the mind, not only for high school kids but for grown-ups. My experience with Jeff Vilencia has led me to wonder about who reads the books I've written and what motivated them to buy the books in the first place. On the one hand, it's almost gratifying to think that insect pest management can arouse people's interests to such an extreme extent. On the other hand, it has convinced me not to list my shoe size in the biographical sketch of my next book.

Berenbaum, May. 1996. Fatal Attractions. *American Entomologist,* Summer, 70–71.

No Sex, Please

Tim Hunkin

Hunkin on nonsexual reproduction: The Observer *cartoonist makes an admirable stab at sorting out parthenogenesis, a popular, though not terribly social, method of insect reproduction. For the record,* parthenos *is the Greek word for "virgin," and* genesis *means "origin."*

Hunkin, Tim. 1988. *Almost Everything There Is to Know by Hunkin.* London: Pyramid Books.

THE RUDIMENTS OF WISDOM
PARTHENOGENESIS
(REPRODUCTION WITHOUT SEX)
COMPILED & DRAWN BY HUNKIN
LIFE WITHOUT SEX
SOME ANIMALS APPEAR TO LIVE COMPLETELY WITHOUT SEX. IN CERTAIN SPECIES OF PLANKTON & GALL WASP NO MALES HAVE EVER BEEN FOUND.
WATER FLEAS
IN SPRING & SUMMER, THE ENTIRE POPULATION OF WATER FLEAS IS FEMALE, REPRODUCING FROM UNFERTILISED EGGS. COLDER AUTUMN WEATHER STIMULATES THE PRODUCTION OF A FEW MALES & LARGE 'WINTER EGGS' WHICH THE MALES FERTILISE. ALL THE ADULT WATER FLEAS DIE OFF IN THE WINTER & THE NEXT YEAR'S POPULATION STARTS FROM THE FERTILE EGGS.
BEES
HONEY BEES & SAW FLIES HAVE BOTH SEXUAL & ASEXUAL REPRODUCTION. UNFERTILISED EGGS PRODUCE MALES & FERTILISED EGGS PRODUCE FEMALES (MOST FEMALE BEES ARE STERILE WORKER BEES, ONLY A FEW ARE FERTILE QUEENS).
GREENFLY
GREENFLY HAVE A SIMILAR CYCLE TO WATER FLEAS. THE SUMMER PLAGUES ARE ALL FEMALES.
MICE?
EGGS REMOVED FROM FEMALE MICE CAN BE PROMPTED TO DIVIDE & DEVELOP AS IF FERTILISED WHEN EXPOSED TO HEAT OR ALCOHOL. NONE OF THESE EMBRYOS HAS SO FAR SURVIVED UNTIL BIRTH.
ADVANTAGES
THE ABSENCE OF MALES & SEXUAL ACTIVITY SAVES ENERGY & ALLOWS ANIMAL SPECIES TO MULTIPLY MORE RAPIDLY IN THE SHORT TERM.
DISADVANTAGES
IN THE LONG TERM, ONLY SEXUAL REPRODUCTION, WHICH CONTINUALLY MIXES UP THE GENES, ENABLES A SPECIES TO ADAPT & EVOLVE TO SUIT CHANGING CONDITIONS.
STICK INSECTS
PARTHENOGENESIS IS COMMON AMONG STICK INSECTS & IS PROBABLY AN ADAPTATION FOR THEIR 'STATIC' LIFESTYLE. THEY ARE EXTREMELY WELL CAMOUFLAGED AS LONG AS THEY STAY STILL, REMOVING THE NEED TO GO IN SEARCH OF A MATE HELPS THEIR SURVIVAL.
VIRGIN BIRTH?
A TEST-TUBE BABY CLINIC IN MELBOURNE, AUSTRALIA, HAS RECENTLY REPORTED A HUMAN EGG SPONTANEOUSLY DIVIDING AS IF FERTILISED (IT DID NOT SURVIVE BEYOND THIS STAGE). SOME SCIENTISTS SUGGEST THIS RAISES A FAINT POSSIBILITY THAT WOMEN MAY EXIST WHO ARE CLONES OF THEIR MOTHERS (FEW WOMEN HAVE EVER BEEN TESTED).

8

Metamorphosis

Metamorphosis is a powerful metaphor. Franz Kafka explored the surreal connotations of metamorphosis in his bizarre human fantasy-drama, in which a man wakes up one day as a giant beetle. And countless poets, songwriters, and novelists have symbolized transcendence with the image of an earthbound caterpillar transformed into a delicate butterfly.

Not all insects undergo such extensive changes, however. In the "*hemi*metabolous" insects—dragonflies, mayflies, grasshoppers, aphids, cockroaches, termites, and others—the baby is just a small, wingless version of the adult. It goes through successive molts, getting larger each time, until it is fully grown. In the "*holo*metabolous" insects—flies, wasps, beetles, moths and butterflies, and others—the baby, called the larva, usually looks like a worm (caterpillar, grub, maggot) and, when it is fully grown, enters a resting stage called a pupa (chrysalis, cocoon). We now know that during this phase, the body is extensively remodeled; in fact, large portions of the body are entirely resorbed and replaced by the adult body, which grows out of pockets of cells that have been set aside for this purpose during the entire life of the larva.

Metamorphosis has allowed both hemimetabolous and holometabolous insects to lead two separate lives (sometimes more!), and, in many cases, evolution has taken those lives down completely different paths. For instance, cicada nymphs live underground, feeding on plant roots for many years, then emerge into tree-dwelling adults to sing and mate for a brief month before they die. The larvae of caddisflies live in streams, breathe through gills, and prey on other animals for food, but the adults breathe air and drink nectar.

As we will see, humans have known that maggots change to flies and caterpillars change to butterflies since at least the time of Aristotle, but we are only beginning to unravel the deeper mysteries of how and why metamorphosis takes place.

Of Eggs, Grubs, Nymphas, and Wings

Aristotle

For the first view of metamorphosis, we turn to the father of biology.

In Book 5 of Historia Animalium, *Aristotle (384–322 B.C.) recounts the mating habits of the various then-known kinds of insects. In some cases, the details are so explicit that he even relays the favorite mating positions of this or that species. But some of the information is clearly second-hand and mistaken, or interpreted in view of one or other conventions of the day. Instead of "queen bees," for example, he calls them "king bees." The idea that the social wasps, ants, and bees were essentially all-female societies might well have amazed Aristotle. The limitations of having no microscope led Aristotle to the conclusion that some of the smallest "insects are not derived from living parentage, but are generated spontaneously."*

British scientist Sir D'Arcy Wentworth Thompson (1860–1948) (author, On Growth and Form*) spent many years off and on translating* Historia Animalium *and making detailed footnotes, five or more per page; in the preface he says he "felt constrained to omit much that I had written, especially on the zoological side of my commentary. To annotate, illustrate, and criticize Aristotle's knowledge of natural history is a task without an end."*

Still, it remains extraordinary how much Aristotle was able to glean from nature and through his inquiring mind about biology, especially considering his other defining contributions in philosophy and other branches of science. He reports on metamorphosis in considerable detail, and on the discovery that fabric could be made from the cocoons of certain caterpillars, some eight hundred years before the true silkworm came to Greece. As the first natural scientist, or biologist, Aristotle created a body of work that would not be superceded for more than two thousand years after his death. No other natural scientist can make that claim.

The so-called psyche or butterfly is generated from caterpillars which grow on green leaves, chiefly leaves of the raphanus, which some call crambe or cabbage. At first it is less than a grain of millet; it then grows into a small grub; and in three days it is a tiny caterpillar. After this it grows on and on, and becomes quiescent and changes its shape, and is now called a chrysalis. The outer shell is hard, and the chrysalis moves if you touch it. It attaches itself by cobweb-like filaments, and is unfurnished with mouth or any other apparent organ. After a little while the outer covering bursts asunder, and out flies the winged creature that we call the psyche or butterfly. At first, when it is a caterpillar, it feeds and ejects excrement; but when it turns into the chrysalis it neither feeds nor ejects excrement.

The same remarks are applicable to all such insects as are developed out of the grub, both such grubs as are derived from the copulation of living animals and such as are generated without copulation on the part of parents. For the grub of the bee, the anthrena, and the wasp, whilst it is young, takes food and voids excrement; but when it has passed from the grub shape to its defined form and become what is termed a 'nympha', it ceases to take food and to void excrement, and remains tightly wrapped up and motionless until it has reached its full size, when it breaks the formation with which the cell is closed, and issues forth. . . .

From one particular large grub, which has as it were horns, and in other respects differs from grubs in general, there comes, by a metamorphosis of the grub, first a caterpillar, then the cocoon, then the *necydalus;* and the creature passes through all these transformations within six months. A class of women unwind and reel off the cocoons of these creatures, and afterwards weave a fabric with the threads thus unwound; a Coan woman of the name of Pamphila, daughter of Plateus, being credited with the first invention of the fabric. After the same fashion the carabus or stag-beetle comes from grubs that live in dry wood: at first the grub is motionless, but after a while the shell bursts and the stag-beetle issues forth.

From the cabbage is engendered the cabbage-worm, and from the leek the prasocuris or leekbane; this creature is also winged. From the flat animalcule that skims over the surface of rivers comes the oestrus or gadfly; and this accounts for the fact that gadflies most abound in the neighbourhood of waters on whose surface these animalcules are observed. From a certain small, black and hairy caterpillar comes first a wingless glow-worm; and this creature again suffers a metamorphosis, and transforms into a winged insect named the bostrychus (or hair-curl). . . .

Thompson, D'Arcy Wentworth. 1910. *Historia animalium* (Vol. IV of *The Works of Aristotle,* [ed.] J. A. Smith and W. D. Ross). Oxford, England: Oxford University Press.

The Wondrous Transformation of Caterpillars

Maria Sibylla Merian

Maria Sibylla Merian (1647–1717) was a pioneer entomologist and artist who drew, painted, and published handsome books on the insect life of Europe and the tropical Americas, focusing on metamorphosis. Such was her skill that Linnaeus, when he came to designating insect species some decades later, used her illustrations and descriptions as primary sources, considering them as good as actual specimens. As an artist, she has been celebrated as the spiritual ancestor of Audubon, Gould, and the finest zoological artists of later centuries.

A native of Frankfurt, with a Swiss father, Merian grew up in the late seventeenth century in a family of artists. In school Merian became an expert engraver and embroiderer and came under the influence of the Dutch school of painting. But from her childhood, she nourished an interest in rearing caterpillars, starting with silkworms and extending to many other species.

At age eighteen Merian married (probably an arranged marriage) and had two daughters, but she kept up her painting, producing her first book in 1679, the year after the birth of her second daughter, when she was thirty-four. The title, as was the style in those days, says it all. Here's a translation from the German: The wondrous metamorphosis and peculiar plant-nourishment of caterpillars, wherein, by means of an entirely new enquiry, the origin, food and changes of caterpillars, worms, butterflies, moths, flies and other similar little creatures, together with their times, place and special characteristics, for the service of naturalists, artists and lovers of gardens, are diligently investigated, shortly described, depicted from life, engraved on copper and personally published by Maria Sibylla Gräffin, daughter of the late Matthäeus Merian the Elder.

After twenty years of marriage, she drifted away from her husband, resumed her maiden name, Merian, and moved to the

Netherlands. In June 1699, having obtained a stipend through scientists in Amsterdam who loved her work, Merian and her twenty-one-year-old daughter Dorothea boarded a Dutch ship for the two-month journey to Paramaribo, Surinam. Working under difficult conditions, she collected and painted the extraordinary tropical world all around her, again focusing on insects and their food plants and on metamorphosis. After two years, she returned to Holland and prepared her masterwork, The Metamorphosis of the Insects of Surinam. *Published in 1705, it was one of the finest examples of bookmaking in its day, with superb engraving, and some editions hand-colored by Merian and her daughter.*

This fine and prettily-striped caterpillar I have come across in August on fennel. . . . It uses this plant for its food. These caterpillars are a beautiful green in colour and have black stripes like velvet and on the stripes goldish yellow flecks. On any sharp contact they immediately put out two yellow horns from the front of their heads like a snail. Also at the front, in this case on the underpart, they have on each side three small pointed feet (pedicels) or claws followed by two blind segments without pedicels. There follow four segments underneath which, on each side, there are four rounded pedicels. These are followed by two more blind segments, then right at the back a further two rounded pedicels with which they can hold on to things very tightly. If they have no fennel they can also eat turnips. Gardeners call this caterpillar the 'Oebser' [fruit picker] because they think it does much damage to fruit though I have not . . . found it on any other plant than fennel and turnips. But it has a peculiar smell like fruit when many different kinds are stored together. When the caterpillar has reached its full size it sheds its coat or skin entirely and this it leaves hanging above it as I have illustrated. It fastens itself to a wall with its head downwards and attaches the hind part of its body as firmly as if it were glued on. In the middle of its body it spins a white thread round itself in order to stay firmly suspended. Then within half a day it turns into a date stone [a pupa], in shape like a baby wrapped in swaddling clothes so that one can almost detect a human face in it. I have pictured one here hanging on the fennel. These 'date stones' are grey, in part also green, in colour. In this form they hang until April or May. It is true that I have had some emerging in December but I put this down to my having kept them in a warm room. The butterfly, here shown already emerged from the pupa, has four wings. The upper two wings are fine yellow and black and so are the lower two also, except that on the dotted areas or fields they are fine blue as is also the lowermost egg-shaped field, though some red can also be seen there. The body remains

The metamorphosis, by Maria Sibylla Merian (from *Erucarum Ortus,* 1718).

black and yellow and has six black feet and in front on the head it has a long proboscis which it coils right up and if one puts a piece of sugar next to it then it places the proboscis on the sugar as if it meant to eat it; also with the proboscis it sucks nectar from flowers, as I have also observed.

Merian, Maria Sibylla. 1718. *Erucarum ortus,* in *The Wondrous Transformation of Caterpillars: Fifty Engravings Selected from Erucarum Ortus (1718),* with an introduction by William T. Stearn. (Reprinted 1978) Scolar Press.

Everyday Miracles

William Kirby and William Spence

No comprehensive reference work summarizing insect biology existed prior to the 1846 publication of An Introduction to Entomology, or, Elements of The Natural History of Insects: Comprising an Account of Noxious and Useful Insects, of Their Metamorphoses, Food, Stratagems, Habitations, Societies, Motions, Noises, Hybernation, Instinct, Etc. Etc. *by the Reverend William Kirby and William Spence. Their book filled a gaping vacuum and went through many editions, each more comprehensive than the last. Most of the chapters still provide excellent reading, summarizing widespread accounts of fascinating biology in a wonderfully nineteenth-century literary style.*

Here are Kirby and Spence trying to explain the mysteries of metamorphosis for their mid-nineteenth-century readers.

Were a naturalist to announce to the world the discovery of an animal which for the first five years of its life existed in the form of a serpent; which then penetrating into the earth, and weaving a shroud of pure silk of the finest texture, contracted itself within this covering into a body without external mouth or limbs, and resembling, more than any thing else, an Egyptian mummy; and which, lastly, after remaining in this state without food and without motion for three years longer, should at the end of that period burst its silken cerements, struggle through its earthy covering, and start into day a winged bird,—what think you would be the sensation excited by this strange piece of intelligence? After the first doubts of its truth were dispelled, what astonishment would succeed! Amongst the learned, what surmises!—what investigations! Amongst the vulgar, what eager curiosity and amazement! All would be interested in the history of such an unheard-of phenomenon; even the most torpid would flock to the sight of such a prodigy.

But, you ask, "To what do all these improbable suppositions tend?" Simply to rouse your attention to the *metamorphoses* of the insect world, almost as strange and surprising, to which I am now about to direct your view,—miracles which, though scarcely surpassed in singularity by all that poets have feigned, and though actually wrought every day beneath our eyes, are, because of their commonness, and the minuteness of the objects, unheeded alike by the ignorant and the learned.

The butterfly which amuses you with his aërial excursions, one while extracting nectar from the tube of the honeysuckle, and then, the very image of fickleness, flying to a rose as if to contrast the hue of its wings with that of the flower on which it reposes, did not come into the world as you now behold it. At its first exclusion from the egg, and for some months of its existence afterwards, it was a worm-like caterpillar, crawling upon sixteen short legs, greedily devouring leaves with two jaws, and seeing by means of twelve eyes so minute as to be nearly imperceptible without the aid of a microscope. You now view it furnished with wings capable of rapid and extensive flights: of its sixteen feet ten have disappeared, and the remaining six are in most respects wholly unlike those to which they have succeeded; its jaws have vanished, and are replaced by a curled-up proboscis suited only for sipping liquid sweets; the form of its head is entirely changed,—two long horns project from its upper surface; and, instead of twelve invisible eyes, you behold two, very large, and composed of at least seventeen thousand convex lenses, each supposed to be a distinct and effective eye!

Were you to push your examination further, and by dissection to compare the internal conformation of the caterpillar with that of the butterfly, you would witness changes even more extraordinary. In the former you would find some thousands of muscles, which in the latter are replaced by others of a form and structure entirely different. Nearly the whole body of the caterpillar is occupied by a capacious stomach. In the butterfly this has become converted into an almost imperceptible thread-like viscus; and the abdomen is now filled by two large packets of eggs, or other organs not visible in the first state. In the former, two spirally-convoluted tubes were filled with a silky gum; in the latter, both tubes and silk have almost totally vanished; and changes equally great have taken place in the economy and structure of the nerves and other organs.

What a surprising transformation! Nor was this all. The change from one form to the other was not direct. An intermediate state not less singular intervened. After casting its skin even to its very jaws several times, and attaining its full growth, the caterpillar attached itself to a leaf by a silken girth. Its body greatly contracted: its skin once more split asunder, and disclosed an oviform mass, without exterior mouth, eyes, or limbs, and exhibiting no other symptom of life than a slight motion when touched. In this state of death-like torpor, and without tasting food, the insect existed for several months, until at length the tomb burst, and out of a case not more than an inch long, and a quarter of an inch in diameter, proceeded the butterfly before you, which covers a surface of nearly four inches square.

Almost every insect which you see has undergone a transformation as singular and surprising, though varied in many of its circumstances. That active little fly, now an unbidden guest at your table, whose delicate palate selects your choicest viands, one while extending his proboscis to the margin of a drop of wine, and then gaily flying to take a more solid repast from a pear or a peach; now gamboling with his comrades in the air, now gracefully currying his furled wings with his taper feet, was but the other day a disgusting grub, without wings, without legs, without eyes, wallowing, well pleased, in the midst of a mass of excrement.

The "grey-coated gnat," whose humming salutation, while she makes her airy circles about your bed, gives terrific warning of the sanguinary operation in which she is ready to engage, was a few hours ago the inhabitant of a stagnant pool, more in shape like a fish than an insect. Then to have been taken out of the water would have been speedily fatal; now it could as little exist in any other element than air. Then it breathed through its tail; now through openings in its sides. Its shapeless head, in that period of its existence, is now exchanged for one adorned with elegantly tufted antennae, and furnished, instead of jaws, with an apparatus more artfully constructed than the cupping-glasses of the phlebotomist—an apparatus, which, at the same time that it strikes in the lancets, composes a tube for pumping up the flowing blood.

The "shard-born beetle," whose "sullen horn," as he directs his "droning flight" close past your ears in your evening walk, calling up in poetic association the lines in which he has been alluded to by Shakespeare, Collins, and Gray, was not in his infancy an inhabitant of air; the first period of his life being spent in gloomy solitude, as a grub, under the surface of the earth. The shapeless maggot . . .—but it is needless to multiply instances, a sufficient number has been adduced to show, that the apparently extravagant supposition with which I set out may be paralleled in the insect world; and that the metamorphoses of its inhabitants are scarcely less astonishing than would be the transformation of a serpent into an eagle.

Kirby, William, and William Spence. 1846. *An Introduction to Entomology; or, Elements of the Natural History of Insects,* 6th ed. Philadelphia: Lea and Blanchard.

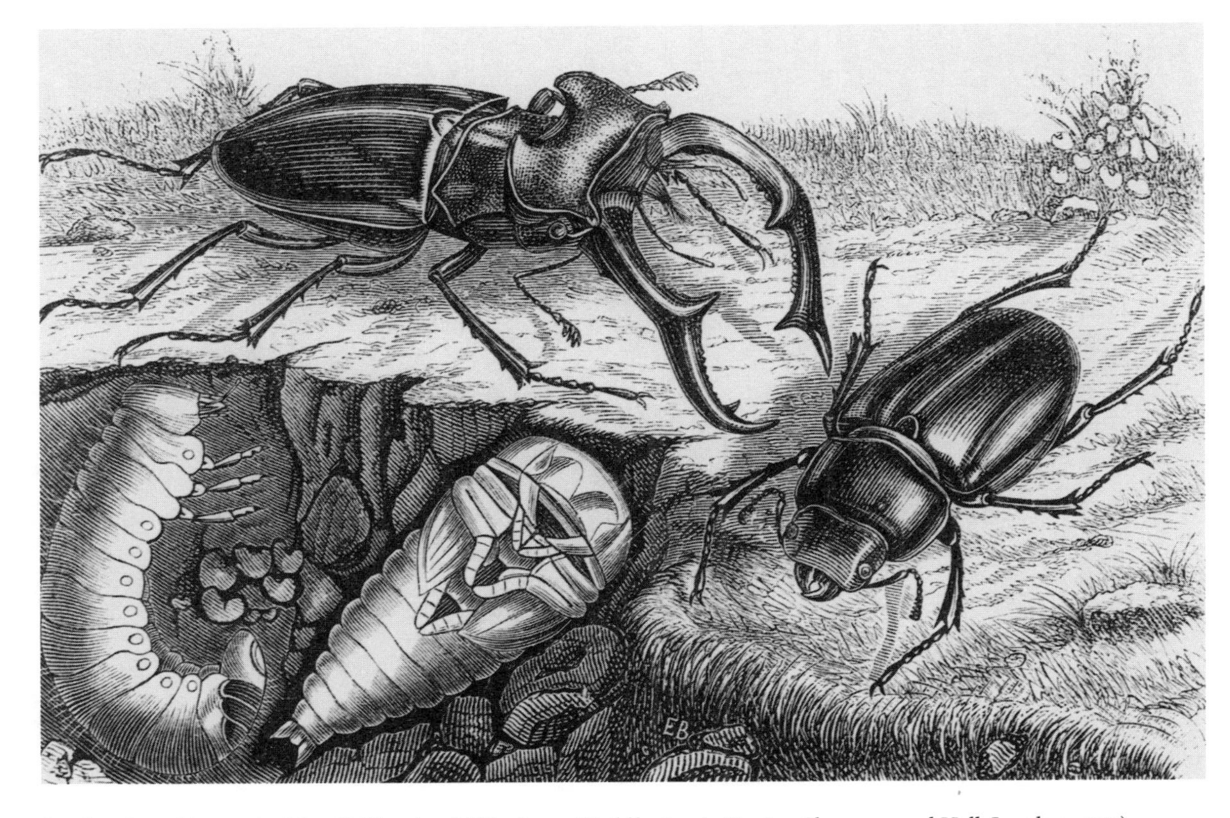

Stag beetle and larvae, by Mm. E. Blanchard (*The Insect World* by Louis Figuier, Chapman and Hall, London, 1869).

The Mediterranean Worm Lion

William Morton Wheeler

In his day, William Morton Wheeler (1865-1937) of Harvard University became one of the best-known biologists in the United States. An erudite writer and speaker, Wheeler increased the visibility of entomology in American life and letters. Toward the end of his life, Wheeler wrote a book on the ant lion and the worm lion, two odd insects that as larvae build funnel-like traps in the sand and lie in wait for ants and other unsuspecting insects to fall in. From Demons of the Dust, *this is Wheeler's account of the Mediterranean worm lion, which metamorphoses from a fierce predator into a delicate, short-lived, rather meek adult fly.*

During the summer of 1925, as guests of Mr. Allison V. Armour on his yacht, the 'Utowana,' my friend Dr. David Fairchild and I were able to visit the three Balearic Islands, Iviza, Majorca and Minorca, and to explore many of their outlying districts for plants and insects. While on the same expedition I had previously found the larvae of *Lampromyia canariensis* on the Island of Teneriffe and was therefore constantly hoping to encounter those of *V. vermileo.* I could find no traces of them on Iviza, but secured many on Majorca and Minorca. . . .

The *Vermileo* larvae collected in the Balearic Islands were kept in tins of dust and fine sand on board the 'Utowana' and fed with workers of a termite belonging to the genus *Calotermes,* which I had found in old stumps in the laurel forest at Las Mercedes on the Island of Teneriffe. When my supply of these insects was exhausted I had nothing with which to feed them between Aug. 31 and Sept. 22, but by the latter date I had arrived with them in Boston and had established them in pans of fresh sand or of kaolin. They were then fed with various ants (*Tapinoma sessile, Lasius brevicornis* and *Formica subsericea*) till late October when they were consigned to a room kept at a temperature of 40° F. for the winter. They were returned to the heated laboratory the middle of March 1926 and were for some weeks abundantly fed with workers of a species of *Calotermes* from Arizona. They throve on this diet and began to pupate early in April. The first fly emerged April 18, and more and more of them appeared from day to day. June 13 I took the remaining larvae, about one hundred in number, to Colebrook, in the Litchfield Hills of Connecticut. For several weeks the weather was unfavorable to the rapid development of the pupae, because long spells of cold alternated with short, hot spells, and although the number of pupae kept increasing, the emergence of the flies was rather irregular and continued throughout the summer. At Colebrook the larvae had been fed so abundantly with our common termite, *Reticulitermes flavipes,* that even those that were very small when they were collected in the Balearic Islands during the preceding summer, had grown to their full size and produced flies. The last individual emerged Sept. 4. I feel confident, however, that many of the larvae, had they remained in their native environment and been constrained to subsist on the few passing ants and other small insects that fell into their pits, would have matured much more slowly and have lived through a second winter before pupating. In other words, they would have behaved like the biennial larvae described by von Siebold, who fed them with ants and miscellaneous insects, which are much less nutritious than fat worker termites.

The larva and pupa of the Balearic *Vermileo* are . . . similar to the corresponding stages of the other Vermileoninae. . . . One point of interest which has not been sufficiently noticed by previous observers, relates to the ecdysis, or moulting of the larva. This occurs repeatedly during larval growth but apparently at rather irregular intervals, depending on the rate of growth, which in turn depends on the amount of food the larva has been able to capture and assimilate. To determine the precise number of moults and the time of their occurrence would require very careful rearing of a number of isolated larvae from hatching to pupation, and this I have not been able to accomplish in the case of *V. vermileo.* . . .

The pupal period varies with the temperature and extends, according to my estimates, over two to three weeks. At first the pupa lies horizontally in the sand one to three centimeters beneath its surface, but when the time for the emergence of the fly arrives, it wriggles up head foremost and thrusts its head and thorax above the surface of the sand. The dorsal thoracic cuticle then ruptures with the usual 'orthorrhaph,' or T-shaped fissure and the imago slowly escapes. It is during the wriggling of the pupa to the surface that the sandgrains glued so firmly to its surface subserve, I believe, an important function. The abdomens of many Lepidopteran and Dipteran pupae, that before eclosion wriggle through burrows or yielding substances such as soil or decomposing wood to a point where the delicate adult can at once issue unimpeded into the open air, are provided with rows of backwardly directed, chitinous teeth or spines, which facilitate locomotion like so many tiny claws, but the *Vermileo* pupa is furnished with only very feeble and therefore useless vestiges of such structures. Their function is assumed by the tightly adherent sandgrains which obviously afford an admirable substitute. Since the abdomen of the young pupa remains contracted and the intersegmental membranes are not exposed while the layer of sand-grains is being glued to its surface, it is possible for the adult pupa to move its abdominal segments very freely while climbing to the surface. One might maintain that the pupa in thus crudely selecting and fixing to its cuticle the sharp sandgrains of its immediate environment and in employing them as an aid in locomotion, is really behaving as a tool-using animal.

Unless there is some difficulty or abnormality that delays eclosion, the flies emerge only during the night and are found in the morning resting on the perpendicular sides of the dishes in the same position as other Rhagionids (*Rhagio, Atherix,* etc.), that is, with the head directed upward and the wings forming acute angles with the long axis of the abdomen as in the resting housefly. They may also rest, though much less frequently, with one wing overlapping the other, as described and figured by Degeer. This position of the wings is never assumed in many other Rhagionids (*Rhagio, Atherix, Chrysopilus, Symphoromyia,* etc.). The flight of *V. vermileo* when disturbed is like that of some gnats and Empidids, exquisitely light and graceful, and with the legs, especially the long hind pair, dangling. . . .

The *V. vermileo* flies which I reared are of a luteous yellow color. Even the wings are very faintly tinged with yellowish, the eyes in life are dark green, the head behind the insertions of the antennae is black, the dorsal surface of the thorax ornamented with four parallel, longitudinal, brown stripes. . . .

The sexes are easily distinguished by the shape of the abdomen, that of the male being much narrower and more cylindrical than that of the female. Unfortunately, the adult insect, unlike the larva, is very delicate and short-lived. Not one of the numerous specimens which I reared survived more than 36 to 48 hours. Many of them, indeed, perished towards the end of the first day of their adult existence. I sought to prolong their lives by feeding them with honey or sugar-water but none of them would even imbibe water, though they have well-developed mouthparts. Strangely, too, though both males and females often emerged during the same night and for several hours rested or flew about together, they exhibited not the slightest inclination to mate. I failed, therefore, to obtain any eggs or young larvae. This failure was the more disappointing because no European entomologist has ever seen the eggs or early stages of any of the Vermileoninae.

Wheeler, William Morton. 1930. *Demons of the Dust.* New York: W. W. Norton.

Mexican Jumping Beans. Real! Live!

Chaparral Novelties, Inc.

Mexican Jumping Beans, sold as novelties in little plastic boxes in souvenir shops in the southwestern United States, afford unsuspecting buyers their first look at this vernacular insect. These are not really beans; they are hollow seeds that contain live caterpillars. One can only admire the folks at Chaparral Novelties, involved in a business that couldn't be particularly lucrative and selling a product with a limited shelf life that effectively ends when the "jumping" Cydia deshaisiana *caterpillar metamorphoses into an adult moth! Buy a cheap "toy"; get a developmental biology lesson in the bargain.*

MEXICAN JUMPING BEANS
Real! Live!
Genuine Jumping Beans
From Old Mexico

JUMPING BEANS 99¢

Story of the Jumping Bean

The MEXICAN JUMPING BEAN is a unique specimen of a living creature, produced through the cooperation of an insect and plant life. Each bean actually contains a tiny caterpillar about 3/16 of an inch long!

When the Yerba de la flecha tree blooms in Old Mexico, certain moths lay eggs in the blossoms. A pod then forms and the egg, inside, hatches into a caterpillar. He eats the seed in the pod, leaving the shell intact, and this shell is his home for the next few months.

The restlessness of the caterpillar flinging himself against the walls of his prison is what causes the bean to hop and roll. He has consumed all the food he will eat in his lifetime, yet his great energy will keep him alive and jumping for several months.

NON-TOXIC
CAUTION: NON EDIBLE
NOT RECOMMENDED FOR CHILDREN
UNDER 6 YEARS OF AGE

Product description from Chaparral Novelties, Inc., Alamogordo, New Mexico.

The Double Life

Robert Evans Snodgrass

R. E. Snodgrass (1875–1962) was, is, and may long remain the god of insect morphology. In the late nineteenth century, he was educated in biology and worked as a university instructor, but he also worked as a commercial illustrator and cartoonist. Throughout his life, he authored a number of extremely influential books on insect biology and anatomy, the foremost being The Principles of Insect Morphology. *Snodgrass's beautiful, zen-simple illustrations for this book have turned up regularly in successive generations of entomology textbooks and probably will continue to do so far into the future.*

The fascination of mythology and the charm of fairy tales lie in the power of the characters to change their form or to be changed by others. Zeus would court the lovely Semele, but knowing well she could not endure the radiance of a god, he takes the form of a mortal. Omit the metamorphosis, and what becomes of the myth? And who would remember the story of Cinderella if the fairy godmother were left out? The flirtation between the heroine and the prince, the triumph of beauty, the chagrin of the haughty sisters—these are but ingredients in the pot of common fiction. But the transformation of rats into prancing horses, of lizards into coachman and lackeys, of rags into fine raiment—this imparts the thrill that endures a lifetime!

It is not surprising, then, that the insects, by reason of the never-ending marvel of their transformations, hold first place in every course of nature study in our modern schools, or that nature writers of all times have found a principal source of inspiration in the "wonders of insect life." Nor, finally, should it be made a matter of scorn if the insects have attached themselves to our emotions, knowing how ardently the natural human mind craves a sign of the supernatural. The butterfly, spirit of the lowly caterpillar, has thus been exalted as a symbol of human resurrection, and its image, carved on graveyard gates, still offers hope to those unfortunates interred behind the walls.

Metamorphosis is a magic word, in spite of its formidable appearance; but rendered into English it means simply "change of form." Not every change of form, however, is a metamorphosis. The change of a kitten into a cat, of a child into a grown-up, of a small fish into a large fish are not examples of metamorphosis, at least not of what is called metamorphosis. There must be something spectacular or unexpected about the change, as in the transformation of the tadpole into a frog, the change of the wormlike caterpillar into a moth, or of a maggot into a fly. This arbitrary limiting of the use of a word that might, from its derivation, have a much more general meaning, is a common practice in science, and for this reason every scientific term must be defined. Metamorphosis, then, as it is used in biology, signifies not merely a change of form, but a particular kind or degree of change; the kind of change, we might say, that would appear to lie outside the direct line of development from the egg to the adult. . . .

Animals live for business, not for pleasure; and all their instincts and their useful structures are developed for practical purposes. Therefore, where the young and the adult of any species differ in form or structure, we may be sure that each is modified for some particular purpose of its own. The two principal functions of any animal are the obtaining of food for its own sustenance, and the production of offspring. The adult insect is necessarily the reproductive stage, but in most cases it must support itself as well; the immature insect has no other object in life than that of feeding and of preparing itself for its transformation into the adult. The feeding function, however, involves most of the activities and structures of the animal, including its adaptation to its environment, its modes of locomotion, its devices for avoiding enemies, its means of obtaining food. Hence, in studying any young insect, we must understand that we are dealing almost exclusively with characters that are adaptive to the feeding function.

When we observe the life of any caterpillar we soon realize that its principal business is that of eating. The caterpillar is one creature, at least, that may openly proclaim it lives to eat. Whatever else it does, except acts connected with its transformation, is subservient to the function of procuring food. Most species feed on plants and live in the open; but some tunnel into the leaves, into the fruit, or into the stem or wood. Other species feed on seeds, stored grain, and cereal preparations. The caterpillars of the clothes moths, however, feed on animal wool, and a few other caterpillars are carnivorous.

The whole structure of the caterpillar betokens its gluttonous habits. Its short legs keep it in close contact with the food material; its long, thick, wormlike body accommodates an ample food storage and gives space for a large stomach for digestive purposes; its hard-walled head supports a pair of strong jaws, and since the caterpillar has small use for eyes or antennae, these organs are but little developed. The muscle system of the caterpillar presents a wonderful exhibition of complexity in anatomical structure, and gives the soft body of the insect the power of turning and twisting in every conceivable manner. In contrast to the caterpillar, the moth or the butterfly feeds but little, and its food consists of liquids, mostly the nectar of flowers, which is rich in sugars and high in energy-giving properties but contains little or none of the tissue-building proteins.

When we examine the young of other insects that differ markedly from the parent form, we discover the same thing about them, namely, the general adaptation of their body form and of their habits to the function of eating. Not all, however, differ as widely from the parent as does the caterpillar from the moth. The young of some beetles, for example, more closely resemble the adults except for the lack of wings. Most of the adult beetles, too, are voracious feeders, and are perhaps not outdone in food consumption by the young. But here another advantage of the double life is demonstrated, for usually the grub and the adult beetle have different modes of life and live in quite different kinds of places. Each individual of the species, therefore, occupies at different times two distinct environments during its life and derives advantages from each. It is true that with some beetles, the young and the adults live together. Such cases, however, are only examples of the general rule that all things in nature show gradations; but this condition, instead of upsetting our generalizations, furnishes the key to evolution, by which so many riddles may be solved.

The grub of the bee or the wasp gives an excellent example of the extreme specialization in form that the young of an insect may take on. The creature spends its whole life in a cell of the comb or the nest where it is provided with food by the parents. Some of the wasps store paralyzed insects in the cells of the nest for the young to feed on; the bees give their young a diet of honey and pollen, with an admixture of a secretion from a pair of glands in their own bodies. The grubs have nothing to do but to eat; they have no legs, eyes, or antennae; each is a mere body with a mouth and a stomach. The adult bees consume much honey, which, like its constituent, nectar, is an energy-forming food; but they also eat a considerable quantity of protein-containing pollen. Yet it is a great advantage to the bees in their social life to have their young in the form of helpless grubs that must stay in their cells until full-grown, when, by a quick transformation, they can take on the adult form and become at once responsible members of the community. Any parents distracted by the incorrigibilities of their offspring in the adolescent stage can appreciate this.

The young mosquito lives in the water, where it obtains its food, which consists of minute particles of organic matter. Some species feed at the surface, others under the surface or at the bottom of the water. The young mosquito is legless and its only means of progression through the water is by a wiggling movement of the soft cylindrical body. It spends much of its time, however, just beneath the surface, from which it hangs suspended by a tube that projects from near the rear end of the body. The tip of the tube just barely emerges above the water surface, where a circlet of small flaps spread out flat from its margin serves to keep the creature afloat. But the tube is primarily a respiratory device, for the two principal trunks of the tracheal system open at its end and thus allow the insect to breathe while its body is submerged.

The adult mosquito, as everybody knows, is a winged insect, the females of which feed on the blood of animals and must go after their victims by use of their wings. It is clear, therefore, that it would be quite impossible for a young mosquito, deprived of the power of flight, to live the life of its parents and to feed after the manner of its mother. Hence, the young mosquito has adopted its own way of living and of feeding, and this has allowed the adult mosquitoes to perfect their specialties without inflicting a hereditary handicap on their offspring. Thus again we see the great advantage which the species as a whole derives from the double life of its individuals.

The fly will only give another example of the same thing. The specialized form of the young fly, the maggot, which is adapted to the requirements of quite a different kind of life from that of the adult fly, relieves the latter from all responsibilty to its offspring. As a consequence, the adult fly has been able to adapt its structure, during the course of evolution, to a way of living best suited to its own purposes, unhampered as it would be if its characters were to be inherited by the young, to whom they would become a great impediment, and probably a fatal handicap.

A [basic] principle of metamorphosis, then, we may say, is that *the species as a whole has acquired an advantage by a double mode of existence, which allows it to take advantage of two environments during its lifetime, one suited to the functions of the young, the other to the functions of the adult.*

We noted, in passing, that the young insect is free to live its own life and to develop structures suited to its own purposes under one proviso, which is that it must eventually revert to the form of the adult of its species. At the period of transformation, the particular characters of the young must be discarded, and those of the adult must be developed.

Insects such as the grasshoppers, the katydids, the roaches, the dragonflies, the aphids, and the cicadas appear in the adult form when the young sheds its skin for the last time. The change that has produced the adult, however, began at an earlier period, and the apparently new creature was partially or almost entirely formed within the old skin before the latter was finally shed. After the molt, only a few last alterations in structure and some final adjustments are made while the wings and legs of the creature that had been confined in the closely fitting skin expand to their full length. The structural changes accomplished after the molt, however, vary with different species of insects, and with some they involve a considerable degree of actual growth and change in the

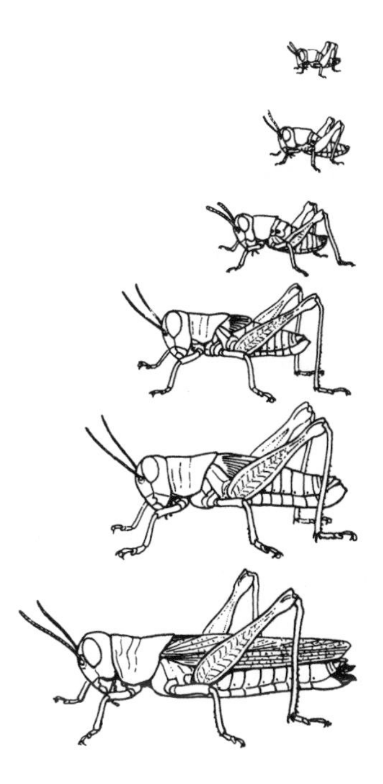

The metamorphosis of a grasshopper, showing its six stages of development from the newly hatched nymph to the fully winged adult. Illustration by R. E. Snodgrass (*Insects: Their Ways and Means of Living* by Robert Evans Snodgrass, Smithsonian Institution, Washington, D.C., 1930).

form of certain parts. The true transformation process, then, is really a period of rapid reconstructive growth preceding and following the molt, in which the shedding of the skin is a mere incident like the raising of the curtain for a new act in a play. During the intermission the actors have changed their costumes, the old scenery has been removed, and the new has been set in place. Thus it is with the insect at the time of its transformation—the special accouterments of the young have been removed, and those of the adult have been put on.

The life of the insect, however, would not make a good theatrical production; it is too much of the nature of two plays given by the same set of actors. The young insect is dressed for a performance of its own in a stage setting appropriate to its act; the adult gives another play and is costumed accordingly. The actor is the same in each case only in the continuity of his individuality. His rehabilitation between the two acts will differ in degree according to the disparity between the parts he plays, that is, according to how far each impersonation is removed from his natural self. . . .

Where the structural disparity between the young and the adult is not great, or is mostly in the external form of the body, the young insect changes directly into the adult, as we have seen in the case of the grasshopper and the cicada. But with many insects, either because of the degree of difference that has arisen between the young and the adult, or for some other reason, the processes of transformation are not accomplished so quickly and require a longer period for their completion. In such cases, the creature that issues at the last shedding of the skin by the young insect is in a very unfinished state, and must yet undergo a great amount of reconstruction before it will attain the form and structure of the fully adult insect. This happens in all the groups of the more highly evolved insects, including the beetles; the moths and butterflies; the mosquitoes and flies; the wasps, bees, ants; and others. The newly transformed insect must remain in a helpless condition without the use of its legs and wings for a period of time varying in length with different species, until the adult organs, particularly the muscles, are completely formed. . . .

It is one thing to know the facts and to see the meaning of metamorphosis; it is quite another to understand how it has come about that an animal undergoes a metamorphic transformation, and yet another to discover how the change is accomplished in the individual. Metamorphosis can be only a special modification of general developmental growth, and growth toward maturity by the individual goes over the same field that the species traversed in its evolution. Yet, the individual in its development may depart widely from the path of its ancestors. It may make many a detour to the right or the left; it may speed up at one place and loiter along at another; and, since the individual is rather an army of cells than a single thing, certain groups of its cells may forge ahead or go off on a bypath, while others lag behind or stop for a rest. Only one condition is mandatory, and this is that the whole army shall finally arrive at the same point at the same time. In each species, the deviations from the ancestral path, traveled for many generations, have become themselves fixed and definite trails followed by all individuals of the species. The development of the individual, therefore, may thus come to be very different from the evolutionary history of its species; and the life history of an insect with complete metamorphosis is but an extreme example of the complex course that may result when a species leaves the path of direct development to wander in the fields along the way.

The larva and the adult insect have become in many cases so divergent in structure, as a result of their separate departures from the ancestral path, that the embryo has become almost a double creature, comprising one set of cells that develop directly into the organs of the embryo and another set held in reserve to build up the adult organs at the end of the larval life.

Snodgrass, Robert Evans. 1930. *Insects: Their Ways and Means of Living.* Washington, D.C.: Smithsonian Institution Series. (Reprinted, New York: Dover.)

9

Symbioses and Mimicry

Some of the most intriguing insect tales are those that involve symbiosis: two unrelated species living together in prolonged and intimate association. The association can be any of three kinds: one in which a species benefits to the detriment of another (parasitism); one in which both species benefit (mutualism); or one in which one species profits from the association without harm or benefit to the other (commensalism). The literature of entomology includes numerous stories of solitary insects that have insinuated themselves into social insect colonies, where, in extreme cases, they are cared for and fed as if they were the host species' own young. So apparently amicable is the relationship between these "symphiles" and their social insect hosts that they are sometimes referred to as "pets" and "guests."

Mimicry is the phenomenon in which one organism has evolved to resemble another, sometimes in astonishing detail. There are, for example, moths with wings that look like owls' eyes and lantern bugs that look like snakes, both, presumably, for the purpose of scaring off hungry birds and lizards. There is a species of firefly in which the female has the ability to mimic a variety of (usually) species-specific flashing signals, luring males of other species in for what they expect to be a romantic encounter. Instead, they get eaten. In the most common kind of mimicry, groups of unrelated insects share similar "warning coloration," for example, the black-and-yellow-striped pattern common among stinging wasps and bees. Birds and other predators (and humans!) learn to avoid such a pattern following an unpleasant encounter (getting stung, sprayed with a chemical, etc.), and all of the participants in the "mimicry complex" benefit as a result. In Müllerian mimicry (named after Fritz Müller, see chapter 10, p. 322), both the "model" and the "mimic" are noxious. But in Batesian mimicry—for example, perfectly harmless and tasty flower flies that also have black and yellow stripes—a nonnoxious, perfectly tasty mimic has evolved to resemble a noxious model, thus benefiting itself by fooling predators that have had past bad experiences with the real thing.

Tiny Pollinator; Big Job

Stephen L. Buchmann and Gary Paul Nabhan

In the early 1990s, entomologist Steve Buchmann and cultural economic botanist Gary Nabhan organized a workshop to explore the overlooked importance of the interrelationship of plants and their pollinators. In the 1980s push to fund biodiversity studies and to conserve habitat, this crucial relationship, upon which most of our food and the health of biodiversity depend, was somehow overlooked. Buchmann and Nabhan's book, The Forgotten Pollinators, *tells their story.*

The pollination services of insects have long been taken for granted. But not by flowering plants! In this passage, Steve Buchmann, who specializes in bees, looks at the extraordinary variety of pollinators.

Some bees are so tiny that it's a wonder I even get to see them in flight. Fortunately, I have a few friends who are faster with a sweep net than I am. I was once in a field of low-growing matlike spurges, wondering what their minuscule flowers could ever muster to offer a pollinator. A moment later, I watched as one of my colleagues swept his net across the tops of the spurge mats, then thrust his head into the gossamer aerial net bag. Strange behavior? Yes, unless you are Jerome Rozen of the American Museum of Natural History, and you are seeking a rather diminutive bee. I walked up to Rozen, baffled, and peered at him through the fine-mesh weave of the net. "Looks like you didn't get anything much."

"You just can't see small bees," Rozen grumbled. He had, in fact, captured the tiniest of the tiny panurgine bees, known as *Perdita minima.* Its name alludes to the fact that the smallest of its kind stretch out to a mere 3/50 of an inch. It is the world's smallest bee, and it rivals a fig wasp and some thrips for being the world's smallest pollinator. And yet its pollination services are dependably provided to the matlike spurges which stuck that day to the soles of Rozen's feet. What the spurge provides in pollen and nectar to this bee would be starvation food for other kinds of pollinators. But for this lilliputian *Perdita,* it does the trick.

At the opposite extreme from the *Perdita minima* is a black giant, an immense leafcutter bee from the Moluccas islands halfway around the world. Alfred Russel Wallace, the codiscoverer of evolution by natural selection, spent time in these islands in the nineteenth century and found remarkable bees about 2 inches long living within the outer ramifications of termite "carton" nests. *Chalicodoma pluto* is the world's largest bee. This sexually dimorphic species has females with grotesquely enlarged mandibles and other mouthparts. They use custom tools to collect plant resins and transport them back to the nest. Until the late 1970s this amazing bee was known only from the type specimen in the British Museum of Natural History collected by Wallace, when it was rediscovered by Adam Messer, then a graduate student at the University of Georgia. You could balance dozens of *Perdita minima* upon the antennae of Wallace's lost bee.

Of course, there are many pollinating insects smaller than the smallest bee. There are over 500 species of flower thrips active in pollination, but they are not typically oriented toward pollen transport from one flower to another of the same species. They are, however, effective pollinators of many flowers. In fact, the diminutive thrips have recently been demonstrated to be highly effective obligate pollinators of certain dipterocarp canopy trees in the rainforests of Indonesia. Thrips typically range in size between 0.04 and 0.12 inch in length.

In this chapter, we can only glimpse some of the bewildering and fascinating shapes, sizes, and behavior of the myriad insects that visit and pollinate flowers. To fully do justice to insect pollinators other than bees, we would need pages to explore each major group of pollinating insects (beetles, butterflies and moths, flies and wasps) merely to highlight their biodiversity, their relative importance to pollination, and their astonishing lifestyles. Nevertheless, we offer a few case histories here to place these organisms in a conceptual framework within the vast worldwide pollinator bestiary. Ever since their origins more than 100 million years before the present, various families and genera of the largest and most successful insect orders (Coleoptera, Diptera, Hymenoptera, and Lepidoptera) have diversified to exploit a wide range of flowering plants utilizing varied strategies. Some specialize on a few related plants; others are generalists that draw on a wider dietary base.

If we were to kneel among the brilliantly colored and fragrant wildflowers of an alpine meadow, our attention would soon be diverted by the guests invited to feed at the banquet. The air is filled with thousands of flying insects of all possible sizes, colors, and forms. The combined noise of their beating wings is especially loud; the sound from a low-pitched passing bumblebee careens past our heads. Smaller insects are everywhere—from tiny straw-colored thrips invisibly feasting upon pollen inside flowers, to acrobatic flower flies, to bee flies, buprestid beetles, spider wasps, bees, and a winged gallery of gaudy butterflies.

All of these insects are floral visitors, but not all will acquire pollen that can be passed along to the next flower on their visits. Some have come to slit "floral throats," robbing them of sweet nectar or stealing away with pollen without fulfilling the implied pact with the flowers. A few of the insects—especially bees, flies, and butterflies—are excellent and faithful pollinators upon which the local flora "entrusts" its same-day pollen delivery service. Let's depart now from the flowers of the meadow and examine the diversity of these pollinators.

More than all the rest combined, the order Coleoptera (with over 350,000 named species worldwide and many yet to be discovered) is the largest extant insect order and probably always was so. From the sap beetle pollinators of western spicebush to the specialized scarab beetles that enter and pollinate the flowers of the giant Amazonian waterlilies, beetles are the customers and pollen vectors of choice for thousands of flowering plants on most continents.

Approximately 30 families of beetles are today engaged in the pollination trade, often acting as what has been termed "mess-and-soil pollinators." While the label is not terribly flattering to this ancient lineage of inordinately successful insects, it does indicate their mode of entry and gustatory pursuits. Thus, sap and rove beetles attracted to the fragrance of the western spicebush merrily chomp on special food tissues and on modified petals, in addition to the pollen grains. In so doing—and amidst a good deal of copulating and defecating—they effectively move the spicebush's and their own gametes around.

Whenever a bloodthirsty female mosquito peskily buzzes around our heads in a darkened room, we aren't likely to thank the mates of its species for the pollination of rare orchids in Wisconsin peat bogs. Yet male mosquitoes seek out nectar-producing orchids and other plants and are dependable pollinators in many parts of the world. In temperate alpine meadows, there are often dozens of species clambering over the open, broad clusters of blossoms on plants like the giant cow parsnip. Our attention is diverted by the high-pitched whine and darting motions of a fuzzy golden bee fly with a black beaklike set of mouthparts used for extracting nectar from nearby blossoms. Thus the flies are exceedingly diverse and important pollinators the world over. The order to which they belong, the Diptera, contains over 150,000 described species. And of those species with a taste for food on the half petal, there are at least 45 families of flies that routinely visit flowers.

Tubular flowers that are often pink or yellow in color with a sweet scent and abundant nectar at their base attract those scaly winged beauties sought out by "butterfliers" (a new breed of butterfly hunter who do their hunting with binoculars, notebook, and pencil). Butterflies are active by day and are found in about 16 families that regularly visit flowers in search of nectar. The order to which moths and butterflies belong, the Lepidoptera, contains at least 100,000 living species according to current estimates by modern taxonomists. It may surprise the nonentomologist to learn that moths, the butterfly's nocturnal cousins (actually butterflies are likely derived evolutionarily from distant moth ancestors), outnumber the butterflies by about ten to one. And yet moths are extremely important pollinators of night bloomers including the sacred datura and many cacti.

Although not so numerous as bees, their "colleagues" in the order Hymenoptera, wasps, also pollinate certain flowers. In the American Southwest, many spider wasps (like the giant tarantula hawk) are important floral visitors and pollinators of native milkweed plants. Similarly, figwort blossoms are especially adapted for visitations by wasps. Many wasps have bodies that are too smooth—especially when compared to their hairy cousins the bees—to pick up much pollen. Some wasps do, however, have legs with coarse hairs that are adequate for picking up and transferring pollen from flower to flower while they go about their business of searching for sweet nectar within blossoms. There are about 10 to 15,000 species of wasps that do function to some degree as pollinators of flowering plants. . . .

All told, we conservatively estimate there may be between 130,000 and 200,000 invertebrate and vertebrate species that regularly visit the flowers of those higher plants which depend on animals to assure crosspollination. This number of animals is at least half the magnitude as the number of flowering plants (other than grasses) described in the floras of the continents of the world. How many are dependable, effective pollinators remains to be seen. And only our grandchildren will be able to grasp how many of these animals and plants survive the next 50 years, for the biological diversity of the entire planet is facing unprecedented threats.

Buchmann, Stephen L., and Gary Paul Nabhan. 1996. *The Forgotten Pollinators.* Washington, D.C.: Island Press, Shearwater Books.

Jerry's Botfly

Adrian Forsyth

This piece, by Adrian Forsyth, a Canadian entomologist and superb natural history writer, gives a new meaning to the phrase "scientific detachment." The author, who is at least physically detached, is not the one who is challenged; it is Forsyth's friend on an expedition who returns to Harvard with a botfly buried in his scalp. He sees it as an opportunity to study the development of the animal, regardless of its chosen place of residence. Forsyth meantime seizes the opportunity to write about symbiosis.

Forsyth's friend's relationship with the botfly might at first glance look like classic parasitism, but the intellectual insights gained may put it squarely in the category of mutualism.

For those of us who dwell in large cities, direct interactions with other species may be limited to encounters with dogs and cats and occasional battles with cockroaches. Modern urban life has pushed us to a distant final link in a disrupted ecosystem. We live our lives far removed from the food chains that support us, perched myopically atop a trophic pyramid of complex and interconnecting relationships that we rarely, if ever, appreciate. But there are still places and occasions when an urban North American can step back into a food chain and experience firsthand the ecological relationships between himself and other species. My friend Jerry Coyne had such an experience on his first visit to the tropical forests of Central America.

Jerry is a biologist. At the time, he was a graduate student working at Harvard's Museum of Comparative Zoology. Well versed in evolutionary logic, genetical theory, Ivy League ecology and the use of biometrical tools, he was also aware that his actual contact with living creatures was "limited to unexciting fruit flies crawling feebly around food-filled glass tubes." Working in the museum had done little to change that. The museum was no longer what it had been in the days of its founder, the celebrated Swiss naturalist Louis Agassiz, whose constant exhortation was to "study nature, not books." Jerry's biological interactions continued to take place in a crowded, sterile laboratory, and the only animals he saw, aside from his fellow graduate students and the ubiquitous dogs of Cambridge, were the stuffed mammals that resided in the display cases between his office and the Pepsi machine. Finally, after a winter and spring of listening to fellow students urge him to get out of the laboratory, he enrolled in a field course in tropical ecology. Soon, he was jetting to Costa Rica, determined to experience for himself the riches of tropical nature.

Jerry's introduction to the tropics was a revelation. It not only confirmed his misgivings about his previous training but also changed his entire approach to his science. No longer would he trust "the naive and simple generalizations about nature produced by so-called theoretical ecologists," as he put it, and no longer would he search for slick hypotheses while glossing over the rich natural historical details of life. But he came away with more than these intellectual revelations.

A few weeks before Jerry was due to return to the museum, his head began to itch. This was hardly remarkable. Skin fungus, chigger and mosquito bites and a wealth of other pruriginous rot are the lot of field biologists in the lowland tropics, as he and his fellow students were by then well aware. Their field station was situated next to a large marsh, and hordes of mosquitoes descended on them as they listened to lectures after dinner. At first, Jerry assumed that the itch on his scalp was a mosquito bite, as indeed it was. But unlike the usual mosquito bite, this one did not subside. It grew larger, forming a small mound, and besides scratching, Jerry began worrying. After several days of private fretting, he sought help. One of his fellow students, a medical entomologist, agreed to examine the wound. Her diagnosis sent a chill of fear through poor Jerry. Poking out of a tiny hole in his scalp was a wiggling insect spiracle. A hideous little botfly maggot was living inside the skin on his head and eating his flesh! This intimacy with nature was too much for Jerry, and he ran around in circles crying for the removal of the maggot.

Unfortunately, removal of a botfly maggot is no simple task. This botfly, *Dermatobia hominis,* has existed as an unwanted guest in the skins of mammals and birds for countless generations. Its larvae have evolved two anal hooks that hold them firmly in their meaty burrows. If you pull gently on the larva, these hooks dig in deeper and bind it tightly to your flesh. If you pull harder, the maggot will eventually burst, leaving part of its body inside, which can lead to an infection far more dangerous than the original bot. A botfly larva secretes an antibiotic into its burrow, a tactic that prevents competing bacteria and fungi from tainting its food supply. A single bot in a nonvital organ thus poses little danger to an adult human, aside from mild physical discomfort and possible psychological trauma.

Occasionally, a bot sets up residence in a particularly tender or private patch of flesh that cries out for immediate removal. In Costa Rica, the locals once used a plant called the *matatorsalo* (bot killer) to kill the embedded larva. Although the acrid white sap of this milkweed kills the larva, the task of removing the corpse remains. The most appropriate course of action then is a deft slice of a sterile scalpel, but surgeons are few and scattered in most tropical forests. Under these conditions, many unwilling botfly hosts choose the meat cure.

This treatment, which is far from perfect, takes advantage of the biology of the botfly larva. The maggot is an air breather and must maintain contact with the air through its respiratory spiracle, a snorkel-like tube that it pokes through the host's skin. If a piece of soft raw meat is sandwiched over this air hole tightly enough, the larva must eventually leave its hole and burrow up through the meat in search of fresh air. When this happens, both meat and botfly are discarded. A student in Jerry's course who was afflicted with a bot in the buttock did this successfully. But the dense mat of hair that was Jerry's pride and joy would have to be shaved off in order for this to work, and toiling in the sweaty tropical heat with a slab of raw meat strapped onto his head was not something he relished. Faced with such a choice, Jerry decided to live and let live for the time being and to seek professional help when he returned to Harvard.

After his initial hysterical revulsion, Jerry learned to accept his guest. It was relatively painless most of the time. Only when the larva twisted would it cause sharp twinges of pain. Swimming made the larva squirm, presumably a reaction to having its air supply cut off temporarily, and Jerry felt this as a grating against his skull. These inconveniences were not enough to blind him to the wonder of it all. The bot was taking Jerry's "own body substance" and rendering it into more botfly flesh. This transmogrification of one creature into another is a miracle easily observed but difficult to experience. The brief bite of a flea or sudden death at the jaws of a large carnivore do not provide the victim the opportunity to reflect on the transmutative nature of predation and parasitism. But for the minor expense of a few milligrams of flesh, Jerry could both contemplate and feel the process at his leisure. He was inside a food chain, rather than at its end. Jerry grew fond of his bot, and the bot grew fat on Jerry.

When Jerry returned to New England, his bot had produced a goose-egg-sized swelling on his head. It hurt more, and he immediately sought medical advice at the Harvard Health Services Clinic. Although he was quickly surrounded by a crowd of physicians and nurses, none of them had seen a botfly before, and they regarded Jerry more as a medical curiosity than as a suffering patient. Chagrined, he abandoned thoughts of a medical

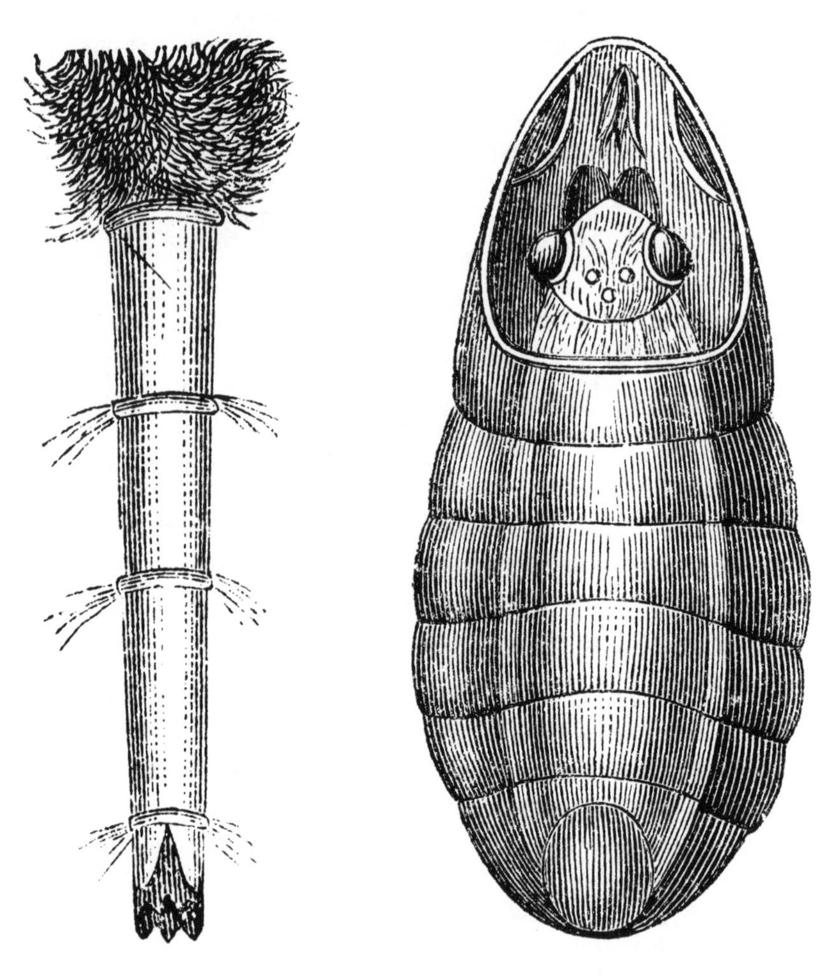

The ovipositor of a botfly and the imago of a botfly as it emerges. Illustration by Mm. E. Blanchard (*The Insect World* by Louis Figuier, Chapman and Hall, London, 1869).

solution and decided to let nature take its course. Despite the discomfort, the bot continued to provide some pleasure. Jerry took great delight in the looks of horror he could produce by telling acquaintances of his guest as he dramatically brushed aside his hair.

While sitting in the bleachers at Fenway Park one evening watching the Red Sox fall prey to the Yankees, Jerry felt the beginning of the end. Protruding from the goose egg atop his scalp was a quarter inch of botfly larva. Over the course of the evening, this protrusion grew, and eventually, the bristly inch-long larva fell free. Jerry filled a glass jar with sterilized sand to act as a nursery for his pupating bot, but despite his tender ministrations, the larva dried out and died before it could encase itself in a protective pupal sheath.

The saga of Jerry's maggot is but one small example, albeit a gripping one, of the interrelatedness of the natural world. The relationship, furthermore, was comparatively straightforward. Jerry's feelings aside, a more ecologically intriguing story was how the maggot came to Jerry in the first place, a story starting with an egg or, more precisely, with a means of getting an egg onto a suitable host.

Although there are many species of botfly, which parasitize a wide range of animals, the species that laid the egg on Jerry favours dexterous, perceptive hosts, such as monkeys and humans. It should thus be, ideally, a quiet, stealthy breed. But adult botflies, large day-flying creatures, are easy to see and hear and therefore to avoid, obvious disadvantages that have brought out an adaptive ingenuity in several species. Those that parasitize rodents, for example, glue their eggs to the root hairs of plants that stick out along the sides of rodent burrows. When a rodent walks by, its body heat causes the eggs to hatch; the larvae wriggle onto the animal's fur and thence into its flesh.

Dermatobia solves the problem of egg placement with a remarkable adaptation: An egg-laden female captures a female mosquito, glues her fertile eggs onto the mosquito and then releases her. The smaller, sneakier night-flying mosquito is well suited for feeding on a suitable host, such as a human. When the mosquito begins her meal, the body heat of the host triggers the hatching of the botfly egg, and the tiny larva falls off its carrier and burrows into the host. It is hard to imagine a more surreptitious strategy than this egg laying by proxy.

The intermediate evolutionary steps that led to such egg-laying behaviour are baffling. Indeed, it is an adaptation that evolutionary biologists could never anticipate. If we assume that the ancestral condition of the botfly was to deposit its eggs directly on the host, or even on grass or leaves, the number of steps between this and egg laying via mosquito is difficult to contemplate. It is not enough to grab any small fly, because this would be wasteful. The botfly must be able to discriminate between suitable vectors and unsuitable ones, and this discrimination must be sophisticated because *Dermatobia* usually uses mosquitoes of the genus *Psorophora* as egg carriers. The recognition cue may be simple; but even if this is the case, the transitional stages necessary to evolve the mechanism would not be apparent.

Whatever the history, however, the relationship today is a complicated one. By the time the egg reached Jerry, much of the story was over, and the subsequent relationship is easily expressed: the bot gained and Jerry lost.

This is parasitism, and it fits neatly into the classification scheme of interspecific interactions found in most textbooks. The following chart is typical of such classifications, where the effect of one species on another is given a sign indicating positive (+), negative (-) or neutral (o) associations:

SPECIES A		SPECIES B
+	–	parasitism or predation
+	0	commensalism
+	+	mutualism

Predation is readily defined: When a spider eats an insect, there is no doubt about who wins and who loses in the relationship. Parasitism is simply a slower, less extreme form of predation. When a spider wasp stings and then lays an egg on a spider, she initiates a process that may take months to complete, but the end result is the same—the egg hatches, and the larva consumes the paralyzed spider.

Equally clear-cut are those relationships, called mutualism, in which both species gain. A classic example is the interaction between various kinds of ants and aphids. Aphids, defenceless creatures that spend their lives sucking plant sap, produce a sugary secretion that is highly desirable to certain ant species. The ants milk the aphids to harvest the sugar but then repay the favour by protecting the aphids from predators.

For an example of commensalism, we need look no further than our own eyelashes, where, in most of us, a species of tiny mite crawls. Although it is known that the mites live on whatever they can glean from the follicles of our eyelashes, no one knows if they work for or against us. Perhaps they keep our eyelashes groomed, or perhaps they browse on a bit of ourselves. Whatever they do is too slight for us to measure, yet the mites themselves clearly benefit.

Forsyth, Adrian. 1984. "Jerry's Maggot." *Equinox* (July/August), no. 16:93–100.

The Ant and the Acacia Tree

Thomas Belt

Ants live in the hollow thorns of some acacia trees, drink nectar from special glands at the bases of the leaves, and feed on ant-size, proteinaceous morsels that grow on the tips of the leaves. The ants repel plant-feeding insects and keep the area around the acacia free from weeds. In 1874, Thomas Belt, the mining engineer and naturalist we met in chapter 4, theorized for the first time that the ants and acacias participated in a mutually beneficial symbiosis. After ninety-two years of sometimes heated debate about this subject, biologist Daniel Janzen proved by experiment in 1966 the remarkable "coevolution" that Belt had first observed.

Clambering down the rocks, we reached our horse and mule, and started off again, passing over dry weedy hills. One low tree, very characteristic of the dry savannahs, I have only incidentally mentioned before. It is a species of acacia, belonging to the section Gummiferae, with bi-pinnate leaves, growing to a height of fifteen or twenty feet. The branches and trunk are covered with strong curved spines, set in pairs, from which it receives the name of the bull's-horn thorn, they having a very strong resemblance to the horns of that quadruped. These thorns are hollow, and are tenanted by ants, that make a small hole for their entrance and exit near one end of the thorn, and also burrow through the partition that separates the two horns; so that the one entrance serves for both. Here they rear their young, and in the wet season every one of the thorns is tenanted; and hundreds of ants are to be seen running about, especially over the young leaves. If one of these be touched, or a branch shaken, the little ants (*Pseudomyrma bicolor*, Guer.) swarm out from the hollow thorns, and attack the aggressor with jaws and sting. They sting severely, raising a little white lump that does not disappear in less than twenty-four hours.

These ants form a most efficient standing army for the plant, which prevents not only the mammalia from browsing on the leaves, but delivers it from the attacks of a much more dangerous enemy—the leaf-cutting ants. For these services the ants are not only securely housed by the plant, but are provided with a bountiful supply of food, and to secure their attendance at the right time and place, the food is so arranged and distributed as to effect that object with wonderful perfection. The leaves are bi-pinnate. At the base of each pair of leaflets, on the mid-rib, is a crater-formed gland, which, when the leaves are young, secretes a honeylike liquid. Of this the ants are very fond; and they are constantly running about from one gland to another to sip up the honey as it is secreted. But this is not all; there is a still more wonderful provision of more solid food. At the end of each of the small divisions of the compound leaflet there is, when the leaf first unfolds, a little yellow fruit-like body united by a point at its base to the end of the pinnule. Examined through a microscope, this little appendage looks like a golden pear. When the leaf first unfolds, the little pears are not quite ripe, and the ants are continually employed going from one to another, examining them. When an ant finds one sufficiently advanced, it bites the small point of attachment; then, bending down the fruit-like body, it breaks it off and bears it away in triumph to the nest. All the fruit-like bodies do not ripen at once, but successively, so that the ants are kept about the young leaf for some time after it unfolds. Thus the young leaf is always guarded by the ants; and no caterpillar or larger animal could attempt to injure them without being attacked by the little warriors. The fruit-like bodies are about one-twelfth of an inch long, and are about one-third of the size of the ants; so that an ant carrying one away is as heavily laden as a man bearing a large bunch of plantains. I think these facts show that the ants are really kept by the acacia as a standing army, to protect its leaves from the attacks of herbivorous mammals and insects. . . .

The thorns, when they are first developed, are soft, and filled with a sweetish, pulpy substance; so that the ant, when it makes an entrance into them, finds its new house full of food. It hollows this out, leaving only the hardened shell of the thorn. Strange to say, this treatment seems to favour the development of the thorn, as it increases in size, bulging out towards the base; whilst in my plants that were not touched by the ants, the thorns turned yellow and dried up into dead but persistent prickles. I am not sure, however, that this may not have been due to the habitat of the plant not suiting it.

These ants seem at first sight to lead the happiest of existences. Protected by their stings, they fear no foe. Habitations full of food are provided for them to commence housekeeping with, and cups of nectar and luscious fruits await them every day. But there is a reverse to the picture. In the dry season on the plains, the acacias cease to grow. No young leaves are produced, and the old glands do not secrete honey. Then want and hunger overtake the ants that have revelled in luxury all the wet season; many of the thorns are depopulated, and only a few ants live through the season of scarcity. As soon, however, as the first rains set in, the trees throw out numerous vigorous shoots, and the ants multiply again with astonishing rapidity.

Both in Brazil and Nicaragua I paid much attention to the relation between the presence of honey-secreting glands on plants, and the protection the latter secured by the attendance of ants attracted by the honey. I found many plants so protected; the glands being specially developed on the young leaves, and on the sepals of the flowers. Besides the bull's-horn acacias, I, however, only met with two other genera of plants that furnished the ants with houses, namely the *Cecropiae* and some of the *Melastomae*. I have no doubt that there are many others. The stem of the *Cecropia*, or trumpet-tree, is hollow, and divided into cells by partitions that extend across the interior of the hollow trunk. The ants gain access by making a hole from the outside, and then burrow through the partitions, thus getting the run of the whole stem. They do not obtain their food directly from the tree, but keep brown scale-insects (Coccidae) in the cells, which suck the juices from the tree, and secrete a honey-like fluid that exudes from a pore on the back, and is lapped up by the ants. In one cell eggs will be found, in another grubs, and in a third pupae, all lying loosely. In another cell, by itself, a queen ant will be found, surrounded by walls made of a brown waxy-looking substance, along with about a dozen Coccidae to supply her with food. I suppose the eggs are removed as soon as laid, for I never found any along with the queen-ant. If the tree be shaken, the ants rush out in myriads, and search about for the molester. This case is not like the last one, where the tree has provided food and shelter for the ants, but rather one where the ant has taken possession of the tree, and brought with it the Coccidae; but I believe that its presence must be beneficial. I have cut into some dozens of the *Cecropia* trees, and never could find one that was not tenanted by ants. I noticed three different species, all, as far as I know, confined to the *Cecropiae*, and all farming scale-insects. As in the bull's-horn thorn, there is never more than one species of ant on the same tree.

In some species of *Melastomae* there is a direct provision of houses for the ants. In each leaf, at the base of the laminae, the petiole, or stalk, is furnished with a couple of pouches, divided from each other by the midrib. . . . Into each of these pouches there is an entrance from the lower side of the leaf. I noticed them first in Northern Brazil, in the province of Maranham; and afterwards at Pará. Every pouch was occupied by a nest of small black ants, and if the leaf was shaken ever so little, they would rush out and scour all over it in search of the aggressor. I must have tested some hundreds of leaves, and never shook one without the ants coming out, excepting on one sickly-looking plant at Pará. In many of the pouches I noticed the eggs and young ants, and in some I saw a few dark-coloured Coccidae or aphides; but my attention had not been at that time directed to the latter as supplying the ants with food, and I did not examine a sufficient number of pouches to determine whether they were constant occupants of the nests or not. My subsequent experience with the *Cecropia* trees would lead me to expect that they were. If so, we have an instance of two insects and a plant living together, and all benefiting by the companionship. The leaves of the plant are guarded by the ants, the ants are provided with houses by the plant, and food by the Coccidae or aphides, and the latter are effectually protected by the ants in their common habitation.

Amongst the numerous plants that do not provide houses, but attract ants to their leaves and flower-buds by means of glands secreting a honey-like liquid, are many epiphytal orchids, and I think all the species of *Passiflora.* I had the common red passion-flower growing over the front of my verandah, where it was continually under my notice. It had honey-secreting glands on its young leaves and on the sepals of the flower-buds. For two years I noticed that the glands were constantly attended by a small ant *(Pheidole),* and, night and day, every young leaf and every flower-bud had a few on them. They did not sting, but attacked and bit my finger when I touched the plant. I have no doubt that the primary object of these honey-glands is to attract the ants, and keep them about the most tender and vulnerable parts of the plant, to prevent them being injured; and I further believe that one of the principal enemies that they serve to guard against in tropical America is the leaf-cutting ant, as I have observed that the latter are very much afraid of the small black ants.

On the third year after I had noticed the attendance of the ants on my passion-flower, I found that the glands were not so well looked after as before, and soon discovered that a number of scale-insects had established themselves on the stems, and that the ants had in a great measure transferred their attentions to them. An ant would stand over a scale-insect and stroke it alternately on each side with its antennae, whereupon every now and then a clear drop of honey would exude from a pore on the back of the latter and be imbibed by the ant. Here it was clear that the scale-insect was competing successfully with the leaves and sepals for the attendance and protection of the ants, and was successful either through the fluid it furnished being more attractive or more abundant. I have, from these facts, been led to the conclusion that the use of honey-secreting glands in plants is to attract insects that will protect the flower-buds and leaves from being injured by herbivorous insects and mammals, but I do not mean to infer that this is the use of all glands, for many of the small appendicular bodies, called "glands" by botanists, do not secrete honey. The common dog-rose of England is furnished with glands on the stipules, and in other species they are more numerous, until in the wild *Rosa villosa* of the northern counties the leaves are thickly edged, and the fruit and sepals covered with stalked glands. I have only observed the wild roses in the north of England, and there I have never seen insects attending the glands. These glands, however, do not secrete honey, but a dark, resinous, sticky liquid, that probably is useful by being distasteful to both insects and mammals.

If the facts I have described are sufficient to show that some plants are benefited by supplying ants with honey from glands on their leaves and flower-buds, I shall not have much difficulty in proving that many plant-lice, scale-insects, and leaf-hoppers, that also attract ants by furnishing them with honey-like food, are similarly benefited. The aphides are the principal ant-cows of Europe. In the tropics their place is taken in a great measure by species of Coccidae and genera of Homoptera, such as *Membracis* and its allies. My pineapples were greatly subject to the attacks of a small, soft-bodied, brown coccus, that was always guarded by a little, black, stinging ant *(Solenopsis)*. This ant took great care of the scale-insects, and attacked savagely any one interfering with them, as I often found to my cost, when trying to clear my pines, by being stung severely by them. Not content with watching over their cattle, the ants brought up grains of damp earth, and built domed galleries over them, in which, under the vigilant guard of their savage little attendants, the scale-insects must, I think, have been secure from the attacks of all enemies.

Many of the leaf-hoppers—species, I think, of *Membracis*—were attended by ants. These leaf-hoppers live in little clusters on shoots of plants and beneath leaves, in which are hoppers in every stage of development—eggs, larvae, and adults. I believe it is only the soft-bodied larvae that exude honey. It would take a volume to describe the various species, and I shall confine my remarks to one whose habits I was able to observe with some minuteness. The papaw trees growing in my garden were infested by a small brown species of *Membracis*—one of the leaf-hoppers—that laid its eggs in a cottony-like nest by the side of the ribs on the under part of the leaves. The hopper would stand covering the nest until the young were hatched. These were little soft-bodied dark-coloured insects, looking like aphides, but more robust, and with the hind segments turned up. From the end of these the little larvae exuded drops of honey, and were assiduously attended by small ants belonging to two species of the genus *Pheidole*, one of them being the same as I have already described as attending the glands on the passion-flower. One tree would be attended by one species, another by the other; and I never saw the two species on the same tree. . . . A third ant, however—a species of *Hypoclinea*—which I have mentioned before as a cowardly species, whose nests were despoiled by the *Ecitons*, frequented all the trees, and whenever it found any young hoppers unattended, it would relieve them of their honey, but would scamper away on the approach of any of the *Pheidole*. The latter do not sting, but they attack and bite the hand if the young hoppers are interfered with. These leaf-hoppers are, when young, so soft-bodied and sluggish in their movements, and there are so many enemies ready to prey upon them, that I imagine that in the tropics many species would be exterminated if it were not for the protection of the ants.

Belt, Thomas. 1874. *The Naturalist in Nicaragua.* London: John Murray.

To a Louse

Robert Burns

Scotland's much loved national poet Robert Burns (1759–1796) was the son of a poor farmer, and even after he was fêted by society, he retained his healthy disregard for those with airs. Here he takes the attempts of a louse to parasitize an upperclass lady as a chance to make a bit of mischievous social commentary. Burns calls the louse "ye ugly, creepin, blastit wonner, detested, shunn'd by saunt an' sinner," but he's really having fun and celebrating the louse and its arrival on the lady's hat, though concerned that the louse may not have found the best home!

To a Louse

On seeing one on a lady's bonnet at church

HA! whaur ye gaun, ye crowlin ferlie?
Your impudence protects you sairly;
I canna say but ye strunt rarely,
   Owre gauze and lace;
Tho', faith! I fear ye dine but sparely
   On sic a place.

Ye ugly, creepin, blastit wonner,
Detested, shunn'd by saunt an' sinner,
How daur ye set your fit upon her—
   Sae fine a lady?
Gae somewhere else and seek your dinner
   On some poor body.

Swith! in some beggar's haffet squattle;
There ye may creep, and sprawl, and sprattle,
Wi' ither kindred, jumping cattle,
   In shoals and nations;

Whaur horn nor bane ne'er daur unsettle
  Your thick plantations.
Now haud you there, ye're out o' sight,
Below the fatt'rels, snug and tight;
Na, faith ye yet! ye'll no be right,
  Till ye've got on it—
The verra tapmost tow'rin height
  O' Miss's bonnet.

My sooth! right bauld ye set your nose out,
As plump an' grey as ony groset:
O for some rank, mercurial rozet,
  Or fell, red smeddum,
I'd gie you sic a hearty dose o't,
  Wad dress your droddum.

I wad na been surpris'd to spy
You on an auld wife's flainen toy;
Or aiblins some bit duddie boy,
  On's wyliecoat;
But Miss's fine Lunardi! fye!
  How daur ye do't?

O Jeany, dinna toss your head,
An' set your beauties a' abread!
Ye little ken what cursed speed
  The blastie's makin:
Thae winks an' finger-ends, I dread,
  Are notice takin.

O wad some Power the giftie gie us
To see oursels as ithers see us!
It wad frae mony a blunder free us,
  An' foolish notion:
What airs in dress an' gait wad lea'e us,
  An' ev'n devotion!

Burns, Robert. 1786. "To a Louse." In *The Complete Illustrated Poems, Songs & Ballads of Robert Burns.* 1990. London: Chancellor Press/Reed Consumer Books.

An Earful of Mites

Asher E. Treat

Asher Treat—longtime research associate of the American Museum of Natural History and a former professor of biology at the City University of New York—devoted many years to the study of mites that live on butterflies and moths. He was especially interested in a curious group of mites that live communally in the ears (located on the sides of the body) of night-flying moths in the family Noctuidae. Years spent peering down a microscope watching mites in action revealed a wonderful fact: These mites, which "come aboard" the moth at different times and in different places, always aggregate in only one of the moth's two ears, though whether this is the right or left ear varies from moth to moth. The mites' activities cause the moth to go deaf, but since the moth is deaf in one ear only, it can still hear bats and, along with its mite passengers, avoid being eaten.

At various times several authors have recorded erythraeids, trombidiids, and other mites, often unrecognized, from moths and butterflies. In 1871, 1889, and again in 1947–1951 there were minor eruptions of interest among British amateurs in "red mites" on Lepidoptera and other insects. *Hardwicke's Science Gossip* for 1871 contained notes by Joseph Anderson Jr. and by T.W. Wonfor concerning scarlet mites ("insects" or "little red things") that they had seen on various moths and butterflies. . . . [but] until a few years ago the mites that had been found on or with Lepidoptera, larval or adult, were either of undetermined species or of species known also in association with other insects or from other ecological contexts as well as from Lepidoptera. None of these acarines could properly be called "moth mites" or "butterfly mites" in an exclusive sense. *Dicrocheles phalaenodectes*, described in 1954 as *Myrmonyssus phalaenodectes*, differed from previously known species in being able to complete its entire life cycle on adult moths, and in having no other known hosts. Its discovery, in Tyringham, Massachusetts, in 1952, was the chance result of work on the physiology of noctuid tympanic organs. Since the previous summer I had been examining the ears of almost every kind of tympanate moth that came to my collecting light. On the night of 5 July 1952, I found a "volunteer" that had somehow got into the attic of the country house where we spent our summers; it was flying about the lamp on my laboratory table. I had finished work for the night, but couldn't resist the temptation to inspect the ears of one more moth. My visitor was a male of *Leucania pseudargyria* Guenée, a tan-colored cutworm moth with a conspicuously hairy body and legs. In the tympanic recess at the base of its left hind wing were several pearly white objects that could only have been eggs of some kind. I transferred one of these to a coverslip inverted over a depression slide and was rewarded a couple of days later by finding a larval mite creeping about inside the glass cell. By the end of the summer several more infested moths had appeared. My ignorance of mites being at that time complete, I wrote to G. W. Wharton, then of Duke University, whose book (with E. W. Baker, 1952) *Introduction to Acarology* had just appeared. On learning from Dr. Wharton that the mites were of potential interest and probably undescribed, I embarked upon the study that has occupied most of my working life ever since.

The surge of interest in acarology that followed the Second World War created an atmosphere and a literature that encouraged the study of mites in general but that did not immediately yield new knowledge of moth or butterfly mites. The recent literature on these is still the work of but few writers, and the immense reservoirs of potential discovery represented by the world's great collections of Lepidoptera remain almost entirely untapped. . . .

The magic of the microscope is not that it makes little creatures larger, but that it makes a large one smaller. We are too big for our world. The microscope takes us down from our proud and lonely immensity and makes us, for a time, fellow citizens with the great majority of living things. It lets us share with them the strange and beautiful world where a meter amounts to a mile and yesterday was years ago.

Let us shrink to the height of a moth ear mite, creep under the wing of a sleeping noctuid, and roam for a little while through the sculptured caverns of the insect's ear. The illustration (page 270) can serve as a map for the tour. We pause at first, within the hood and beneath the alula, in the deep hollow of the outer recess. Between the crescent of the epaulette and the pale stigma in the tympanic membrane, we peer through this frail window into the dim depths of the tympanic air sac, where the white cradle of the scoloparium hangs like a hammock overhead. Within that cradle lie the twin nerve cells of the acoustic sense, tuned to sounds too high for human hearing. Through the thin tracheal veil that lines this cavity, we see strange forms—the long, thin tongue of the scutal phragma, the curling lip of the *Bügel,* the sinuous shelf of the tendon plate, and the receding cord of the tympanic nerve.

Let us pierce the membrane and enter this inner chamber. The soft carpet yields to our feet at every step, and through the transparent inner wall we can see the moth's great muscle-engines resting in readiness for the evening flight. Beside us, one above another, yawn the mouths of far dark grottoes, the four pockets of the tympanic frame. Partitioned like the cells of a cloister, but each of a singular shape, they form a hollow, four-storied pillar between the thorax and the abdomen. Above us is the closed skylight of the countertympanic cavity. Let us broach this membrane and climb up into the gleaming vault above it.

The dim light of this chamber comes from a window in its outer wall, like a broad mouth, slightly open to the outer air. Turning inward, we see in the distance another such opening, as though the nearer one were mirrored in glass. It is not a reflection; it is real. But as we grope toward it our way is blocked by a firm yet totally transparent wall, unnoticed until we stumble into it. It is the countertympanic septum, the median barrier between this ear and the one of the opposite side. We cannot enter that looking-glass ear; the septum is too stout to be pierced. We turn and creep through the narrow exit and find ourselves upon the jutting roof of the tympanic frame, once more beneath the alula and overlooking the hollow where we entered. Our journey is over, and it is nearly dark now. The moth spreads its wings and begins to tremble. It is ready for flight.

This, then, is the world of the moth ear mite from the time of its emergence as a larva to its maturity as a fertile female, perhaps to its death as a worn-out male, or to its destruction, with ill luck, in the jaws of a hungry bat. The mite, of course, whatever its aesthetic sensibilities, does not perceive this world as we have described it, because, if for no other reason, the mite is blind. But the scene is there, the stage is set, the actors are ready, and anyone with an entomological microscope can draw apart the curtain. . . .

Early speculation that certain moths might respond to the cries of insectivorous bats has been abundantly confirmed and amplified by Schaller and Timm, Roeder and Treat, and others. . . . More recent studies by Roeder and others have greatly extended our knowledge of the tympanic organs as ultrasonic receptors and have revealed much of their role in the eluding of predatory bats by flying noctuoids. . . .

That a niche as attractive as the noctuid ear should occasionally harbor a vagrant mite is certainly not surprising, and in fact mites of about a dozen species have been taken from the outer ears of these moths at one time or another. Only a few are frequent or regular occupants, however, and only

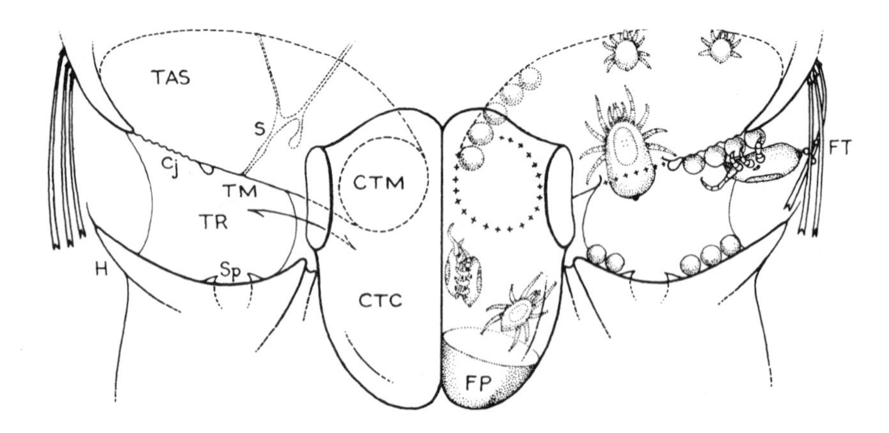

Schematic view of moth ears (noctuid tympanic organs) showing occupancy in one ear by the mite *Dicrocheles phalaenodectes.* Cj, conjunctiva; CTC, countertympanic cavity; CTM, countertympanic membrane; FP, fecal plug; FT, fecal hair thatch; H, hood; S. scoloparial sensillum; Sp, spiracle; TAS, tympanic air sac; TM, tympanic membrane; TR, tympanic recess (*Mites of Moths and Butterflies* by Asher E. Treat, Cornell University Press, Ithaca, N.Y., 1975).

those of the genus *Dicrocheles* are known to invade the inner chamber. Perhaps this, too, is not surprising, for the invasion of the tympanic air sacs can scarcely be accomplished without the destruction of the scoloparia and, with them, of one of the moth's best means of escaping total destruction, in which the mites must inevitably share. We shall see how the mites offset this hazard. . . .

Dicrocheles phalaenodectes (Treat, 1954)

The name *phalaenodectes* comes from the Greek *phalaena,* a moth, plus *dectes,* one who bites, pricks, or annoys. This is the best known and perhaps the commonest of New World moth mites, at least in North America. . . .

Larvae of the preferred hosts are mainly grass feeders. The armyworm, *P. unipuncta,* is among our most important economic pests of cereals and forage crops. At present there is no direct evidence that the mites significantly affect the populations of their host. Under uncontrolled laboratory conditions the life span of infested moths does not appear to be appreciably shortened, and moths are occasionally seen with "ghost colonies," in which no mites remain, though signs of previous occupancy are unmistakable. Although the acoustic sense is doubtless important in the evasion of insectivorous bats, infested moths have been seen to "junk" and dive effectively under attack by these predators, even with one ear totally incapacitated. Evasion at short range, however, is not always successful even with both ears functional, and when a moth is attacked at close range the chance of its being caught may be as high as one in ten. Roeder, however, has shown that binaural localization of faint ultrasonic pulses like the hunting cries of a distant bat can cause a flying noctuid to turn away from the source of sound and thus, perhaps, avoid the hunting area. The destruction of one ear by mites may therefore increase the likelihood that the host will be exposed to short-range predation, and this may well be a factor, though perhaps a minor one, in the natural regulation of the moth population. There are opportunities for valuable field observations and experimental work in this connection.

At whatever point one picks up the story of the moth ear mite, he is almost sure to be fascinated by what he sees. From the boarding of the host by a fertile female to the dispersal of her fertile daughters, every stage is marked by curious and distinctive behavior. In midsummer, the cycle may be completed in only five or six days, although the life of a colony may continue for a couple of weeks or longer, with a first and second filial generation living together in a rudimentary social organization. Although freshly or recently invaded hosts may appear throughout the summer and early autumn, in New England at least, large and mature colonies are commonest in late July and early August, suggesting that reproduction does not proceed at a uniform rate throughout the season.

Perhaps the most noteworthy aspect of a *D. phalaenodectes* colony is its unilaterality. The mites normally invade and occupy only one of their host's two ears, leaving the other ear intact and fully functional as a detector of ultrasound. The adaptive value of this behavior is obvious, for if an infested moth is caught by a hungry bat, not only does the moth perish, but the mites as well. A one-eared moth may have its acoustic defenses somewhat impaired, but it is still a safer mite vehicle than a moth that is deafened altogether. The chosen ear may be either the left one or the right, but regardless of which it is, and regardless of how crowded it may become, the other ear is almost invariably left wholly undisturbed. This seems the more remarkable when one considers that most mites that travel in numbers on insects distribute themselves quite symmetrically upon their hosts. In some such instances, where the mites become attached and only a small odd number are present, the odd mite will occupy a median position. Mites that move about freely, like the gamasine deutonymphs that commonly infest sexton beetles, manage nevertheless to maintain approximately equal numbers on the two sides of the host. Even among the tympanicolous mites of moths, those that spare the ears from damage use both ears equally or nearly so. It is only in certain species of *Dicrocheles* that unilaterality is the rule and not the exception. In these species the restriction of the colony to one ear appears to be the result of a highly evolved behavioral pattern perhaps including pheromonal and other types of communication.

Almost nothing is known about the selection, detection, and boarding ("infaunation") of acceptable hosts in nature. It is noteworthy that among the numerous host species on record, at least one—the armyworm moth—is a strong, long-lived flier and a known or suspected migrant with a wide geographical range, with several annual broods, and with flight periods in temperate latitudes extending from April or May until October or November. Similar attributes characterize hosts of at least two of the other three species of *Dicrocheles.* The effect of host migration upon the distribution of the mites, and upon their seasonal recurrence in latitudes where the moths cannot winter as adults, are at present matters for speculation. . . .

If a fertile female of *D. phalaenodectes,* engorged or unengorged, is placed upon the body of an acceptable mite-free host such as an armyworm moth, at rest and unrestrained, her actions can often be followed in some detail under stereoscopic magnification. Her first response to the new situation is usually a brief "freeze"—a cessation of all motion—lasting from a few seconds to perhaps half a minute. The thaw is gradual, with the first leg movements slow and tentative. If the vestiture is deep at the point where the mite was placed, she soon begins a succession of short, jerky forward thrusts or lunges that sink her body deeper and deeper among the hairs and scales until she is well down among their bases, partly or completely hidden from surface view. But even when she herself cannot be seen, her stirrings can be followed readily by the twitching movements that they impart to the moth's pelage. The jerky progress continues, guided perhaps by the characteristic "lay" of the vestiture, as the mite slowly makes her way, not toward the ears of the moth as might be expected, but toward the patagial region—the "collar."

Having arrived at the collar, she proceeds to explore this region, moving first to or toward one shoulder and then to the other, jerkily pushing aside the hairs as she rummages among them. At times she may disappear entirely beneath or between the patagia, possibly puncturing their soft articular membranes for a taste of hemolymph, but soon the hairs begin to twitch again as her "rummaging" is resumed. In some instances this activity lasts for only a couple of minutes; in others it goes on for an hour or more. Usually both shoulders are visited several times.

If one watches continuously during this rummaging period, he becomes conscious of a certain tempo in the mite's behavior. Her movements, or the twitching of the hairs that mark them when she herself is not directly visible, while not exactly rhythmical, seem to occur at fairly regular intervals, say about once a second. At length there comes a time when the tempo is suddenly increased. As a rule this occurs when the mite is somewhere near the midline of the thorax, near the anterior thoracic tuft if one is present. Whatever the cause of the change, the impression of excitement is inescapable. The jerky movements are vigorous and in rapid succession. Soon a funnel-like gap appears among the thoracic hairs. The mite has turned rearward and is parting the long hairs with her forelegs much as one might push aside the tall stalks in a canebrake. Now she emerges from the neck of the funnel and creeps rearward in the midline, palpating the tiny flooring scales with her mouthparts as she goes. The jerky movements have ceased, and her progress is steady, for she is going "downstream" now, and the hair tips give way easily before her. Usually there is no hesitation and no turning back. Within some ten or twenty seconds she has traversed the thoracic disc and has reached the hairs projecting rearward over the base of the abdomen. She pushes her way down among them and into the anterior abdominal tuft, bordered and almost concealed by the inner margins of the folded forewings.

This is the "crossroad," the midpoint on the transverse path between the two ears. Here she hesitates, probing a bit in all directions. Which will she choose? On a few occasions I have seen a mite turn back at this stage, return to the collar and repeat the whole performance, but usually a "choice" is made within a few minutes, and in some instances the hesitation is very brief. She creeps under the margin of a resting wing, either the left one or the right, and disappears. A few moments later, she may be found, upon examination, in the tympanic recess.

It might be imagined that these actions would evoke some response on the part of the moth, but apparently they do not. Neither during the rummaging period nor later at any stage does the host exhibit any sign of distress or annoyance attributable to its guests, unless by chance one of them wanders on to a leg or an eye, in which case it may be brushed off. The mites, on their part, appear equally indifferent to activities of the host. If the moth is fastened with wax to a pedestal beneath its leg bases, one may tune a stroboscope to its wing beats and observe its mites going about their business undisturbed by vibrations that in ordinary light will blur the entire body of the host.

Newly arrived in the ear of her choice, the female *D. phalaenodectes* does not immediately set up housekeeping. During the first hour or so she returns from time to time to the midpoint or "crossroad" between the two ears, pausing there on each visit to probe again in various directions and then returning to the ear first entered. The significance of these back trips can perhaps be surmised, but it has not been fully clarified. Is she laying a trail or posting a sign of some kind to guide the next mite to come? I think that she is, although I cannot read the trail or find the sign myself. In any event, the next traveler and all that follow will normally take the same route as the first mite and with little hesitation at the midpoint will lodge with her in the same inn, never visiting the empty tavern across the street. Yet if mites are persistently placed and replaced in that opposite ear, they will sometimes remain there and establish a second normal colony. Very rarely a bilateral infestation is found in nature. Evidently it is no compelling asymmetry of the host that dictates the habitual unilaterality of these parasites. True, the wings of resting moths overlap some one way, some the other, but this is not constant in any one moth, and in any event the mites plainly disregard this factor in making their "choice" of ears. Are the mites themselves congenitally right- or left-handed? No—a mite that on one host has chosen the right ear, upon experimental transfer to another host, may choose the left.

If the first comer does indeed make a trail to the chosen ear, what can be the nature of that trail? Is it a chemical one—a pheromone—or is it some physical disturbance of the path traversed? I have tried to answer this question experimentally but with equivocal results. By removing hairs and scales of a mite-free moth from the path between the "crossroad" and an ear, one can make an artificial trail that a mite will follow when placed on the insect. One can even mislead the mites occasionally in this way, and produce a bilateral colony on a previously infested moth. Yet such an artificial trail is grossly unlike any physical disturbance that could be created by the mites themselves. To me, the circumstances suggest a pheromone, but there is no proof.

Within a few hours after her arrival, the mother of the future colony has removed both the tympanic and countertympanic membranes and has become fully engorged. Her soft, transparent interscutal integument is stretched smooth and glossy over her distended body, its contents yellow with the ingested hemolymph of her host. By puncturing the articular membranes above or beside the ear she could feed at any of several places, but probably the one of easiest access is through the tracheal epithelium of the tympanic air sac, which in a mature colony is always dotted and discolored with feeding scars. By this time, of course, the scoloparium, containing the two acoustic cells, has been destroyed or irrevocably damaged and the moth is deaf—in one ear.

The first egg is usually laid on or near the ledgelike shelf of the anterior tendon plate within the tympanic air sac. The next few eggs are placed close by. If the countertympanic cavity is opened by dissection at this time, one can see these eggs through the open frame of the countertympanic membrane, which the mite herself has previously destroyed. They are round, glossy, and water-white, like miniature pearls. When several eggs have been deposited in the air sac, and perhaps a few in the anterior part of the countertympanic cavity, the mite begins to pave the conjunctival membrane with eggs. The conjunctiva is the soft, white integument lateral to the epaulette, in the outer ear or tympanic recess. Together with the tympanic membrane (which has been destroyed) it forms the outer, posterior wall of the tympanic air sac; it is bounded anterolaterally by the subalar sclerite. Here, the act of oviposition can be watched again and again.

At summer temperatures an egg is laid about every two or three hours. The process is quite ceremonious. The gravid mother first selects and prepares the site, choosing a spot in close proximity to a previous egg if one is present. She applies her chelicerae to the soft substrate, punctures it, and scarifies a small area, which she works over repeatedly, lifting and slightly lacerating the cuticle. Having finished this business, she may leave the spot for a while and withdraw into the air sac to await her time of delivery. As the moment approaches, she returns to the spot she has made ready and stands facing it. Through her transparent body wall her internal organs can be seen in frequent churning motion, and from time to time dimples appear on the sides and rear of her hysterosoma as the anal and metapodal muscles contract. At length the egg appears, pressed forward beneath her from the genital orifice between her fourth coxae. As it passes under her gnathosoma she grasps it with her palpi and presses it against the scarified cuticle. She rocks it about a bit and nestles it into place, then strokes it with her palpi from apex to base. This done, she leaves the spot and goes off for a rest and a meal of moth juice. The eggs adhere lightly to the substrate, perhaps cemented there in some way, though they can be pried free without damage to their viability. Their "mother" visits them occasionally, but gives them no special care. She does not explicitly defend them, though she may vigorously resist the entrance of another mite into the area that they occupy. Before her reproductive life is over she may produce as many as eighty offspring.

At midsummer room temperatures, the eggs hatch in about two days. The developing larva can be seen clearly through the transparent cuticle. The larva too is transparent and colorless. . . . In most of the eggs, which would produce females, the chromosome number seems to be six, although the form of the stainable elements leaves some doubt about this. Among the first eggs laid there are usually one or a few in which only four chromosomes can be seen, and in which each embryonic interphase cell contains a small comma-shaped chromatic mass. These are the eggs of males.

In hatchings that I have witnessed, the larva has emerged rear end first, leaving the shriveled egg cuticle at its site of attachment. From eggs laid in the outer recess, the newly hatched larvae grope their way unattended through the open frame of the tympanic membrane and into the air sac. Placed in an intact ear they seem unable to penetrate the membrane and do not survive. There is some evidence that feeding may take place in all developmental instars, but whether feeding is required for transformation in every instance, I do not know. . . .

Within a few days the offspring of even a single brood female can crowd the ear with mites and eggs, and when several females are ovipositing the whole tympanic area may literally overflow with their offspring. Yet despite this population explosion, the opposite ear remains inviolate, except in very rare instances, even when both sides of the collar and neck are swarming with young females ready to seek a new home.

In the close and somewhat confined quarters of a moth's ear, contamination with acarine wastes could become a serious hazard to a growing community of mites, but in fact public sanitation is well regulated in a *Dicrocheles* colony. Defecation is performed in either of two places: at the outer periphery of the tympanic recess or in the rear part of the countertympanic cavity. I have witnessed defecation by adult females only. The anus is at the tip of the body rather than on the ventral surface as it is in many mite species. A mite about to use the outer "performery" creeps to the edge of the recess, turns around and backs up until the anal region touches one of the long hairs that roof this cavity. She then extrudes a droplet of turbid yellow liquid. When she moves forward, this is left clinging to the hair, where it quickly shrinks to perhaps half its initial volume. With repeated defecations the fecal substance gradually accumulates in a waxy brown- or orange-colored mass that eventually entangles many of the roofing hairs and forms a mat that may completely cover and conceal the recess and its occupants. The fecal material is hygroscopic, swelling and softening in moist air, reversibly shrinking, darkening, and hardening when the atmosphere is dry. One wonders how this affects the microclimate of the colony. . . .

In a normal colony the tympanic air sac serves as the dining room for the developing mites. Here they are separated from their food supply by nothing more than a thin tracheal epithelium, which is easily pierced. Although eggs and young are sometimes found in the countertympanic cavity, feeding cannot take place there because the walls are sclerotized and impenetrable even by the adults. It is here, however, that mating usually takes place. As the deutonymphs mature they evidently become attractive to the males, who comprise only about 7 percent of the population. Some time before her final molt occurs, a marriageable deutonymph is embraced, piggyback, by her prospective mate. As the time for ecdysis approaches, the male maneuvers himself into a venter-to-venter position, with his mouthparts at the level of the genital orifice of the pharate female. When the molt begins he may even help to remove the deutonymphal cuticle and expose the genital area. . . .

The low ratio of males to females suggests that the males must mate repeatedly, but in a large and crowded colony it is possible that some females nevertheless escape insemination. This may account for the occasional failure of filial females to establish colonies when transferred experimentally to fresh hosts. Parthenogenesis probably does not occur, at least not regularly, for although virgin females can produce eggs, these in my experience are never viable.

Some of the daughters may stay at home and contribute a second generation to the colony, but as the population increases many leave the car and move forward to the collar and neck of the moth. Here feeding sites are still available, neither side being "off limits," and here also conditions are favorable for debarkation in search of a fresh host. The exodus of young females may perhaps be stimulated or encouraged by a curious kind of behavior often displayed in crowded colonies and seen occasionally when a lone female with her eggs is confronted by an intruding mite of her own species. This behavior consists of repeated series of rapid, side-to-side oscillations, usually with a yawing component reminiscent of the "wagging dances" of honey bees. I have often witnessed such actions when I have transferred a second female mite to a moth whose tympanic recess is already occupied by a previously established female and her eggs. If, as usually happens, the second female attempts to enter the air sac via the external recess, she is met, in many instances, with strong resistance. At every approach the defending mite interposes her body between the intruder and the open tympanic frame and on contact performs the jostling oscillatory movements just described. This may go on for as much as an hour before her resistance weakens and the newcomer is at last able to enter. Late arrivers sometimes gain access to the ear through the "back door"—the external orifice of the countertympanic cavity—where they seem to be unopposed.

When I first observed this peculiar jostling performance, I imagined that it must represent a kind of territorial defense, and in a sense it may in fact be something of this sort. In some instances the blocking and jostling actions are supplemented by direct though apparently harmless attack with the mouthparts. Jostling movements can sometimes be evoked experimentally by prodding with a bristle, but only when the mite is in her own "territory." And yet it is common to find incipient colonies comprising several gravid females in frequent contact with one another and with no such show of resistance or "hostility." Jostling sometimes occurs, however, in response to other members of the same colony and when no active intrusion is taking place. Its effect upon other mites is mildly repellent, causing them to move aside at least for the duration of the oscillations.

Shaking movements of various kinds have been noted in other mites, in termites, in roaches, and in crickets, as well as in bees, wasps, and flies. They are usually regarded as having some communicative function. It is possible that in *Dicrocheles* this behavior aids in driving young females out of the overcrowded ear and in initiating their search for a new host. Jostling in a prospective feeding area might make it difficult for the young females to achieve engorgement and might thus enhance the stimulus of hunger as a drive toward dispersal.

As more and more emigrants leave the ear, their number increases in the patagial region. Here they feed and defecate, using both sides of the host indiscriminately. They do not become engorged, however. During the day they stay mainly on the dorsal side of the collar region, concealed among the long hairs and beneath the patagia. At nightfall, when the moth's evening flight period is at hand, many of the mites move downward to the ventral side of the neck and head. If the proboscis is uncoiled at this time, from one to a dozen or more mites can be seen at its base and between the palpi, all facing forward or climbing over one another and waving their forelegs in an agitated manner. Some, a few at a time, may venture out upon the tongue for some distance and may readily leave the moth for a needle, a stalk of grass, or almost any inert object that is presented to them. If an odorous flower is offered, several mites may quickly swarm on to it. Once there, the urge to wander may be appeased for a time. The mites creep about, palpating the surfaces and entering any cups or crevices that they encounter. Often they will lodge for hours in such places, but I have not seen them take nectar or pollen. A flower thus occupied can be used experimentally to infaunate a fresh moth that feeds from it.

Some flowery or fruity odors seem to attract the mites, while others repel them. The effect may depend upon the nature and concentration of the volatile substance. I have seen mites leave a moth for a bit of filter paper moistened with ethyl acetate. If no flower or other source of food for the moth is provided, the young female mites will eventually leave the host anyway and will wander about, seemingly at random. In the close confinement of a five-ounce glass jar, such mites readily find and board a fresh host, but when (in one experiment) they were required to traverse a distance of ten centimeters in order to reach the moth, they failed to board it even in a trial period of more than twenty-four hours.

At the season of peak incidence it is common to find not one but several newly engorged females in a freshly invaded ear, and when a single moth is confined with a number of wandering mites or with an infested moth carrying a mature colony, as many as a dozen mites may get aboard. Very rarely, however, does this result in a bilateral infestation. In a typical experiment, in August 1955, when many young females had found and boarded a single moth, the left ear was seen three days later closely packed with eggs and larvae, but with no adult mites. The moth had died, though not necessarily as the result of the infestation. Eleven engorged females were found on the head, neck, and prothorax, where they had been feeding. Both sides of this region were occupied, but the right ear was empty and undamaged.

It is sometimes suggested that bilateral colonies are commoner than would appear from light-trap collections, and are simply not observed because the hosts, being totally deafened, are too frequently caught and eaten by bats. This, of course, is exactly the sort of selection pressure that should result in habitual unilaterality. That it has already had this effect is evident from facts such as those just described. Continuing selection of this kind may be reinforcing the unilateral behavior pattern. In moths, the survival value of the acoustic sense may also be reflected in the rarity of defects, anomalies, or aberrations in the tympanic organs. . . .

For me, there is something perennially fascinating in a *Dicrocheles* colony. Few animals and fewer parasites lend themselves better to close and inquiring scrutiny. The strangely interwoven lives of the mites, the moths, and the bats that prey upon them now interweave with my own life, as they will, inevitably, with that of anyone willing to mingle his threads with theirs. . . .

The next time a swooping bat snatches a choice moth from before your eyes just as it is about to settle at your light, swallow the curse that was on your tongue and breathe a word of thanks to this blind marauder. His voice, unheard by you, has called into being that marvel of form and sensitivity the noctuid ear, and thus has made a home for some of the most fascinating creatures in the world of night.

Treat, Asher E. 1975. *Mites of Moths and Butterflies.* Ithaca, N.Y., Cornell University Press.

Murder by Narcosis

Edward Jacobson

Do ants take drugs? There is at least one beetle known to live in ant nests that secretes an "appeasement substance" from a special gland. The ants, which are normally disturbed by the beetle's presence, become placated and friendly after imbibing this irresistible potion. In the account reproduced here, a Dutch naturalist living in Java in the early part of this century describes an assassin bug that kills ants with a kind of gruesome mercy: It takes its victims only after first allowing them to feast on its narcotic secretions and then patiently waiting until the ants have drifted into unconsciousness. As William M. Wheeler pointed out, this is "a flagrant example of appetite perversion."

The very curious species of . . . *Ptilocerus ochraceus* . . . was found by me at Wonosobo, a place in the Kedou Residency (Central Java), situated at a height of about 2600 feet. In the vicinity is a natural spring, used as a bathing place, called Mangli.

In the bamboo sheds surrounding the basin I discovered large numbers of the curious bug. A few of them were flying about, but the majority was to be found in the open ends of the bamboo poles of which the roof was constructed, and on the top of all the wooden posts supporting the structure. Hundreds of the full grown insects as well as their nymphs were crowded together in certain places. They were attended by large numbers of small black ants, which Professor A. Forel had the kindness to determine as *Dolichoderus bituberculatus* Mayr, one of the most common species in this country. These ants usually make their nests in trees where they fasten together two leaves, between which they store the larvae and cocoons; but other sheltered places serve the same purpose, and many dwelling-houses are infested by them. On post and beams countless numbers are busily moving along definite paths. Their hunting grounds extend to the surrounding trees and shrubs, where they keep large herds of Aphids, Coccids and Membracidae for the sake of the sweet excretions these insects and their larvae afford them. I often found with these ants caterpillars of the Lycaenid butterfly *Gerydus Boisduxali* Mre, which make themselves agreeable to the ants by yielding a fluid of which the ants are fond, but in their turn they feed on the scale insects kept by the ants.

This all goes to show that *Dolichoderus bituberculatus* is particularly fond of the sweet excretions of different insects, a taste which sometimes leads to its wholesale destruction, as we shall presently learn.

Most of the ants, which I found . . . near the bugs, appeared to be in a more or less paralysed state and the ground beneath was in some places covered an inch thick with dead ants. These corpses were continually carried off by another kind of small red ant, but new victims, dropped from the roof, incessantly replaced those that were removed. As the spot where the bugs had settled down was a very inconvenient one for closer inspection, I gathered a large number of the insects, adults as well as nymphs, and carried them home alive, together with a section of a bamboo pole from the roof, which being split open revealed a large number of bugs' eggs, fastened to the inner surface.

Soon afterwards I left Wonosobo and returned to my dwelling-place Samarang, where I had a better opportunity for observing the bugs at leisure. The ants I had found with them at Wonosobo also abounded near my house; so I brought a great number of these ants together with the bugs in a small cage with glass windows, specially constructed for the observation of insects.

The bugs had fasted for about a week, the only thing I had given them being pure water, sprinkled in their cage, and which they readily absorbed. They were however none the worse for the fasting, only a few of the many hundreds I had captured having died. . . .

The bug possesses a very curious tuft of yellow hair (a trichome), situated on the under-side of the body, which apparently secretes some substance with a flavour agreeable to the ants.

The way in which the bugs proceed to entice the ants is as follows. They take up a position in an ant-path or ants find out the abodes of the bugs, and attracted by their secretion visit them in great numbers.

On the approach of an ant of the species *Dolichoderus bituberculatus* the bug is at once on the alert; it raises half way the front of the body, so as to put the trichome in evidence. As far as my observations go the bugs only show a liking for *Dolichoderus bituberculatus;* several other species of ants, e.g. *Cremastogaster difformus* and others, which were brought together with them, were not accepted; on the contrary, on the approach of such a stranger, the bug inclined its body forwards, pressing down its head; the reverse therefore of the inviting attitude taken up towards *Dolichoderus bituberculatus.* In meeting the latter the bug lifts up its front legs, folding them in such a manner that the tarsi nearly meet below the head. The ant at once proceeds to lick the trichome, pulling all the while with its mandibles at the tuft of hairs, as if milking the creature, and by this manipulation the body of the bug is continually moved up and down.

At this stage of the proceedings the bug does not yet attack the ant; it only takes the head and thorax of its victim between its front legs, as if to make sure of it; very often the point of the bug's beak is put behind the ant's head, where this is jointed to the body, without, however, doing any injury to the ant.

It is surprising to see how the bug can restrain its murderous intention as if it was knowing that the right moment had not yet arrived.

After the ant has indulged in licking the tuft of hair for some minutes the exudation commences to exercise its paralysing effect. That this is only brought about by the substance which the ants extract from the trichome, and not by some thrust from the bug, is proved by the fact that a great number of ants, after having licked for some time the secretion from the trichome, leave the bug to retire to some distance. But very soon they are overtaken by the paralysis, even if they have not been touched at all by the bug's proboscis. In this way a much larger number of ants is destroyed than actually serves as food to the bugs, and one must wonder at the great prolificacy of the ants, which enables them to stand such a heavy draft on the population of one community.

As soon as the ant shows signs of paralysis by curling itself up and drawing in its legs, the bug at once seizes it with its frontlegs, and very soon it is pierced and sucked dry.

The chitinised parts of the ant's body seem to be too hard for the bug to penetrate, and it therefore attacks the joints of the armour. The neck, the different sutures on the thorax and especially the base of the antennae are chosen as points of attack.

Nymphs and adults of the bug act in exactly the same manner to lure the ants to their destruction, after having rendered them helpless by treating them to a tempting delicacy.

The bugs are very sluggish in their movements, advancing by little jerks, brought about by stretching alternately the right and left hindleg, making brief stops between each movement.

Their flight also is very slow and laboured, and the bugs can easily be captured when on the wing.

The hindwings are much reduced in size and consequently the forewings are chiefly used for the purpose of flying.

In copulating the male does not mount the female, but clings to its side, a position necessitated by the shape of the body.

The eggs are deposited in hidden places, as e.g. the inner wall of a bamboo. They are irregularly spread over the surface, and more or less covered with a white exudation.

Although the bugs occurred in thousands in the locality mentioned . . . I rather think that they are not very common, as I have never met with them before, notwithstanding I often visited localities of the same altitude, and even at Wonosobo I could not find them anywhere else.

Samarang (Java), June 1909.

Jacobson, Edward. 1911. "Biological Notes on the Hemipteron *Ptilocerus ochraceus.*" *Tijdschrift voor entomologie* 54:175–79.

The Ant-Decapitating Fly

Theodore Pergande

What would you do if you noticed lots of disembodied ant heads strewn about? Theodore Pergande (1840–1916), a good naturalist, took notice, stored his observations in memory, correlated those memories with something he noticed years later, and made an important discovery. Theodore Pergande was a self-educated German immigrant who worked as an assistant to the applied entomologist L. O. Howard in the U.S. Department of Agriculture in Washington, D.C., from 1878 to 1916. By all accounts hard-working, eccentric, and a keen observer, Pergande was particularly interested in ants.

Pergande's observation was undeniably fascinating, but, like so much "pure" (rather than "applied") biological information, it turned out to be important for the human species as well. Since Pergande's time, dozens of ant-decapitating phorid fly species have been discovered, as a general rule each one specializing in a single species of ant. Today, almost a century after Pergande's report, scientists in the modern version of Pergande's agency, the USDA, are hard at work studying the phorid fly parasites of the so-called imported fire ants (obtained along with other fire ant parasites during repeated expeditions to the fire ants' native turf in South America). The hope is that these flies may be able to do the job that humans thus far have failed at: controlling this ever-spreading fire ant scourge of the southern United States.

For many years past, when rambling about the woods surrounding the city of Washington, I frequently came across larger or smaller colonies of the so-called carpenter ant (*Camponotus pennsylvanicus* De Geer), the largest and most powerful of our indigenous ants, which, as a rule, prefers to select for its home, dead, or partially decayed, forest trees, stumps, and logs in which it excavates cavities of various sizes and shapes, for the purpose of having a congenial home and safe dormitories for its progeny. Frequently, on finding such a colony, I watched them excavating new chambers, the detached chips of which were either carried patiently to the base of the tree or stump, or simply dropped to the ground. In watching this work and seeing the chips dropped or deposited, I frequently noticed around the base of the tree, stump, etc., numbers of heads of this ant strewn about, which always aroused my curiosity as to the cause of this strange phenomenon. Thinking, however, that the ants to which these heads belonged had succumbed to disease or old age while in the colony, and that their earthly remains had been disposed of in this simple manner, I dropped the subject entirely. Recently, however, this subject of heads without bodies flashed suddenly on my mind, while in the woods near Cabin John Bridge, Md., on the 5th of September, 1900, after concluding some observations on certain insects inhabiting the witchhazels and birches at the edge of the small creek at the bottom of the little valley. I ascended the steep and wooded slope for the homeward trip; getting tired and out of breath when about two-thirds up the hill, I stopped to accumulate enough steam or lung power to enable me to gain the crest of the slope. While standing there to readjust my respiratory organism, I happened to be near a beech tree, which in this locality abounds. Casually looking over the smooth trunk I observed a small worker of *Camponotus pennsylvanicus,* head downwards, about five feet above the ground, which had evidently come down from a foraging expedition, collecting the honey-dew or nectar of *Phyllaphis fagi,* which was very abundant on this tree. As a rule, these ants are very active in ascending and descending trees; this particular specimen, however, had stopped short for some mysterious reason and remained in this position while I was watching it and kept motionless until I touched it. The poor creature appeared to be tired and sleepy, and moved rather aimlessly and laboriously to one side on being touched. Being urged again, it moved a short distance in the opposite direction, seemingly in a trance or having lost the memory of its home. The action of this specimen appeared rather remarkable to me, since this species is very active, especially when disturbed, when it darts along at a rapid gait; it seemed to have lost control of its limbs and movements of its body. The head was drooping, as if of no use to its owner, though the antennae were still moving. In fact, it seemed as if it had lost its head. In order to discover something of the cause of this ailment I transferred it to a large vial and took it home. After an hour or two of resting, after reaching my home, I examined the vial to learn how my little sufferer was getting along, but found to my surprise that the poor thing was minus a head, though still alive and quite as active as before on being urged to move about. Further investigations of the contents of the vial disclosed the head at the farther end of the vial, minus its antennae and mouth parts, which were some distance in front of the head, and, while examining the skull of the ant through the glass, I observed the anterior part of a Dipterous larva protruding from the anterior opening of the shell, swaying back and forward, but soon to retire into the empty shell of the head. Next morning the poor victim was dead.

Having kept the specimen reasonably damp, I was rewarded and delighted to find on the morning of the 21st of September, or 17 days after the head had been cut off, a very handsome, extremely active and agile little fly in the vial, scurrying along with extremely rapid motions, only to stop now and then to clean itself and to exercise its horny genital apparatus in anticipation of the important work to be performed by it.

To continue my observations, and possibly to obtain additional infested ants, I went again to the same locality on the 24th of September, but failed utterly in discovering additional specimens showing symptoms of being infested. Being disappointed, and sitting down near the base of a beech tree badly infested with *Pemphigus imbricaria,* suspecting that there might be a colony of this ant at or near its base, I removed the loose bark, just above the ground, and found that I was not mistaken, for, as soon as it was removed, there tumbled and scrambled forth numbers of this ant in utter consternation, which apparently had been hiding for some unexplained reason, instead of climbing the tree in quest of the abundant honey supply above. Some of these ants, in their headlong haste to escape imminent danger, were captured and placed in a vial in the hope of breeding one or more of the flies, though without success. All others succeeded in hiding themselves under fragments of sticks and dry leaves, which had accumulated around the base of the tree. On removing this accumulated waste gradually, I found that the ants had their formicary under and between the larger roots of the tree, and on being again disturbed they scampered in evident alarm for their underground passages. While this was going on most of the exit holes had been closed with earth, preventing the stragglers from entering their home in time. While watching the frantic efforts to conceal themselves, I observed minute objects flitting rapidly about and hovering above and near the spot where the majority of the ants had disappeared, alighting now and then in quest of some particular object. The light in this particular locality is rather dim and subdued, with a streak of sunlight here and there, which renders it very difficult, especially at the surface of the ground, to distinguish with any certainty the minute insects whirling about near the entrance of the formicary. Being suspicious, however, that some of those animated dots, darting back and forth, might be the enemy of these ants, I managed to capture four of them, and found on examination that they really belonged to the species infesting the head of this ant, though all of them were females. While thus engaged I observed one of the ants which had secreted itself under one of the roots, make its appearance for the obvious purpose of reaching the entrance to the nest, and observed that no sooner had it made its appearance than one of those winged atoms made a dart for it, which frightened the poor creature to such a degree that it almost tumbled over itself when it scrambled in great haste to hide itself. This observation seems to indicate that the ants are in mortal terror of their diminutive foe and deadly enemy.

Another trip was again made to the same locality on the 27th of September. But very few of the ants were observed, while the flies were still more scarce. Only one female was captured, which was still alive and very active the next morning. On placing one of the ants in the tube containing the fly, I soon observed that the ant had become aware of the presence of its enemy and commenced to run restlessly back and forth, the fly watching it rather interestedly; coming accidentally in contact with the fly, the latter darted at the ant, which, enraged at this unsuspected attack, went for it with a furious rush and widely open mandibles. For a moment there was a general mix-up, reminding one strongly of two tom-cats in a fight, during which combat the ant was trying hard to catch the fly with its legs and mandibles, though on account of its extreme agility it rushed between the legs of the ant and escaped unharmed. This battle was kept up for some time, the fly jumping on the back and head of the angry ant till both became tired out, especially the ant, which walked about slowly, cleaning her head, mouth, antennae and legs; the poor creature became at last so completely exhausted that her legs commenced to tremble. The fight between the two had evidently lasted through some part of the night, when the ant at last succeeded in capturing the fly, which it crushed to a shapeless mass. On examining the ant the next morning, I failed to find any eggs on any part of its body.

After rearing this fly I wrote to the distinguished European writer on Formicidae, Professor Charles Emery, and asked him whether he knew of any observations upon the decapitation of ants. He replied that nothing of this nature had been called to his attention, except the fact that *Formica exsecta,* a very savage fighter, in the course of its battles frequently cuts off the heads of its opponents.

Later, mentioning the matter to Dr. L.O. Howard, he called my attention to the observations reported by Dr. W.H. Fox, at the September, 1887, meeting of this society, published in Volume I, pp. 100 and 101. Dr. Fox found the decapitated heads of *Camponotus pennsylvanicus* at Hollis, New Hampshire, in the summer of 1887, and discovered that they contained Dipterous larvae. At a meeting of the Biological Society of Washington, held in October, 1887, Dr. Howard mentioned this observation of Dr. Fox's and suggested that the Dipterous larvae in question might belong to the family Conopidae, the larvae of certain species of which in Europe are parasitic upon Hymenopterous insects. The present observations set the matter at rest and indicate the true nature of the parasite. It is probably identical with or at least congeneric with the species observed by Dr. Fox in New Hampshire. The adult parasites were referred to Mr. Coquillett, who finds that they constitute a new genus of the family Phoridae.

Pergande, Theodore. 1901. "The Ant-Decapitating Fly." *Proceedings of the Entomological Society of Washington* IV:497–502.

For the Love of Nature

Thomas Eisner

As an undergraduate premed student at Harvard in the early 1950s, Tom Eisner (b. 1929) experienced an epiphany when he began to associate with entomology graduate students and realized that some people make their livings studying insects. He signed up for graduate study in entomology at Harvard and never looked back. Entomology is the richer for it. His encounter with lacewing larvae lays bare the scientific method as well as the playful nature of a naturalist's mind.

I do not remember the date, but I will never forget the occasion. It was early autumn in 1971, I believe, and I was spending a few days at one of my favorite hideouts, the Huyck Preserve near Albany, New York. I often go there after completion of my summer's experimental program in search of peace and a chance to explore nature at leisure before onset of the formal academic year at Cornell. The region is indescribably beautiful at that time, with the foliage in the midst of its spectral shift and the weather usually crisp and sunny. I was with Robert Silberglied of Harvard on that day, a close friend and fellow nature enthusiast, strolling about in the field with collecting gear and camera, observing colonies of the wooly alder aphid, *Prociphilus tesselatus.* Our fascination was not so much the aphids themselves, but with the attendant ants that stood guard over the aphids, drinking their honeydew and providing them with protection in return. We knew such "shepherding" behavior to be widespread among ants, but neither of us had spent much time watching it. We poked the ants and noted how they held their ground and attempted to bite whatever instrument we used to provoke them. We saw a more interesting phenomenon: Wasps, which were also attracted to the honeydew, were actively prevented by the ants from feeding on the aphids and forced to restrict their drinking to excess honeydew that had dribbled from the aphids to leaves lower down on the alder plant where there were no guarding ants. While attempting to photograph the aphids at close range, I caught sight of something that I knew full well could not be. Aphids, my own experience had told me, are usually sedentary, and when walking are slow at best. Yet here, clearly apparent in the viewing screen of my camera, was a *running aphid!* A careful second look revealed the actual nature of our find. It was the larva of a green lacewing, a so-called chrysopid larva (Neuroptera: Chrysopidae)—not an aphid at all, but an aphid predator—so similar in appearance to the *Prociphilus* with which it was living that it could easily pass as one of them.

We spent the evening in our makeshift laboratory in a cottage at the preserve, examining the larva and watching its behavior. We saw that it fed on the *Prociphilus* aphids, which it pierced with its sickle-shaped mandibles and sucked dry, as chrysopid larvae typically do with their aphid prey. And we noted again the extraordinary resemblance of the larva to the aphids, rendered all the more striking at the higher magnification of the microscope. The "woolen" investiture of the larva seemed identical to that of its prey. Work done in collaboration with associates at Cornell had shown the aphid wool to consist of tufts of very fine strands of wax, later identified as a long-chain ketoester. The wool of the larva appeared to be made of the same strands, although they were not rooted, as in the aphids, and seemed to be more irregularly arranged. I knew that there were certain chrysopid larvae, called "trash carriers," that cover their backs with debris, and it occurred to me that our larva might be of that type and that it obtained its wool by plucking it from the aphids. A simple experiment confirmed this. I removed the wax from the back of the larva with a brush, and when it was thoroughly denuded, released it again among the aphids. Within minutes it began reloading itself. Using its mandibles as a two-pronged fork, it plucked one tuft of wax after another from the aphids and applied the material to its back. In less than a half hour it had rebuilt its cover. I was fascinated by what I saw and was hooked on the prospects of working with this insect. Some of my Cornell associates, including Karen Hicks and my wife Maria, joined in the project.

There were intriguing questions to be answered. Does the waxy covering protect the larvae against ants? Do the ants actually mistake the larva for an aphid? Vis-à-vis the "shepherding" ants and their aphid "flock," is the larva a true "wolf in sheep's clothing"? We soon learned that the larvae were not at all rare and even relatively easy to collect once we had learned to tell them apart from the aphids. We maintained some in the laboratory, raised them, and had them identified when the lacewings emerged. They turned out to be *Chrysopa slossonae,* known from the adult stage only. The larva had never been described. In its near-perfect disguise it had apparently escaped detection.

Experiments with denuded larvae showed that they give high priority to the reloading procedure. They usually began gathering wax soon after being reintroduced among the aphids and continued doing so until their shield was complete. If starved beforehand and therefore driven by the dual need to reload and feed, they divided their time about equally between both activities.

We spent hours observing larvae in the field and found to our delight that the "wolf in sheep's clothing" analogy really held. The ants seemed truly oblivious to the presence of the larvae. As the latter fed on the aphids, impaling one after the other on the mandibles and sucking them out, they induced little overt reaction in their prey. The ants failed to detect the larvae even when they tread on them and continued drinking honeydew from aphids in the immediate vicinity of the larvae.

Without their shields, the larvae are relatively helpless. We denuded 27 larvae, released them into *Prociphilus* colonies, and followed their fate. All except four were discovered by the ants and removed from the colonies. Individual ants grasped them and dropped them to the ground or carried them to the ground by descending along the branches of the plant. The four larvae that escaped detection made their way to unguarded sites of the colonies and proceeded to rebuild their shields.

Larvae that were released in the near vicinity of ants with their shields intact, in such fashion that they were bound to be encountered by the ants, were sometimes bitten, but the wax proved an effective deterrent. The ants released their hold, and with their mouthparts heavily contaminated with wax, backed away. As the ants then proceeded to cleanse themselves, the larvae made their escape.

The stunning camouflage of the larvae suggested that they might also be protected against other predators, but we did not test for this. Birds, for example, appear not to feed on *Prociphilus* and might therefore also ignore visual mimics of the aphids. And there are other unanswered questions. We know that the female *Chrysopa* lays its eggs in the close vicinity of *Prociphilus* colonies, but do not know how she locates these. We also know little about how the larvae react to one another. Are they, like so many other chrysopid larvae, cannibalistic? Does their resemblance to aphid prey increase their chances of being cannibalized? As is so often the case, discovery leads to followup, and the followup creates its own need for further exploration and discovery.

Eisner, Thomas. 1982. "For Love of Nature: Exploration and Discovery at Biological Field Stations." *BioScience* 32, no. 5:321–26.

Mimics, Aggressive and Otherwise

John Alcock

Not every flying insect with black and yellow stripes is capable of stinging, but they all usually get our respect. This is the essence of Batesian mimicry, a factor that apparently has been important in the evolution of the two Sonoran Desert fly species wonderfully described in this selection by biologist and natural historian John Alcock, whom we met in chapter 7.

It is the season of the robberflies. In a palo verde on Usery Peak three orange robberflies bumble through the branch tips to perch on bent legs that seem far too long for their bodies. Members of this species will occupy the tree for a few short weeks, their flight season squeezed into a niche in July and August. They are part of a succession of insects that began in late February; since that time, everything from tiny midges to giant tarantula hawk wasps has come to the hilltops according to schedule to wait for sexual partners or to feed on the species that rendezvous here.

Robberflies are killers. They ambush their insect prey, darting out from a perch to grasp a victim with their hairy legs. After stabbing it with its stiletto mouthparts, the predatory fly slowly drains the subdued prey of its internal contents, then drops the husk of its meal to the ground. One of the orange robberflies dangles from a branch, suspended by its front legs, while it consumes a small black bee that droops at the end of the fly's piercing proboscis.

The orange robberfly is only one of about a thousand North American members of its family, the Asilidae. The wonderful thing about insects is the diversity in body form and behavior that has arisen among related species, a diversity that often overshadows that found in comparable assemblages of vertebrates. Robberflies are delightfully diverse. They come in all sizes and colors, from those a few millimeters long to giants (for flies) as long as your thumb. Variation in body size is paralleled in variety of diet. The tiny robberflies swoop after minute midges or delicate true bugs; the biggest species embrace and kill big, chunky grasshoppers or robust bumblebees, incapacitating their hefty targets with lethal injections of saliva.

At the peak of the robberfly flight season, representatives of several species coexist on Usery Peak. In addition to the clownish long-legged orange fly, there are other grey-black asilids, with long, sleek, tapering abdomens. Many of these superficially drab animals have patches of silvery white at the tips of their bodies or streaking their abdomens, so that when they fly up light flashes from them. A much more showy asilid, about the length of my thumbnail, sometimes perches in the palo verdes with its larger cousins. This fly's most striking feature is its brilliant green eyes. Golden brown hairs coat its thorax, and its robust, darker abdomen is banded by rings of pale tan hairs. Backlighted on its perch, the creature assumes a radiance not usually associated with flies. One member of the genus has been given a name that might apply to all its close relatives, *Mallophorina pulchra*, the beautiful *Mallophorina*.

When the Usery Peak *Mallophorina* leaves one perch to go to another or to zip after a potential victim, its golden hairs, posture in flight, and buzzing wings combine to create the illusion that it is a small bee. A great many asilids resemble various species of bees and wasps, and entomologists early on recognized that these mimetic robberflies generally specialized in the dispatch of bees and wasps. *Mallophorina* is a fairly small asilid, and it usually captures small to medium-sized solitary bees, of which some manage a living even in July in the Sonoran Desert. Other, much larger asilids elsewhere look very much like paper wasps or bumblebees, and these hefty predators typically hunt and kill their look-alikes.

The pioneering American entomologist C. V. Riley suggested that the bee- and wasp-mimicking flies were "aggressive mimics" that used their mimicry to approach their victims closely without alarming them. A "bumblebee" asilid might, according to this hypothesis, be able to practice a kind of wolf-in-bee's-clothing tactic, the better to approach and grasp bumblebees, its favored meals. The resemblance of *Mallophorina to* certain small desert bees would constitute aggressive mimicry if these bees were more likely to ignore the fly than some other predatory insect enemies.

The aggressive mimicry hypothesis is an intriguing one, but it has not been tested with asilids and their prey, and there are reasons to think such tests would lead to rejection of the hypothesis. E. B. Poulton pointed out in 1904 that asilids almost never approach their prey stealthily or casually, but instead launch short, rapid attacks from perches in the areas where their prey are flying. The nature of these ambushes is such that a resemblance to the victim would hardly advance the success of the attacker.

Thus it seems likely that the bee- and wasp-mimicking robberflies are engaged in what is called Batesian mimicry, after Henry Bates, the English lepidopterist who first proposed that an edible species could gain some protection from its predators by resembling another species that is noxious, poisonous, or stinging. Birds relish flies but are less fond of stinging Hymenoptera, and with good reason; a flycatcher stung in its mouth by a bumblebee or paper wasp receives a huge dose of toxin relative to its weight and can be expected to suffer proportionally. Humans quickly learn to avoid bees and wasps by their color patterns. Insect-eating birds are equally adept at this form of learning, as many experiments have shown. Because bumblebee-hunting asilids occur by necessity in habitats with their prey, birds that have learned through unhappy experience that bumblebees should be left to their own devices may mistakenly also ignore mimetic stingless asilids, which fool their predators.

Batesian mimics rely on deceit to gain a survival advantage. Their trick can work only on "educated" predators, those that have had an unpleasant encounter or two with a genuinely protected species before they come across the mimic. As a general rule, therefore, Batesian mimics are rare relative to their models, a rule that holds for bee-mimicking robberflies. In addition, as G. P. Waldbauer of the University of Illinois has documented for an extensive array of mimetic flies, the deceiving species usually emerge as adults later in the season than do their models. This scheduling tactic gives local bird predators, particularly the young of the year, an opportunity to sample models and get stung or sick before the mimics appear on the scene.

It is possible that the timing of adulthood in *Mallophorina* has been shaped by bird predation, with the fly coming out only after fledgling flycatchers have tried, with unhappy but highly educational consequences, to eat some small stinging bees. On the other hand, it is just as likely that the mimetic robberfly metamorphoses to adulthood slightly later than its bee models because it would be disastrous for a bee hunter to emerge before there were bees to hunt.

A more probable (but still uncertain) case of adaptive developmental scheduling by a mimetic fly on my study site involves a mydid fly, *Mydas ventralis*. This species belongs to a family, the Mydidae, that is closely allied with the Asilidae. They not only look rather like some asilids, they may also employ the asilid tactic of ambushing insect prey. This family, however, does not contain a great diversity of species; there are less than five hundred species worldwide and only about fifty in the United States. The low number of species, the rarity of those that do exist, and their preference for "the hottest climates at the hottest times of the year" have combined to make mydid flies one of the least-studied of dipterans. Indeed, Frank Cole philosophizes in his *Flies of Western North America* that "like many large and ungainly creatures, they have probably found it difficult to survive in a highly competitive world." Simple accidents of evolutionary history could be responsible for the rarity of mydid flies, of course, but it is true that they are large; indeed, the largest of all flies is a Brazilian mydid, *Mydas heros,* which has a wingspan of four inches. This must be a formidable-looking fly, although adults of this species apparently are actually among the Ferdinands of the mydid world; they feed on nectar, not insect prey.

When I see *Mydas ventralis* on Usery Peak in June and July, however, the flies are not sipping flower nectar or the juices of insect prey, but instead are looking for mates. Like many of their insect predecessors on the local peaktops, males gather on the ridge to secure perch territories where they wait for a female to show up and copulate. One of the species in this group that precedes the mydid to the peaks is a big tarantula hawk wasp whose males claim palo verdes, creosotes, and jojoba bushes as their territories. Male tarantula hawks fly about in conspicuous patrols, showing off their jet-black bodies and bright, red-brown wings. Female tarantula hawks have immense stingers and even the stingless males appear to possess some chemical protection; ash-throated flycatchers never seem to attempt to capture them, despite numerous opportunities. The males smell unpleasantly acrid, and so I suspect they are not tasty morsels.

The flight season of the wasp peaks in May and peters out in June, giving fledgling flycatchers ample time to make a mistake and grab a tarantula hawk or two before the mydid flies come to the ridgetops. These flies, which are just as big as tarantula hawks, do a wonderful imitation of the wasp, but only when they are flying. Although they have dark wings and a red abdomen, rather than red wings and a dark abdomen, when circling their territories they maneuver in a dipping, tilting flight that closely matches that of tarantula hawk wasps. Presumably, it is when it is flying that a mydid fly—or an asilid, for that matter—is at special risk of detection and attack by those birds that ambush insect prey. If juvenile ash-throated flycatchers do sometimes sample a tarantula hawk, only to find it repellent, they might be prone to look the other way when they see a *Mydas ventralis* looping about its peaktop territory. Today, in a truly hot place at the hottest time of the year, there are no flycatchers on Usery Peak, no birds at all, no thought of rain, only a few mydid flies and dormant brittlebushes, and a grey robberfly grimly clutching a skeletal creosote limb.

Alcock, John. 1990. *Sonoran Desert Summer.* Tucson: University of Arizona Press.

The walking stick and the leaf mimic (the left-hand leaf is a butterfly at rest). Illustrations by Mary Wellman (*American Insects* by Vernon L. Kellogg, Henry Holt and Co., New York, 1908) and E. A. Smith and J. B. Zwecker, engraved by G. Pearson (*Insects Abroad* by J. G. Wood, Longmans, Green & Co., London, 1874).

10

Lives under the Microscope: Insect Behavior

Set out walking in a wood or a jungle or a desert until you are miles away from the nearest human being. Stoop down to the ground and begin to dig, or climb a tree and tear away some bark, or turn over a rock or a log. There you will find thousands of insect lives, and, had you not ventured this random act, they would have been lived from start to finish without the notice or knowledge of any human mind. And so it has been for countless generations in countless lineages for 400 million years, and so it will be, perhaps, for millions of years after we humans have gone.

This vast evolutionary history of the insects, combined with the immense diversity of environments they inhabit, has produced an array of rich, intricate behavior that defies even the wildest human imagination. The study of this behavior has evolved as well. Following the natural-history approach of the nineteenth and early twentieth centuries, the behaviorist approach dominated the study of insect behavior in the United States until the 1960s, characterized by an interest in learning and memory and experiments in artificial settings. Today, the field is dominated by behavioral ecology, which poses hypotheses about the evolutionary (selective) advantage of certain behaviors over others, and then seeks to test those hypotheses with experiment and observation.

Whatever their approach, behavioral entomologists have been and continue to be driven by the inevitable sense of wonder that accompanies the study of insects. This fascination is not the exclusive province of science, however; as we shall see in this chapter, it is available to anyone who takes the trouble to become acquainted with how insects behave.

Little Crumple-Wing

George D. Shafer

George Shafer, a retired professor of physiology turned wasp researcher whom we met in chapter 1, dedicated the book from which this excerpt was taken to a wasp, writing: "To 'Crumple-Wing', who was true to the tradition of all mud daubers. She honored her Creator *and* published his name *so that 'He who runs may read.'" Obviously, as he became closer to his subjects, Shafer began to appreciate them as individuals with unique personalities and "minds."*

It was the "mud dauber of the window" that first showed me how females of her tribe deliver the paralyzing thrust, with the sting, to spiders which are to be stored as food for their young. She demonstrated the act on one spider only. On three other occasions, spiders were offered, but she refused—really seemed to spurn them. The reason for her refusal was a mystery at the time. Much later, when other individuals had made known their habit of taking very definite little rests and vacations from work, I began to suspect that my offers might have been timed wrongly. Perhaps, during vacation, a proffered spider was not a gift, but an offense. Of course fear is probably the main reason why mud daubers, in general, will take gifts from no man. But between friends, it seemed, there must be some other good reason for refusing an intended gift.

It was not until little "Crumple-Wing" came into my care that the acts of accepting spiders and of stinging them were demonstrated over and over for me. She was, to all appearances, a perfect female of the large species (*S. cementarium*)—perfect except for one wing which was crumpled. For some reason, that wing had not expanded fully before it dried. She had gnawed the mud plug and released herself from the cell, however. I found her on my workbench not long after that, and she ate honey from my hand. But things looked rather hopeless for her. She could not fly. She could not build and care for a nest, it seemed. How could she even care for herself? It was a very busy day; there was little time to think of her. When she finished eating from my hand, a drop of honey was placed beside her, and she was left standing on the workbench.

Two days later, walking through the beehouse, I somehow happened to look down. My foot was within an inch of stepping on a mud dauber—almost too close, but not quite. She was safe. My hand was offered, and she crawled upon it. It was the mud dauber with the crumpled wing. She was hungry and she remembered that hand. Long before this incident, it had become clear that the best way to teach a mud dauber to eat from the hand is to keep her where she cannot fly away for a few hours after she emerges from her mud cell, and then proffer the food. There seems to be less fear at that time than later in life, and hunger is sure to be present then too. Fear seems to be instinctive, of course, but once a mud dauber has voluntarily rested on the hand and eaten there, it never forgets. It will always be timid—some individuals more so than others—but if the proper, careful approach is made after that first surrender, it will always come again to eat when it is hungry. Males and females of both species (*cementarium* and *coerulium*) have been taught to eat from my hand. The few less timid ones may become a nuisance and a care sometimes if one is working constantly where they are. At any moment such a one may fly trustfully to you when you are not looking and endanger its life.

For example, one of these less timid pets came on a certain afternoon to be fed. I was extracting honey and just starting to pour a large pail of it upon the straining cloth when in came a hungry mud dauber. She flashed past my hand and alighted on the broad stream of honey. The surface must have looked solid enough to her, but that surface was both sticky and in motion. Down she went in the stream, and out of sight at once, engulfed in a sea of honey. That was a blundering thing for a mud dauber to do, and had she not been a pet, it would never have happened. For fifteen minutes, she could not be seen. Then she emerged, lifting her feet with difficulty, struggling to crawl up the side of the strainer cloth. The honey did not adhere much to her shiny, smooth body, but wings and feet were sticky. Tissue paper was used to wipe all surplus honey away, and then she was left outside the beehouse to finish cleaning herself.

But to return to the mud dauber with the crumpled wing. There she was, crawling about on my hand, searching for the expected honeymeal, and she could not fly! We walked over to the workbench together, and she soon began to eat. After that, she was placed under a bell jar, and I became her keeper. That is when she received her name, "Crumple-Wing." She soon came to regard the bell jar as home. Sometimes she was given a little freedom to run about on the bench while she was watched. At the end of such a play period, if I began tapping on the bench with my finger, closer and closer to her, she would run under the tipped bell jar and then turn quickly toward the finger, as if to say, "Now you can't get me." It seemed remarkable to me that she should learn to use legs and bell jar as a means of refuge, when if her wings all had been normal, they and "the air about her" would have served instead. It was the one crumpled wing that made all the difference in her life, and (because mud wasps are always so trim and neat in the way they carry their wings when they walk) she was constantly reaching back with her hind leg to push it along the wing—trying to smooth down the offending crinkle—until gradually she wore the whole crumpled part away, leaving only a wing stub.

Not long after that I chanced upon a dead mud wasp. I cut the corresponding wing from this wasp and endeavored to splice it to Crumple-Wing's stub and fasten it in place there with Testor's cement. The wing length was measured accurately, and she was held willy-nilly until the cement was dry, and the splice seemed firm. Then she was released. Indignant at her manhandling, she leaped from my hand and used her wings; but they whirled her in a circle and down she went. We manipulated the wing again and again, trying to achieve a balance, but it was no good. We had to give up, and Crumple-Wing, not pleased with my gift, worked away with her hind leg until by the end of a week she had pushed it off from her wing stub. There seemed to be no other way. She was going to have to continue to be the "little animal in a cage," and I her keeper. A few cells of uncapped honey were placed under the bell jar with her so that she might never suffer from neglect.

Several days later, another type of experiment that might be tried with Crumple-Wing occurred to me. The year before, a nest had been collected from which six male mud daubers and only one female had emerged. That seemed to be a very unusual proportion of males. It caused me to wonder, at the time, if a mud-dauber queen's eggs (like a virgin queen honeybee's eggs) would hatch and produce males in the absence of fertilization. Crumple-Wing had never mated. Perhaps she would answer my question. She could be supplied with mud and a partially darkened box, under the bell jar, for nest building. If she built a nest, spiders would be supplied. Then if she stored the spiders and laid eggs, an opportunity would be given to learn whether the eggs would hatch and, if so, whether the larvae would produce male mud daubers. Forthwith, a shallow little box, open on two sides, and a small tin container of mud were prepared. These were placed under the bell jar. Crumple-Wing examined them curiously. She went into the box and out, again and again. Every day she went through a routine of taking exercise. Climbing up into the top of the jar, she would jump down onto the box and then off it, flipping her wings as if trying to fly. But day after day went by, and she did nothing about the mud. A little water was added to the tin container as often as was necessary to keep the mud at the proper consistency. The days ran into weeks—a little more than three weeks—with no prospect of ever seeing a nest in the box.

Then, one afternoon when I came in, Crumple-Wing seemed unusually excited, and mud had been taken from the top of the container. Sure enough, in the upper corner of the box was a mud cell drawn out to full length. This mud dauber that had been unconcerned about nest building for so long now seemed eager. Her cell was ready; apparently she wanted spiders and at once. There was only one way of supplying them quickly on such short notice. The cell of another mud dauber was opened and its spiders, recently paralyzed, were removed. A choice specimen was picked out with fine-pointed forceps and offered to Crumple-Wing. She seized it and went through the motions, four times, of stinging it. Then she settled down and began to chew at the spider's cephalothorax and to sip at its body fluid. Finally she walked away, leaving the spider's abdomen untouched. It was removed from the bell jar. In about fifteen minutes she was back, turning this way and that with her antennae vibrating, as if hunting. I offered her another small spider. She seized it quickly, used her sting three times, and then began to chew until she had broken the body wall. When she was through this time and had walked away, only a comparatively dry, flattened little spider body was left. It was removed. Crumple-Wing seemed quite well satisfied. She stood quietly by the nest box. Of course she could swallow only fluids through her very narrow esophagus which traverses the thorax and the long, slender, inflexible waist petiole, to the crop. But these fluids were very different from honey, since they were rich in animal proteins. This was Crumple-Wing's first food of that kind, but it was not to be her last.

For twenty minutes she stood and rested. Now and then she cleaned an antenna or brushed an eye, but mostly she just rested. Then, suddenly, rousing herself as if she remembered something left undone, she hastened to the tin of mud, elevated her abdomen on those long hind legs, and began to sing as she rolled up a pellet of mud. Three times she hastened from the tin container to the nest box. Three pellets she gathered in this way, singing as she gathered them and singing as she plastered the wet mud into chinks around her cell, anchoring it more securely to the wood. This done she ceased work as abruptly as she had begun and settled down for another rest. She had no interest in repeated offers of a spider now. About two hours later she became active again, and I offered her another of the paralyzed spiders. She took it readily, stung it twice at least, and after making several false starts, carried it to her mud cell. Entering headfirst, she pushed the spider about halfway back in the cell. Then she backed out, examined the sides of the cell carefully, and finally rested by it for nearly three minutes. Suddenly another impulse of activity possessed her. She rushed out of the box, gathered a pellet of mud, and skillfully spread it as a thin curtain over the entrance to the cell. Following that she walked deliberately, with an air of assurance, out to the side of the box. Her spider in the cell was safe from marauders.

The next day she did not disturb the seal on the cell. She showed no interest all morning, but in the afternoon when I offered a spider she seized it readily. However, she did not go toward the closed cell. She performed the routine acts of stinging and then chewed the legs and thorax, sipping the body fluids. The next forenoon the cell was still sealed, but she accepted another small spider and dined on its body fluids. In the afternoon she used another small spider in the same way. Later she opened the cell, but she took neither of the chewed spiders to it. She closed the cell in the evening. That was on Saturday. She was not visited on Sunday and not until one o'clock on Monday. She had her cell open then and seemed very eager. As I approached the bell jar she ran to the side to meet me, dancing sideways back and forth with her face toward me. I lifted the jar and offered a spider. She jumped to seize it, performed her stinging act, and then, carrying the spider in her jaws, she began to run. She hopped with all six feet clear of the surface, as if trying to fly. She could not fly, but she tried, over and over. She climbed up the inside of the jar to the top and jumped down to the box; then off onto the bench, only to race through the whole performance again—overjoyed, it seemed to me. For nearly twenty minutes she kept going. Then she put the spider in the cell, got a pellet of mud at once, and sealed the entrance with a protective curtain. That done, she rested for the remainder of the day. After opening her cell the next day, she received a spider, stung it, and then ran for three minutes before she stored it in the cell. These two were the only long runs she made, carrying a spider. Immediately following this last exhibition, she received five spiders, one after the other, stinging them and storing them in the nest at once. When a sixth was given, she chewed its cephalothorax and fed on the body fluids. A few minutes later she gathered mud and sealed her cell with a curtain. She was through work for the day.

The next spiders accepted by Crumple-Wing were two large grass spiders which I captured for her. She was able to paralyze both of them and carry them. That filled her cell, and she began gathering pellets of mud at once to seal it—a permanent seal this time. She carried eleven pellets of mud before she stopped to rest. Three of them, at least, were used to form the cell plug, the others were spread along a crack in the box next to the cell.

Then she went to the bit of honeycomb, kept in the bell jar with her, and sipped honey. A short rest followed, after which she went to work once more, carrying mud. She plastered the whole cell with fresh mud and then began construction of a new cell. The latter was finished before noon next day. It was a longer cell than the first. The nest box was removed temporarily from the jar, inverted, and a photograph obtained of the two cells—one sealed, the other open. Later, color photographs were obtained of Crumple-Wing before her nest box as she manipulated a spider to sting it . . . rising on her toes ready to dip her abdomen under and sting her captive. The wasp stands before the nest box. At the right is the can of fresh mud and the bit of honeycomb.

When Crumple-Wing placed spider No. 2 in the new cell, she did not push it in; she backed into the cell pulling the spider, as she held it beneath her, in her mandibles. She remained in the cell five minutes and then came out, headfirst, without the spider. She pushed all other spiders into the cell and then backed out. It was my belief (and hope) that she had deposited an egg on spider No. 2. Sixteen days elapsed from the time this cell was ready for spiders until Crumple-Wing sealed it with a permanent plug. She had taken many rest periods in that time. Perhaps it might not have taken quite so long if I had always been present with spiders when she had the protective curtain gnawed away and was ready for work. But it was clear as the days passed that she needed longer periods of rest—that she was growing older and weaker. She had waited long, too long, before beginning the nest. On her last day, she sealed the second cell with a permanent plug and finished plastering the surface of the nest. The next morning I found her in the darkest corner of the box, dead. She had lived between two and a half and three months.

Her work may be recapitulated thus: after the first cell was ready, she had been offered nineteen spiders before it was sealed permanently. She placed eight of the proffered spiders (including two of those I captured for her) in the cell; eight were chewed and the body fluids sipped. Three specimens offered had been refused altogether. Seven days passed from the time the cell was ready until it was plugged. However, on three of the seven days no spiders were placed in the cell. In the case of the second cell (the longer one), seventeen days passed from the time it was ready to receive spiders until it was plugged. Eighteen spiders were placed in this cell and, of these, nine were spiders which I had captured alive for her. During the time Crumple-Wing was occupied with the second cell, she chewed only two spiders and dined on the body fluids; this in contrast to eight used for that purpose while she was provisioning the first cell. Without doubt she was greatly in need of animal food when the nest was first started. There was no indication of any egg laying in the first cell, but when spider No. 2 was introduced into the second cell it will be remembered that Crumple-Wing backed in and remained there five minutes. As already described . . . the eggs examined on spiders within cells had been aligned in a certain very definite way. Backing in seemed to be the only technique that could so orient the wasp's long body as to enable oviposition to accomplish this observed result.

It was my mistake that spider No. 2 was not examined at once to see if an egg had been deposited on it. As the matter turned out, other work interfered and the cells were not opened until eight days later. Then an unfortunate situation was revealed. All the spiders that had been already paralyzed when they were given to Crumple-Wing were found to be dead* and completely covered with the mycelium of a fungus. None of the specimens showed any indication of having been fed upon by a larva, and no vestige of an egg could be found. If an egg had been deposited, the hyphae of the fungus had destroyed it completely. It must be confessed, therefore, that the experiment did not prove whether an unmated queen mud dauber may lay eggs, nor whether such eggs, if laid, would produce males.

However, the former surmise that a mud dauber's rest periods are sacred for rest was confirmed. Also confirmed was the fact that any cell partially filled with spiders is always closed with a thin protective curtain of mud when work for the day is over. This temporary seal is left intact until the wasp is ready for work again, and no spider will be accepted by her while the cell is so closed—a single exception being that she may accept a spider during such a period if she desires its body fluid as food for herself. When any spider's body wall (exoskeleton) had been broken by chewing, it was never carried to the cell. There was one large spider whose legs stuck out stiffly after it had been paralyzed. Crumple-Wing bit those legs repeatedly until they relaxed so that she could tuck them down beneath the spider and carry it. That spider she did place in her cell, but its exoskeleton had not been broken. Only by merest chance might one ever witness enactment of the capturing and stinging operations in the field. Rarer still the occasion when he might possibly discover there the fact that a mother mud dauber, herself, feeds on the vital body fluids of spiders. Crumple-Wing revealed

* All the paralyzed spiders which were given to Crumple-Wing were alive when she received them, but it is reasonably certain that they were killed by her additional stings. No spider lives very long after being stung by a mud dauber. Very small spiders may live only a few days. However, it is my observation that most spiders stored in a cell normally live to be fresh meat all through the comparatively short feeding period of the larva.

A solitary wasp drags a large wingless locustid to her nest. Illustration by Mary Wellman (*American Insects* by Vernon L. Kellogg, Henry Holt and Co., New York, 1908).

and repeatedly demonstrated this dining habit, as well as the acts of seizing, subduing, holding, and stinging the prey. Moreover, she had been induced to build a nest and stock it with spiders supplied to her while she was confined in a cage. She proved that the experiment, as first conceived, is perhaps not impracticable. She did all this and more; she was a little solitary wasp—handicapped, dependent, timid—but she accepted the ministrations of a man and became a friend.

Shafer, George D. 1949. *The Ways of a Mud Dauber.* Stanford, Cal.: Stanford University Press.

Brute Neighbors

Henry David Thoreau

After Thoreau (1817–1862) read Kirby and Spence's legendary An Introduction to Entomology, *the first comprehensive reference work on insect biology, he was inspired to make a literary transcription of "war" among the ants in* Walden, or Life in the Woods. *The details as Thoreau reports them are true—only the supposed rarity of the conflict is exaggerated and suggests that Thoreau may not have often watched ants. To an entomologist, ant wars are a day-in, day-out occurrence. Along the American eastern seaboard, where Thoreau lived and wrote, the everyday wars of the pavement ant are there for the watching all summer long—even in the towns and villages of New England.*

It is remarkable how many creatures live wild and free though secret in the woods. . . . You only need sit still long enough in some attractive spot in the woods that all its inhabitants may exhibit themselves to you by turns.

I was witness to events of a less peaceful character. One day when I went out to my wood-pile, or rather my pile of stumps, I observed two large ants, the one red, the other much larger, nearly half an inch long, and black, fiercely contending with one another. Having once got hold they never let go, but struggled and wrestled and rolled on the chips incessantly. Looking farther, I was surprised to find that the chips were covered with such combatants, that it was not a *duellum,* but a *bellum,* a war between two races of ants, the red always pitted against the black, and frequently two red ones to one black. The legions of these Myrmidons covered all the hills and vales in my wood-yard, and the ground was already strewn with the dead and dying, both red and black. It was the only battle which I have ever witnessed, the only battle-field I ever trod while the battle was raging; internecine war; the red republicans on the one hand, and the black imperialists on the other. On every side they were engaged in deadly combat, yet without any noise that I could hear, and human soldiers never fought so resolutely. I watched a couple that were fast locked in each other's embraces, in a little sunny valley amid the chips, now at noonday prepared to fight till the sun went down, or life went out. The smaller red champion had fastened himself like a vise to his adversary's front, and through all the tumblings on that field never for an instant ceased to gnaw at one of his feelers near the root, having already caused the other to go by the board; while the stronger black one dashed him from side to side, and, as I saw on looking nearer, had already divested him of several of his members. They fought with more pertinacity than bulldogs. Neither manifested the least disposition to retreat. It was evident that their battle-cry was "Conquer or die." In the meanwhile there came along a single red ant on the hillside of this valley, evidently full of excitement, who either had despatched his foe, or had not yet taken part in the battle; probably the latter, for he had lost none of his limbs; whose mother had charged him to return with his shield or upon it. Or perchance he was some Achilles, who had nourished his wrath apart, and had now come to avenge or rescue his Patroclus. He saw this unequal combat from afar,—for the blacks were nearly twice the size of the red,—he drew near with rapid pace till he stood on his guard within half an inch of the combatants; then, watching his opportunity, he sprang upon the black warrior, and commenced his operations near the root of his right fore leg, leaving the foe to select among his own members; and so there were three united for life, as if a new kind of attraction had been invented which put all other locks and cements to shame. I should not have wondered by this time to find that they had their respective musical bands stationed on some eminent chip, and playing their national airs the while, to excite the slow and cheer the dying combatants. I was myself excited somewhat even as if they had been men. The more you think of it, the less the difference. And certainly there is not the fight recorded in Concord history, at least, if in the history of America, that will bear a moment's comparison with this, whether for the numbers engaged in it, or for the patriotism and heroism displayed. For numbers and for carnage it was an Austerlitz or Dresden. Concord fight! Two killed on the patriots' side, and Luther Blanchard wounded! Why here every ant was a Buttrick,—"Fire, for God's sake fire!"—and thousands shared the fate of Davis and Hosmer. There was not one hireling there. I have no doubt that it was a principle they fought for, as much as our ancestors, and not to avoid a three-penny tax on their tea; and the results of this battle will be as important and memorable to those whom it concerns as those of the battle of Bunker Hill, at least.

I took up the chip on which the three I have particularly described were struggling, carried it into my house, and placed it under a tumbler on my window-sill, in order to see the issue. Holding a microscope to the first-mentioned red ant, I saw that, though he was assiduously gnawing at the near fore leg of his enemy, having severed his remaining feeler, his own breast was all torn away, exposing what vitals he had there to the jaws of the black warrior, whose breastplate was apparently too thick for him to pierce; and the dark carbuncles of the sufferer's eyes shone with ferocity such as war only could excite. They struggled half an hour longer under the tumbler, and when I looked again the black soldier had severed the heads of his foes from their bodies, and the still living heads were hanging on either side of him like ghastly trophies at his saddle-bow, still apparently as firmly fastened as ever, and he was endeavoring with feeble struggles, being without feelers and with only the remnant of a leg, and I know not how many other wounds, to divest himself of them; which at length, after half an hour or more, he accomplished. I raised the glass, and he went off over the window-sill in that crippled state. Whether he finally survived that combat, and spent the remainder of his days in some Hôtel des Invalides, I do not know; but I thought that his industry would not be worth much thereafter. I never learned which party was victorious, nor the cause of the war; but I felt for the rest of that day as if I had had my feelings excited and harrowed by witnessing the struggle, the ferocity and carnage, of a human battle before my door. . . . The battle which I witnessed took place in the Presidency of Polk, five years before the passage of Webster's Fugitive-Slave Bill.

Thoreau, Henry David. 1854. *Walden, or Life in the Woods.* Boston, Mass.: Ticknor and Fields.

Slave-Making in Ants

Charles Darwin

Darwin was clearly fascinated with slave-making behavior in ants and drew on Pierre Huber's painstaking observations to try to understand more. Here, in On The Origin of Species, *he proposes an evolutionary sequence for slave-making.*

Slave-making instinct. This remarkable instinct was first discovered in the *Formica (Polyerges) rufescens* by Pierre Huber, a better observer even than his celebrated father. This ant is absolutely dependent on its slaves; without their aid, the species would certainly become extinct in a single year. The males and fertile females do no work. The workers or sterile females, though most energetic and courageous in capturing slaves, do no other work. They are incapable of making their own nests, or of feeding their own larvae. When the old nest is found inconvenient, and they have to migrate, it is the slaves which determine the migration, and actually carry their masters in their jaws. So utterly helpless are the masters, that when Huber shut up thirty of them without a slave, but with plenty of the food which they like best, and with their larvae and pupae to stimulate them to work, they did nothing; they could not even feed themselves, and many perished of hunger. Huber then introduced a single slave *(F. fusca),* and she instantly set to work, fed and saved the survivors; made some cells and tended the larvae, and put all to rights. What can be more extraordinary than these well-ascertained facts? If we had not known of any other slave-making ant, it would have been hopeless to have speculated how so wonderful an instinct could have been perfected.

Formica sanguinea was likewise first discovered by P. Huber to be a slave-making ant. This species is found in the southern parts of England, and its habits have been attended to by Mr. F. Smith, of the British Museum, to whom I am much indebted for information on this and other subjects. Although fully trusting to the statements of Huber and Mr. Smith, I tried to approach the subject in a sceptical frame of mind, as any one may well be excused for doubting the truth of so extraordinary and odious an instinct as that of making slaves. Hence I will give the observations which I have myself made, in some little detail. I opened fourteen nests of *F. sanguinea*, and found a few slaves in all. Males and fertile females of the slave-species are found only in their own proper communities, and have never been observed in the nests of *F. sanguinea*. The slaves are black and not above half the size of their red masters, so that the contrast in their appearance is very great. When the nest is slightly disturbed, the slaves occasionally come out, and like their masters are much agitated and defend the nest: when the nest is much disturbed and the larvae and pupae are exposed, the slaves work energetically with their masters in carrying them away to a place of safety. Hence, it is clear, that the slaves feel quite at home. During the months of June and July, on three successive years, I have watched for many hours several nests in Surrey and Sussex, and never saw a slave either leave or enter a nest. As, during these months, the slaves are very few in number, I thought that they might behave differently when more numerous; but Mr. Smith informs me that he has watched the nests at various hours during May, June and August, both in Surrey and Hampshire, and has never seen the slaves, though present in large numbers in August, either leave or enter the nest. Hence he considers them as strictly household slaves. The masters, on the other hand, may be constantly seen bringing in materials for the nest, and food of all kinds. During the present year, however, in the month of July, I came across a community with an unusually large stock of slaves, and I observed a few slaves mingled with their masters leaving the nest, and marching along the same road to a tall Scotch-fir-tree, twenty-five yards distant, which they ascended together, probably in search of aphides or cocci. According to Huber, who had ample opportunities for observation, in Switzerland the slaves habitually work with their masters in making the nest, and they alone open and close the doors in the morning and evening; and, as Huber expressly states, their principal office is to search for aphides. This difference in the usual habits of the masters and slaves in the two countries, probably depends merely on the slaves being captured in greater numbers in Switzerland than in England.

One day I fortunately chanced to witness a migration from one nest to another, and it was a most interesting spectacle to behold the masters carefully carrying, as Huber has described, their slaves in their jaws. Another day my attention was struck by about a score of the slave-makers haunting the same spot, and evidently not in search of food; they approached and were vigorously repulsed by an independent community of the slave species *(F. fusca)*; sometimes as many as three of these ants clinging to the legs of the slave-making *F. sanguinea*. The latter ruthlessly killed their small opponents, and carried their dead bodies as food to their nest, twenty-nine yards distant; but they were prevented from getting any pupae to rear as slaves. I then dug up a small parcel of the pupae of *F. fusca* from another nest, and put them down on a bare spot near the place of combat; they were eagerly seized, and carried off by the tyrants, who perhaps fancied that, after all, they had been victorious in their late combat.

At the same time I laid on the same place a small parcel of the pupae of another species, *F. flava,* with a few of these little yellow ants still clinging to the fragments of the nest. This species is sometimes, though rarely, made into slaves, as has been described by Mr. Smith. Although so small a species, it is very courageous, and I have seen it ferociously attack other ants. In one instance I found to my surprise an independent community of *F. flava* under a stone beneath a nest of the slave-making *F. sanguinea;* and when I had accidentally disturbed both nests, the little ants attacked their big neighbours with surprising courage. Now I was curious to ascertain whether *F. sanguinea* could distinguish the pupae of *F. fusca,* which they habitually make into slaves, from those of the little and furious *F. flava,* which they rarely capture, and it was evident that they did at once distinguish them: for we have seen that they eagerly and instantly seized the pupae of *F. fusca,* whereas they were much terrified when they came across the pupae, or even the earth from the nest of *F. flava,* and quickly ran away; but in about a quarter of an hour, shortly after all the little yellow ants had crawled away, they took heart and carried off the pupae.

One evening I visited another community of *F. sanguinea,* and found a number of these ants entering their nest, carrying the dead bodies of *F. fusca* (showing that it was not a migration) and numerous pupae. I traced the returning file burthened with booty, for about forty yards, to a very thick clump of heath, whence I saw the last individual of *F. sanguinea* emerge, carrying a pupa; but I was not able to find the desolated nest in the thick heath. The nest, however, must have been close at hand, for two or three individuals of *F. fusca* were rushing about in the greatest agitation, and one was perched motionless with its own pupa in its mouth on the top of a spray of heath over its ravaged home.

Such are the facts, though they did not need confirmation by me, in regard to the wonderful instinct of making slaves. Let it be observed what a contrast the instinctive habits of *F. sanguinea* present with those of the *F. rufescens.* The latter does not build its own nest, does not determine its own migrations, does not collect food for itself or its young, and cannot even feed itself: it is absolutely dependent on its numerous slaves. *Formica sanguinea,* on the other hand, possesses much fewer slaves, and in the early part of the summer extremely few. The masters determine when and where a new nest shall be formed, and when they migrate, the masters carry the slaves. Both in Switzerland and England the slaves seem to have the exclusive care of the larvae, and the masters alone go on slave-making expeditions. In Switzerland the slaves and masters work together, making and bringing materials for the nest: both, but chiefly the slaves, tend, and milk as it may be called, their aphides; and thus both collect food for the community. In England the masters alone usually leave the nest to collect building materials and food for themselves, their slaves and larvae. So that the masters in this country receive much less service from their slaves than they do in Switzerland.

By what steps the instinct of *F. sanguinea* originated I will not pretend to conjecture. But as ants, which are not slave-makers, will, as I have seen, carry off pupae of other species, if scattered near their nests, it is possible that pupae originally stored as food might become developed; and the ants thus unintentionally reared would then follow their proper instincts, and do what work they could. If their presence proved useful to the species which had seized them—if it were more advantageous to this species to capture workers than to procreate them—the habit of collecting pupae originally for food might by natural selection be strengthened and rendered permanent for the very different purpose of raising slaves. When the instinct was once acquired, if carried out to a much less extent even than in our British *F. sanguinea,* which, as we have seen, is less aided by its slaves than the same species in Switzerland, I can see no difficulty in natural selection increasing and modifying the instinct—always supposing each modification to be of use to the species—until an ant was formed as abjectly dependent on its slaves as is the *Formica rufescens.*

Darwin, Charles R. 1859. *On the Origin of Species by Means of Natural Selection, or the Preservation of Favoured Races in the Struggle for Life.* London: John Murray.

The Daintiness of Ants

Reverend Henry C. McCook

Most ants live in rotting wood or soil, and their eggs and larvae are kept in the open rather than in closed cells like those of bees and wasps. To exclude the various fungi and bacteria with which they are in close contact, ants have evolved a variety of methods for keeping themselves extraordinarily clean, including a special antibiotic-secreting gland unknown in McCook's day. The first American popularizer of ants, the Rev. Henry C. McCook, (1837–1911), wrote this as one of a series of articles about ants for Harper's Monthly Magazine, *and one can imagine the housewives of a century ago reading and reacting with newfound respect to the news that ants might well be the cleanest of the insects encountered in their kitchens.*

If there be truth in the old saying, cleanliness is next to godliness, insects are but one remove from piety. As tidy as an emmet—is more truthful than most proverbial comparisons. Who ever saw an untidy ant, or bee, or wasp? The writer has observed innumerable thousands of ants, has lived in his tent in the midst of their great communities, and watched them at all hours of day and night, under a great variety of conditions, natural and artificial, unfavorable to cleanliness, and has never seen one really unclean. Most of them are fossorial in habit, digging in the ground, within which they live; are covered with hair and bristles, to which dirt pellets easily cling; they move habitually in the midst of the muck and chippage and elemental offal of nature—yet they seem to take no stain and to keep none.

This is true of other insects. Take, for example, the interesting families of wasps. Many burrow in the earth to make breeding-cells for their young. Others, like the mud-daubers, collect mortar from mud-beds near brooks and pools to build their clay nurseries and storehouses. Some, like the yellow-jackets, live in caves which they excavate in the ground. They delve in the dirt; handle and mix and carry it; mould and spread it, moving to and fro all day long, and day after day, at work in surroundings that would befoul the most careful human worker—yet do not show the least trace of their occupation.

Of course there is much in temperament and training. There are women who remind us of insects in their faculty of moving unmarred amidst the current defilements of daily duty. They will pass to the parlor from kitchen, nursery, or sewing-room with no adjustment of toilet but a discarded apron or turned-down sleeves, yet quite sweet and presentable. But there are women, high and low, and men innumerable, of a different pattern. With insects, however, the type of dainty tidiness is the absolute rule. There are no exceptions; no degenerates of uncleanness, as with men. Temperament is wholly and always on the side of cleanliness; and training is not a factor therein, for it is inborn, and as strong in adolescents as in veterans. How has nature secured this admirable result?

If the reader were told that ants possess brushes, fine and coarse tooth combs, and other toilet articles quite after the pattern of our own, he would probably think he was being gulled. Yet it is even so. Let us take an inventory of these. To begin with, the body is covered more or less closely with fine pubescence, corresponding somewhat with the fur of beasts. This is interspersed with bristles and spines, which are sometimes jointed, and are so arranged as to aid materially in keeping the body clean. Particles of soil cling to this hairy covering, but it is a protective medium, holding the dirt aloof and isolated from the skin surfaces, so that it can be readily shaken off or taken off. The brushing, washing, and combing of this hairy coat constitute the insect's toilet-making.

One of the efficient toilet articles is the tongue. Around the sides of this organ curves a series of ridges covered with hemispherical bosses. The ridges are chitinous, and thus by greater hardness are fitted for the uses of a brush. When eating, this structure rasps off minute particles of solid foods, so fitting them for the stomach. For toilet uses it serves as both sponge and brush, and takes up bits of dirt not otherwise removed. In short, ants use their tongues as dogs and cats do, for lapping up food and licking clean the body. One is continually reminded, as he watches the tiny creatures at their toilet, of the actions of his cat and dog at the fireside.

The tibial comb or fore-spur is another toilet implement, unique in form and function. This is a real comb, which might well have served the inventor of our own combs for a model, its chief difference being that it is permanently attached to the limb that operates it. It has a short handle, a stiff back, and a regularly toothed edge. It is set into the apical end of the tibia of the fore legs, upon which it articulates freely, thus giving the owner the power to apply it to various organs. Placed along the edge are sixty-five teeth of equal length, except towards the apex, where they are shorter. They are pointed at the free end and enlarged at the base, are stiff but elastic, and spring back when bent, as do the teeth of a comb.

The efficiency of this instrument is greatly increased by an arrangement of the tarsus, opposite whose base it is placed. That part of the leg is so shaped that the curved outlines of the tibial spur when pushed up against it fit into it. It is furnished with forty-five teeth, coarser and more open than those just described. Thus ants have the useful arrangement of fine and coarse toothed combs which for toilet uses we unite in one instrument. A further contribution to the toilet paraphernalia is a secondary spur, a simpler form of that on the fore legs, set upon the tibiae of the second and third pairs of legs. Moreover, the mandibles, or upper jaws, which are palm shaped and serrated, are used freely, especially in cleaning the legs, which are drawn through them. In this action there is a salivary secretion that moistens the members, and furnishes a good substitute for those "washes" which are valued by men and women as softening the hair and making it more pliable. Indeed, one might almost conjecture that it is also the emmet equivalent for our toilet soaps!

There are no pastes and powders among these toilet articles—at least as far as known,—but the repertoire, it will be seen, is tolerably complete: fine-tooth combs, coarse or "reddin'" combs, hairbrushes and sponges, washes and soap!—and all so conveniently attached to the body and working-limbs, which are arms as well as legs, that they are always literally "on hand" for service.

Ants have no set time for brushing up. But certain conditions plainly incite thereto—as when they feel particularly comfortable, as after eating, or after awaking from or before going to sleep. The keen sense of discomfort aroused by the presence of dirt incites to cleansing. Often one may see an ant suddenly pause in the midst of the duties of field or formicary and begin to comb herself. Here is a mountain mound-maker driven by the passion of nest-building to the utmost fervor of activity. Suddenly she drops out of the gang of fellow workers, and mounting a near-by clod, poses upon her hind legs and plies teeth, tongue, and comb. For a few moments the aim of being is centred upon that act. Around her coign of vantage sweeps to and fro the bustling host of builders with all their energies bent upon reconstructing their ruined city. She combs on unconcernedly. From top of head to tip of hind legs she goes, smoothing out ruffled hairs and removing atoms of soil invisible to human eyes. Her toilet is ended at last. A few leisurely finishing-strokes and she rises, stretches herself, calmly climbs down her pedestal, and is immediately infected with the fervor that lashes on the surging throng around her, and is lost in the crowd. Meanwhile other workers have dropped out of the lines, and may be seen here and there at their ablutions. Thus it goes in the field, as one may easily see if he have tact and patience.

But artificial nests give the best opportunity for careful observation, although one must allow for the unnatural surroundings. No doubt with ants, as with man, artificial conditions of society induce greater attention to personal appearance. Thus the writer's imprisoned ants would invariably be drawn out from their underground lodgings by the light and heat of lamps at night. They would gather in clusters against the glass of the formicary next to the lamp, and after some preliminary jostling and skirmishing for position, would begin to wash themselves. Slight elevations, afforded by irregularities in the surface, were favorite seats. The modes of operating are so various that it is difficult to describe them, much more to fix the attitudes with the pencil. But typical poses at least may be described.

In cleaning the head and fore parts of the body the insect often sits upon the two hind legs and turns the face to one side. Then the fore leg is raised and passed over the face from the vertex to the mandible—that is, from the top of the head to the mouth. Meanwhile the head is slowly turned to expose both sides to manipulation; and if this is not satisfactory the position is reversed and the opposite leg brought into play. In "doing up the back hair"—as one may say—the head is further dropped and the leg with its movable spur-comb, which has free play like a comb in a human hand, is thrown quite behind the vertex, and moved forward again and again through the tuft of hairs growing there. In these and other cleansing movements the leg will be drawn through the jaws at intervals, to moisten it or to wipe off the dust caught in the comb. The action reminds one of the alternations of pussy's paw between mouth and neck when washing the back of her head and ears.

Cleaning the abdomen and the stinging organs at the apex, which is surrounded by circles of hairs, places the ant in grotesque attitudes; although herein also one notes a miniature of the ways of domestic animals. For example, the hind legs will be thrown backwards and well extended; the middle pairs set nearly straight outward from the thorax and less extended, so that the body is nearly erect. The abdomen is then turned under the body and deflected upward towards the head, which at the same time is bent over and downward. The body of the ant thus forms a letter C, or nearly a circle. Meanwhile, the fore feet have clasped the abdomen, the tarsus passing quite around and beneath it, and the brushing has begun. The strokes are directed toward the tip of the abdomen, which is also sponged off by the tongue. Occasionally the leg is rubbed over the head after being drawn through the mouth, and so again to the abdomen. One ant was seen cleansing its abdomen while hanging by the hind legs from the roof of the formicarium. The abdomen was thrown up and between the legs, as a gymnast on the turning-bar throws his body upward between his arms. The head was then reached upward, and tongue and fore feet were engaged as above described. Another emmet acrobat was caught in the act of cleansing its legs while hanging by one foot, the under part of the body being toward the observer.

During these toilet actions the formicarium presented a most interesting view, especially in the evening, when the table-lamps were lit and the ants had been fed, and a general "washing-up" was in progress. But one of the most interesting features was the part which the insects took in cleansing one another. This was a new and pleasing revelation in life habit. It was unexpected, but after-experience showed that nature has taught these little creatures the value of cooperation in such matters among fellow communists. Ants are particularly liable to attack of parasites—a danger increased by imprisonment. As these enemies pass from one to another, and thus become a common peril, every individual has an interest in the personal health and habits of his neighbors. This is shown in the friendly offices here described. We may easily think of men as saying, "My neighbor's premises are untidy; he lacks the means and the disposition to keep clean; he is infected—what is that to me?" But citizens of an emmet commune are apt to be superior to such selfishness, and seem to feel instinctively—at least so to act—that the pernicious habits and personal misfortunes of the individual highly concern his fellows and the public. Perhaps this is fortified by a natural amiability that delights to give pleasure. And what a pleasure most animals feel in manipulation of the hair and body! The now popular art of massagerie appears to be naturally practised by ants, doubtless antedating by ages the habit of men.

McCook, Henry C. 1905. "The Daintiness of Ants." *Harper's Monthly Magazine* CIX(652): 604–610.

The original caption for this illustration, written by the early American myrmecologist the Rev. H. C. McCook, was "Winged female ants at play on the plaza," but these virgin females are actually preparing to take off for their once-in-a-lifetime mating flight (*Ant Communities and How They Are Governed* by Henry C. McCook, Harper & Brothers, New York, 1909).

The Tenderness of Earwigs

Charles Degeer and George John Romanes

Earwigs are members of the order Dermaptera, usually have pincers at the rear end of the body, and do not, as popular belief would have it, favor habitation in human ears. That earwig mothers tenderly care for their young was first noted in 1773 by the Swedish entomologist Charles Degeer (1720–1778). In fact, maternal (and sometimes paternal) care is the rule in many insects. These two short passages about earwig behavior come from Degeer's original report in the mid–eighteenth century followed by Romanes's Animal Intelligence *in the late nineteenth century. George John Romanes's (1848–1894) book is a fascinating collection of anecdotes documenting complex behavior throughout the animal kingdom.*

At the beginning of June I found under a stone a female earwig with several small insects, which were quite obviously her progeny. They did not leave her, and even placed themselves under her body as chickens under a hen. So insects of this kind take care, in a way, of their offspring after they are born, and stay near them as if wishing to protect them.

Except in a few points the young resemble their parents. . . . I placed them with their mother in a box, wherein I had put a little fresh earth, and it was curious to see how they crept under the body and between the legs of their mother, who remained very quiet and allowed them to do so. She, as it were, covered them as a hen does her chicks, and they often remained in this position for hours together. . . .

Again at the beginning of April 1759 I found some female earwigs under stones, with a mass of eggs, on which the mother was seated, and of which she took the greatest care, never going a step away from them. (M. Frisch has observed this before me.) I took the mother with her eggs and placed them in a box half-filled with fresh earth, the eggs being scattered here and there; but she soon picked up the eggs, carrying them in her jaws. After a few days I saw that she had collected the eggs into one place on the surface of the earth in the box, and that she remained constantly seated upon them in such a way that she really seemed to be covering them.

Maternal care in earwigs. A male is above and two females are below. At left is the heap of eggs that a mother is collecting together. In the foreground, the other female has hatched her eggs and is now brooding her young progeny. Illustration by Theo. Carreras (*Marvels of Insect Life,* edited by Edward Step, Robert M. McBride & Co., New York, 1916).

I must devote a short division of this chapter to the earwig. M. Geer [Charles Degeer] describes a regular process of incubation as practised by the mother insect. . . .

A young lady, who objects to her name being published, informs me that her two younger sisters (children) are in the habit of feeding every morning with sugar an earwig, which they call 'Tom', and which crawls up a certain curtain regularly every day at the same hour, with the apparent expectation of getting its breakfast. This resembles analogous instances which have been mentioned in the case of spiders.

Degeer, Charles. 1773. *Mémoires pour servir à l'histoire des insectes 3.* Stockholm: P. Hesselberg. (Translated and quoted in William John Lucas, 1920, *A Monograph of the British Orthoptera.* London: The Ray Society)

Romanes, George John. 1899. *Animal Intelligence.* New York: D. Appleton and Company.

Letter from Brazil: Termites and Stingless Bees

Fritz Müller

Introduced by Charles Darwin

As contemporary American myrmecologist Jack Longino has pointed out, Fritz Müller (1822–1897) was the field biologist's field biologist. A nineteenth-century photograph of Müller (full German name: Johannes Friedrich Theodor Müller, but he published as "Fritz") reveals a long-bearded, barefooted hermit figure carrying a small shoulder bag and leaning on a rough-hewn walking stick. During the years 1852 to 1897, the year of his death, Müller sent numerous dispatches back to Europe from his stomping grounds in southern Brazil, and published about 150 papers, describing for the first time all sorts of important natural phenomena. Here Müller combines natural history observation with astute reasoning in a discussion of social evolution in termites and stingless bees. This letter from the field arrived at Nature *just fifteen years after Darwin's publication of* On the Origin of Species. *Müller's 1863 book* Für Darwin *(*Facts and Arguments for Darwin *was the English title) made him one of the first defenders of Darwin.*

The accompanying letter, just received from Fritz Müller, in Southern Brazil, is so interesting that it appears to me well worth publishing in *Nature.* His discovery of the two sexually mature forms of Termites, and of their habits, is now published in Germany; nevertheless few Englishmen will have as yet seen the account.

In the German paper he justly compares, as far as function is concerned, the winged males and females of the one form, and the wingless males and females of the second form, with those plants which produce flowers of two forms, serving different ends, of which so excellent an account has lately appeared in *Nature* by his brother, Hermann Müller.

The facts, also, given by Fritz Müller with respect to the stingless bees of Brazil will surprise and interest entomologists.

Charles Darwin

Feb. 11

For some years I have been engaged in studying the natural history of our Termites, of which I have had more than a dozen living species at my disposition. The several species differ much more in their habits and in their anatomy than is generally assumed. In most species there are two sets of neuters, viz., labourers and soldiers; but in some species (*Calotermes* Hg.) the labourers, and in others (*Anoplotermes* F.M.) the soldiers, are wanting. With respect to these neuters I have come to the same conclusion as that arrived at by Mr. Bates, viz., that differently from what we see in social Hymenoptera, they are not modified imagos (sterile females), but modified larvae, which undergo no further metamorphosis. This accounts for the fact first observed by Lespès, that both the sexes are represented among the sterile (or so-called neuter) Termites. In some species of *Calotermes* the male soldiers may even externally be distinguished from the female ones. I have been able to confirm, in almost all our species, the fact already observed by Mr. Smeathman a century ago, but doubted by most subsequent writers, that in the company of the queen there lives always a king. The most interesting fact in the natural history of these curious insects is the existence of two forms of sexual individuals, in some (if not in all) of the species. Besides the winged males and females, which are produced in vast numbers, and which, leaving the termitary in large swarms, may intercross with those produced in other communities, there are wingless males and females, which never leave the termitary where they are born, and which replace the winged males or females, whenever a community does not find in due time a true king or queen. Once I found a king (of a species of *Eutermes*) living in company with as many as thirty-one such complemental females, as they may be called, instead of with a single legitimate queen. Termites would, no doubt, save an extraordinary amount of labour if, instead of raising annually myriads of winged males and females, almost all of which (helpless creatures as they are) perish in the time of swarming without being able to find a new home, they raised solely a few wingless males and females, which, free from danger, might remain in their native termitary; and he who does not admit the paramount importance of intercrossing, must of course wonder why this latter manner of reproduction (by wingless individuals) has not long since taken the place through natural selection of the production of winged males and females. But the wingless individuals would of course have to pair always with their near relatives, whilst by the swarming of the winged Termites a chance is given to them for the intercrossing of individuals not nearly related. I sent to Germany, about a year ago, a paper on this subject, but do not know whether it has yet been published.

From Termites I have lately turned my attention to a still more interesting group of social insects, viz., our stingless honey-bees (*Melipona* and *Trigona*). Though a high authority in this matter, Mr. Frederick Smith, has lately affirmed, that "we have now acquired almost a complete history of their economy," I still believe, that almost all remains to be done in this respect. I think that even their affinities are not yet well established, and that they are by no means intermediate between hive- and humble-bees, nor so nearly allied to them, as is now generally admitted. Wasps and hive-bees have no doubt independently acquired their social habits, as well as the habit of constructing combs of hexagonal cells, and so, I think, has *Melipona*. The genera *Apis* and *Melipona* may even have separated from a common progenitor, before wax was used in the construction of the cells; for in hive-bees, as is well known, wax is secreted on the ventral side: in *Melipona* on the contrary, as I have seen, on the dorsal side of the abdomen; now it is not probable, that the secretion of wax, when once established, should have migrated from the ventral to the dorsal side, or *vice versa*.

The queen of the hive-bee fixes her eggs on the bottom of the empty cells; the larvae are fed by the labourers at first with semi-digested food, and afterwards with a mixture of pollen and honey, and only when the larvae are full grown, the cells are closed. The *Meliponae* and *Trigonae*, on the contrary, fill the cells with semi-digested food before the eggs are laid, and they shut the cells immediately after the queen has dropped an egg on the food. With hive-bees the royal cells, in which the future queens have to be raised, differ in their direction from the other cells; this is not the case with *Melipona* and *Trigona*, where all the cells are vertical, with their orifices turned upward, forming horizontal (or rarely spirally ascending) combs. You know that honey is stored by our stingless bees in large, oval, irregularly clustered cells; and thus there are many more or less important differences in the structure, as well as in the economy, of *Apis* and *Melipona*.

My brother, who is now examining carefully the external structure of our species, is surprised at the amount of variability, which the several species show in the structure of their hind legs, of their wings, etc., and not less are the differences they exhibit in their habits.

I have hitherto observed here 14 species of *Melipona* and *Trigona*, the smallest of them scarcely exceeding 2 millimetres in length, the largest being about the size of the hive-bee. One of these species lives as a parasite within the nests of some other species. I have now, in my garden, hives of 4 of our species, in which I have observed the construction of the combs, the laying of the eggs, etc., and I hope I shall soon be able to obtain hives of some more species. Some of our species are so elegant and beautiful and so extremely interesting, that they would be a most precious acquisition for zoological gardens or large hot-houses; nor do I think that it would be very difficult to bring them to Europe and there to preserve them in a living state.

If it be of some interest to you I shall be glad to give you from time to time an account of what I may observe in my *Melipona* apiary.

Believe me, dear Sir, etc.
Fritz Müller

Müller, Fritz. 1874. "Recent Researches on Termites and Honey Bees." Letter, with introduction by Charles Darwin, *Nature* 9:308–309.

The Social Behavior of Burying Beetles

Lorus J. Milne and Margery Milne

The mother and father perch on the edge of the nest, admiring their brood: six babies straining upward, mouths agape. The mother leans close to one of the brood and opens her mouth; the baby eagerly feeds. A nest of birds? No. A nest of burying beetles.

Lorus and Margery Milne tell us about these insects that often work together as a mated pair to bury the carcasses of small mammals that provide food for them and their young.

An observant person who sees the carcass of a small animal lying on the ground is likely to find, if he returns to the spot the next day, that the object has disappeared. The chances are that it has been buried, either there or nearby, by a pair of *Nicrophorus* (or *Necrophorus*) beetles. They will use it as food for their young during the larval stages. The feat of these small insects in rapidly interring a carcass that is many times their size is remarkable enough, but it is only a step toward the most advanced form of parental cooperativeness known among the Coleoptera (the beetles). We have spent much time watching these burying beetles (also called carrion beetles and sexton beetles) and putting them to various tests, which demonstrate an impressive plasticity in the behavior of the insects.

The patient French naturalist Jean-Henri Fabre set out fleshy bait of several kinds to lure burying beetles to where he could watch them. He admired these little gravediggers of the animal world, describing them as being "elegantly attired" in black, with a "double, scalloped scarf of vermilion" across their shining wing covers. The observer cannot watch for long. Unlike the scarab beetle of Mediterranean countries, which walks in plain view while rolling a ball of dung to some still undiscovered place of burial, a burying beetle quickly slides out of sight below the carcass of a mouse or a bird it has found. There, lying on its back, the insect uses all six of its powerful legs as levers to shift its prize. From time to time it rights itself and bulldozes headfirst into the earth to loosen the soil and push it away.

Inconspicuously, a fraction of an inch at a time, the carcass moves horizontally or disappears into the ground.

Nicrophorus beetles are by no means the only insects that sequester food for their larvae before they lay the eggs that will give rise to the larvae, but they work as a team, whereas the others (the scarab and its relatives and various solitary bees and wasps) work alone. Either a male or a female *Nicrophorus* will initiate the flexible behavior that gets the larval food into a safe place. At any time during the operation a mate is likely to arrive. The partner is accepted with no time off for courtship. The two labor together at intervals and also separately in a loose cooperation that advances the common effort. Yet either member of the pair may also creep into a more or less concealed place and appear to sleep for as much as half an hour or depart on feet or wings to some unknown destination for a comparable period, thereafter returning and resuming the work. Ordinarily copulation is deferred until the beetles are securely in possession of their carrion in a chamber of their own making, an inch or more below the surface of the ground.

At this juncture the male might be expected to perform his brief sexual duty and depart. The inseminated female would then carry on alone to the end of the sequence of behavior specified by her inheritance. Occasionally this pattern is followed, but usually both parents remain. Together they work the mass of food into a compact ball. They free it of fur or feathers, perhaps adding secretions that modify the course of decomposition.

As the insects clamber around the carcass, which will provide food for them as well as for their young, the floor, the walls and the roof of the earthen chamber become firmly packed. The female constructs a short vertical extension of the chamber above the carrion and lays her eggs in the side walls of the passageway. She returns to the carcass, and by a combination of selective feeding and clawing at the upper surface prepares a conical depression. Both beetles regurgitate into the depression droplets of partly digested tissue. The fluid accumulates as a pabulum for the larvae that will soon hatch.

This much Fabre or any other persistent observer could discover by exhuming the beetles and their food supply at the proper time, just before the young hatch. Erna Pukowski, studying species native to her Polish countryside, managed to learn more. She made captive burying beetles so much at home, notwithstanding the unnatural condition she created by illuminating their burial chamber, that she could follow the next steps.

One beetle (perhaps the female, although the members of a pair are too much alike externally for an observer to distinguish sex) stood beside the pool of liquid nourishment and began to stridulate. The sound brought hatchling larvae (some two or three millimeters long and almost like maggots in appearance) to the parent's side. The parent sipped from the pool and then transferred the fluid food to one larva after another. The larvae lifted their mouth ends, the better to receive the food. Sometimes both parents shared in the feeding operation.

The British entomologist R. L. Morley discovered in 1902 that the sound of stridulation arises when twin plectrums on the inner surface of the cover of a beetle's wing fret against crosswise ridges on the fifth segment of the abdomen. The sound is clearly audible to human cars. Pukowski noticed it also during the three or four seconds when copulation is in progress. We have heard it when burying beetles are under stress, as they are in repelling an insect of another species or a smaller member of the same sex and in confronting an obstruction that impedes the movement of a carcass.

In 1972 Carsten Niemitz of Justus Liebig University in West Germany discovered that very young larvae will orient themselves to the sound of an adult's stridulation recorded on tape. This response disappears, however, after the larvae have molted for the first time. Even so, older larvae renew their solicitation of regurgitated food for a few hours after each molt by approaching any adult that is close to the pool of food and pressing their mouthparts against its jaws or palps. This action stimulates regurgitation as before. Otherwise the growing larvae feed directly from the pool or pull fragments from the surface of the carrion.

The larvae receive parental care all through their period of feeding growth. The parents may even prepare a horizontal passageway into which the fully grown larvae can crawl to pupate. Only then, when the adults can contribute nothing more to their brood, do they force their way upward through the soil and fly away.

We have not yet marked and followed the departing parents to see whether they do the same thing all over again. They probably do, since adult beetles live from three to 15 months, depending on the species. They search widely for the odor of recent death and are remarkably efficient at finding carrion. Frantisek Petruska, a Czechoslovak ethologist, has found, by capturing beetles with carrion bait, marking them and releasing them at various distances, that they will return to the carrion within 24 hours from as much as four kilometers away.

During one period of four hours, beginning just 35 minutes after we had laid out a newly dead mouse on birchleaf litter, nine burying beetles arrived. Each beetle followed the guidance of the olfactory organs in its antennae. It dropped to the ground within three meters of the mouse, quickly folded its flying wings under its wing covers and came crashing through the litter to the carcass. There, after only a moment's hesitation, the beetle turned over onto its back, slid under the body and lifted the mouse slightly from the ground, apparently to test whether the body was movable. Emerging on the other side of the mouse and righting itself, the beetle began testing the soil.

We had placed our bait on hard ground. Each beetle rejected the site for burial and began to explore, seemingly at random, for softer earth. This was our cue to remove the active beetle and wait for the next one. Each one of the nine beetles followed essentially the same routine, even though four of them were members of one species and five were members of another.

Competition for small carcasses is frequently intense. Ants and flies (particularly blowflies, which deposit active maggots) tend to take over during the day. Burying beetles of the species that are most active by day succeed only if they can inter a body quickly. For species that are active at night the competition is mainly from other species. The largest beetle generally repels all the others except a mate. That is probably why one more often finds a large male cooperating with a small female (or a large female cooperating with a small male) than one finds two beetles of the same size cooperating.

Burying beetles have other ways of reducing competition. Each species has a preferred combination of temperature range and relative humidity. This pattern, as Jean Théodoridès of the University of Paris showed in his laboratory, keeps certain beetles in woodland and others in open fields. Burying beetles that are active in the spring belong to species that go through the winter as adults, whereas the beetles found competing in the summer are likely to represent species that spend the cold months dormant as pupae or full-grown larvae.

Animals that eat insects are likely to constitute a hazard for burying beetles that are active by day. At least one diurnal species of burying beetle in Europe and one in North America may escape being eaten as a result of their resemblance to a small bumblebee. Unlike most burying beetles, these species have golden hair over some of their hard black surfaces. Color, sound and style of flight combine so convincingly that the British biologists Charles Lane and Miriam Rothschild have suggested that this is an example of mimicry, at least with respect to sound. Even a superficial resemblance might have survival value for the beetles.

The most spectacular feature of the activity of burying beetles is the way they transport a carcass from hard ground to soft, in one steady direction, a fraction of an inch at a time. A beetle that has yet to acquire a mate may identify a suitable burial site several meters away from the carrion. The beetle will alternate between loosening the earth at the burial site and rushing back to the carcass. There it performs its lifting feat, starting under the body at the end closest to the burial site. The dead weight is progressively shifted until the beetle emerges from under the opposite end. The insect may run around the carcass and repeat the process time after time. If a mate arrives, the progress is more nearly continuous.

A measure of the success of this way of life can be seen in the fact that the genus *Nicrophorus* includes almost 100 species, with some overlap in distribution. About half of the species are Asiatic. Almost the only areas where burying beetles have not been found are the West Indies, Africa south of the great deserts, Australia and New Zealand.

Adult burying beetles range in length from 10 to 35 millimeters, with considerable variation within a species. All of them appear able to transport the compact body of a bird or a mammal weighing up to 100 grams—up to the size of a rat or a big robin. Anything heavier is usually abandoned unless it is only slightly overweight and can be interred where it is found. A dead snake, however, can weigh more and still be buried expeditiously. Its carcass is subdivided into two or more zones of operation. One pair of beetles attends to each zone.

As Fabre observed, burying beetles show considerable plasticity in behavior. Noting the number of obstacles a typical environment is likely to present to beetles trying to move or bury a body, he wrote that the insect therefore "cannot employ fixed methods in performing its task. Exposed to fortuitous hazards, it must be able to modify its tactics within the limits of its modest discernment. To saw, to break, to disentangle, to lift, to shake, to displace—these are so many means that are indispensable to the gravedigger in a predicament. Deprived of these resources, reduced to uniformity of procedure, the insect would be incapable of pursuing its calling."

Fabre's evaluation rested on watching the beetles at work rather than on experiments. He was reluctant to disturb the carrying and burying operations because so few *Nicrophorus* beetles came to his bait. Our studies have centered in countryside (New Hampshire and Ontario) where richer woodlands and more varied fields support a larger population of beetles. Simple tests confirm the versatility encompassed within the insects' programmed patterns of behavior.

To create a reasonable facsimile of the type of obstacle a burying beetle might encounter naturally, we place a dead mouse close to a clover plant and then tie the carcass down by arching the stalk of a leaf over the torso and fastening the stalk to the ground with a hairpin. The first burying beetle to arrive discovers that the front and rear ends of the carcass can be raised but that the middle cannot. The beetle promptly climbs over the mouse, discovers the tight leaf stalk, forces its head under the stalk and pushes forward. The stalk does not break, but it stretches enough to release the carcass for transport and burial. Repetitions of the experiment with other beetles all have the same result.

Once we drove a good-sized stake into the ground at a 45-degree angle and tied a strong cotton string around its upper end. We tied the dangling end of the string around a hind leg of a dead mouse lying on soft ground. A pair of *Nicrophorus* beetles pushed away the soil below the body until the mouse hung from the tethered leg over a cup-shaped depression. The insects cleared a space the thickness of their bodies between the mouse and the soil and then kept swiveling the carcass in wide arcs. The tail of the mouse dragged on the rim of the depression until one of the beetles chewed it off.

That did not solve the problem, and so both beetles explored the surface of the carcass. Only about six hours after they had begun to work did one of them discover the tether. In less than a minute the insect settled down to gnaw through the cotton fibers. By dawn the carcass had been liberated and buried.

To test the strength of *Nicrophorus* beetles we rested one end of a flat rock on the body of a 50-gram mole. The rock applied about half a kilogram of unyielding weight to the body. Two beetles were nonetheless able to work the body free. First they took up positions side by side with their back against the rock and their legs against the body. They shifted the body about a centimeter in relation to the rock and then repeated the performance with respect to the hard soil below the body. Alternating between these two areas of contact, they freed the carcass in less than half an hour, whereupon they transported it to soft ground and quickly buried it.

In tests of the memory of burying beetles we have found that if a beetle has had 15 or 20 minutes of experience with a suitable carcass, it can be removed and held captive for at least 16 hours without losing its readiness to return to the body within minutes of being released. After 24 hours of separation from its trophy the beetle is more likely to fly off. Two beetles of the species *Nicrophorus orbicollis* that had shifted a mouse about six inches were picked up and put in separate boxes with moist earth. Two hours later, while two smaller beetles of the species *N. tomentosus* were working on the mouse, we released the male *orbicollis* six inches to the east of the body and the female six inches to the north. They both feigned death for a few seconds and then set out almost directly for the body of the mouse. They repelled the *tomentosus* beetles and resumed their normal activities.

In this test the female was the larger *orbicollis.* If the disparity had been the other way, the behavior probably would have been different. We find that females are much more aggressive than males in ousting rivals from carrion. A male is more likely to allow other *Nicrophorus* beetles, particularly those of other species, to work for a while before repelling them. In the end each manageable carcass serves as food for the adults and larvae of only one pair of *Nicrophorus* beetles—the ones that bury it.

Beetles need both memory and some special sense (probably olfaction) to recognize a particular trophy. If we move a carcass a meter or less to one side while the beetles of a pair are momentarily away from it, they immediately begin exploring on their return to the vacant site. In a few minutes, aided no doubt by scent, they find the carcass and resume work as though nothing had happened.

Possibly burying beetles mark a carcass with a chemical secretion, which would explain what happens when the beetles return to a site where they have been working on a carcass only to find a different carcass there. They examine the substitute and then go off exploring. If they find the original carcass within a meter or less, they resume work on it. If they fail to locate their prize, they are as likely to fly away as they are to accept the substitute. A volatile substance that conferred a distinctive odor on a carcass, as a message to be read later by the same insect or its mate, might serve also as a pheromone. Pukowski noticed that, a lone *Nicrophorus* beetle, after laboring for a long time without being joined by a mate, would climb on top of a plant or a stone, elevate its abdomen obliquely and extend it as though emitting a secretion.

The social behavior of burying beetles fits between extremes in the behavior of other insects. In the most primitive insect social behavior the parent or parents attend only to their own offspring. The most advanced social insects have a female at least providing care for the offspring of other females, often as a sterile surrogate parent. Burying beetles often show some altruistic behavior in that small members of the same species or a different one may contribute significantly to the rapid burial of a carcass and then leave, taking no part in reproduction. The dominant, mated pair take over the food supply. The female, at least, remains to care for the larvae, but she will not tend the larvae of any other female.

Parental interactions that promote the survival of further generations have evolved independently in more than two dozen families of insects. Among the beetles, which are the most varied order of animals, *Nicrophorus* is unique in extending maternal care so far and in having the aid of the male so often until the larvae are ready to pupate. No other members of the same superfamily show social behavior of any kind. Indeed, entomologists regard the superfamily (Staphylinoidea) and the family (Silphidae) to which the *Nicrophorus* beetle belongs as being made up of rather unspecialized beetles. It is odd that behavior of such plasticity should have arisen at all and then should have succeeded so widely.

Milne, Lorus J., and Margery Milne. 1976. "The Social Behavior of Burying Beetles." *Scientific American* (Aug.) 235:84–89.

INSECT CONSCIOUSNESS

Donald R. Griffin

Donald Griffin's books The Question of Animal Awareness, Animal Minds, *and* Animal Thinking *are all in the service of asking: Are animals aware? Griffin (b. 1915), best known as the co-discoverer of bat echolocation, obviously believes they are, and he believes so for evolutionary reasons. Here he considers bees and some of the behavioral research most revealing of animal "minds." It is a superb analysis of the importance of Karl von Frisch (1886–1982) and his successors' work on the waggle dance of the honeybees. In 1973, Karl von Frisch shared the Nobel Prize with Niko Tinbergen and Konrad Lorenz for deciphering the "language" of bees—one of the great pieces of detective work on insects.*

The Symbolic Dances of Honeybees

The most significant example of versatile communication known in any animals other than our own species is the so-called "dance language" of honeybees. This type of communicative behavior is so strikingly different from all other known kinds of animal communication that it has been difficult for inclusive behaviorists to integrate it into their general understanding of animal behavior. The difficulty is exemplified by the behavioral ecologist Krebs, who called these dances an "evolutionary freak." The versatility of honeybee learning has been reviewed by Gould and Towne; and many of the complex effects of overlearning and extinction studied by psychologists in rats and pigeons have also been found in honeybees.

Beekeepers and students of bee behavior had noticed for centuries that worker honeybees sometimes move about over the surface of the honeycomb in agitated patterns called dances. It was also well known that once a single foraging worker has discovered a rich source of food, such as flowers that have just come into bloom, many other bees from the same colony may arrive a few minutes later, so rapidly that they could not all have found the food by individual searching. This suggested that some sort of recruiting communication occurred, but how this was achieved remained almost totally unknown until the work of Karl von Frisch. . . . Quite early in his career he proved by elegant simple experiments that bees were capable of discriminating hues. He was led to this discovery by the simple naturalist's belief that the striking colors of flowers must be perceptible to the insects that visit them to obtain nectar and pollen. Early in the twentieth century, prevailing scientific opinion was strongly negative about color vision in invertebrates. But von Frisch developed simple and ingenious experiments demonstrating conclusively that honeybees have excellent color vision.

In the 1920s, when studying the sensory capabilities of honeybees, von Frisch noticed that the agitated dances were carried out by workers that had visited sources of food. In order to see what bees did on returning to the hive, he constructed specialized beehives with glass windows that allowed a clear view of their behavior as they crawled over the honeycomb. To study foraging behavior he set out dishes containing concentrated sugar solutions, which bees visit and take up eagerly just as they take nectar from flowers. To identify individual bees he marked them with small daubs of paint on the dorsal surface while they were sucking up sugar solution. The bees he had marked at his dishes of sugar solution danced in circles, alternating clockwise and counterclockwise motions. At the same time he noticed that others returning with loads of pollen were carrying out a very different type of dance. He could tell that they had gathered pollen because honeybees carry substantial amounts of pollen grains packed between hairs on their legs. These pollen gatherers performed what are called *Schwanzeltanzen* in German, customarily translated as waggle dances. In a waggle dance the bee walks rapidly in a straight line while moving her abdomen back and forth laterally at about thirteen or fourteen times per second. Then at the end of this straight wagging run she circles back and repeats the straight part of the dance, followed by alternating clockwise and counterclockwise returns to the starting point of a series of straight wagging runs. Although the bee may move a short distance between successive cycles of the waggle dance, the basic pattern is relatively constant under given conditions. Von Frisch thus concluded that the two types of dance were somehow related to the sort of food being brought back to the colony, a reasonable interpretation of the observations he was able to make at the time.

Only much later, during World War II, when his laboratory in Munich had been seriously damaged and he was studying bees at his country estate in the Austrian Tyrol, did he have occasion to move the artificial feeding dishes to a considerable distance from the observation hive. When he did this he discovered that the bees gathering sugar solution from more than about one hundred meters performed waggle dances rather than the round dances he had always associated with sugar gathering. This had escaped his notice previously, because for reasons of simple convenience he had set out his dishes of sugar solution relatively close to the observation hive so that he or his assistants could remain in touch while marking bees at the feeder or observing them in the hive.

It is very important to appreciate that these dances occur only under rather special conditions. They are part of an elaborate nexus of social communication that goes on almost continuously in a beehive, as described by Lindauer and Seeley. The workers move about a great deal and interact with their sisters by feeling them with their antennae and being felt in return. There is a sort of mutual palpitation in which the two bees face each other and feel each other's antennae and head region. At this time one of the bees often regurgitates a small portion of her stomach contents, which is taken up by the other. This is very similar to the trophallaxis of weaver ants and other social insects. It is very widespread among social insects and serves to convey not only food material but also the odors that accompany it. In colonies of specialized social insects such as honeybees the queen, the larvae, and younger workers obtain their food in this way. Trophallaxis is so widespread that a given molecule of sugar ordinarily passes through several stomachs before it is finally regurgitated into one of the cells in the honeycomb. By this time the original nectar gathered from the flowers has been modified into honey. Pollen grains transported in specialized pollen baskets formed from stiff hairs on the legs are also transferred to other workers before being stored. During round and waggle dances other bees cluster around the dancer and follow her movements. From time to time they make a brief sound that seems to cause the dancer to stop and engage in trophallaxis with one or more of her sisters.

In spite of all this activity, worker honeybees seem to be doing nothing at all much of the time, but it is difficult to determine by simply watching them whether their leisurely moving about the hive is idle loafing or whether they are sampling odors and other conditions in ways that will later affect their behavior. The workers often take food from partially filled cells, into which other workers are still adding honey or pollen. Thus the food stores of a beehive are in a constant state of flux, with new material being added after foragers have brought it back, but also constantly being drained by workers, which obtain much of their food in this way.

When a forager returns to the hive with a stomach full of nectar, she ordinarily finds other workers ready, after a brief period of mutual palpitation, to take her stomach load by trophallaxis. These other workers then store the somewhat modified nectar in partly filled cells, or they may transfer it to third parties before it is finally stored. This widespread process of mutual palpitation and trophallaxis serves also as a sort of communication, because the ease or difficulty with which a returning forager can transfer her load provides information about the general situation in the colony. This is particularly important when conditions are not optimal and when something is in short supply. The most common shortage is of carbohydrate food. The workers are informed of this by the relative emptiness of storage cells, and when conditions are truly severe, capped cells may be opened and the honey consumed. This results in an eagerness to receive regurgitated nectar during trophallaxis with returning foragers. When, on the other hand, honey is abundant but pollen is in short supply, foragers returning with stomachs full of nectar have more difficulty finding a sister to whom they can transfer their load. Whether there is some chemical or other signal that conveys more than a reluctance to receive one type of material is not clear. But somehow foragers are induced to change what they seek outside the hive.

This distinction becomes all the more clear under special conditions of overheating. When the hive temperature rises above approximately 35 degrees C, workers returning with either sugar or pollen have difficulty unloading. But workers that have gathered water regurgitate it in small droplets, and other workers fan vigorously with their wings, producing a circulation of air that cools the hive by evaporation. Under these conditions foragers shift from gathering nectar or pollen and visit places where they can take up water. It is not clear just how this transfer occurs, whether the high temperature directly stimulates the returning forager, along with her difficulty in unloading whatever she was bringing in previously, or whether some sort of information is transferred from other workers.

The important point is that the older workers that fly outside the hive searching for things needed by the colony shift their searching behavior between different commodities according to the needs of the colony. This network of social communication conveys to the older workers not only what is required by the colony but how badly it is needed. Thus when a forager leaves the hive, she has been induced to search for some particular thing, and this motivation clearly varies in intensity. The most common need is for carbohydrate food, so that the nectar of flowers is the usual target of this searching activity. But sometimes it may be pollen grains or water. Under other special conditions the need may be for waxy material used in building honeycomb. This need is relatively rare in agricultural beekeeping practice because beehives have been built to provide a waxy foundation, but under natural conditions honeybees and other bees must build their own wax foundation or plug holes in a natural cavity. Dances do not occur at all when everything is going nicely and nothing is in short supply. Foragers return with nectar or pollen, these substances are transferred to other workers, and the net stores of both carbohydrate and protein food are either constant or slowly increasing. Under these favorable conditions it seems that workers can find adequate supplies of nectar and pollen by individual searching efforts.

Under the special conditions when something has been in short supply, older workers ready to fly outside the hive somehow receive the message that sugar, or something else, is very badly needed. When they have discovered a rich source of nectar they return to the hive and engage in communicative dances. After walking a short distance in from the entrance of the hive, and usually after antennal contact with other workers, the returning forager begins either round dances or waggle dances. Dancing is not something the bees do mechanically and automatically, but only as part of the larger social nexus of communication. This point is often overlooked in elementary discussions of the bee dances. The dances are entirely dependent on the presence of an audience of other bees.

With this background we can better appreciate the symbolic nature of the honeybee communication system. It is partly a chemical communication system, for the odors of flowers are conveyed along with the nectar or pollen. And when bees visit desirable things, they often mark their location with secretions from certain glands that produce a long-lasting odor which serves to attract searching foragers. But in addition to these specific odors that are transmitted, the dances convey the direction and distance to the food by a sort of geometrical symbolism.

Round dances are performed when food has been discovered relatively close to the hive, and waggle dances are used for desirable things located at a greater distance. The transition from round to waggle dances is gradual, and it occurs at different distances ranging from two or three meters in the Indian species *Apis dorsata* and *A. florea* to 50–100 meters in the widely used Carniolan strain of European honeybees. In a typical round dance a bee circles alternately clockwise and counterclockwise, although occasionally two cycles may be executed in the same direction. When von Frisch originally discovered that round dances were used for food sources at relatively short distances and waggle dances for those farther from the hive, he concluded that the round dances contained no directional information and simply informed recruited workers to search in all directions close to the hive. He also found that bees stimulated by round dances arrived in approximately equal numbers at experimental feeders located in different directions within a few meters of the hive. Many interested scientists have observed and made motion pictures of round and waggle dances over a forty-year period, but only recently has it been noticed that even at very short distances the point at which the circling reverses does contain directional information of the same kind conveyed by the waggle dances discussed below. The fact that it took so long to appreciate this simple fact indicates how easily we overlook matters that do not readily fall into our preconceived patterns of expectation.

The transition from round to waggle dances begins by a very brief lateral vibration of the bee's body just at the moment when she has completed one circle and is about to begin circling in the opposite direction. As the distance to the food increases this lateral wagging motion lasts longer and the bee walks in a straight line for a gradually increasing distance. This can be observed with the same bees by inducing them to gather concentrated sugar solution from an artificial feeder which is gradually moved away from the hive. This experimental procedure is effective only when the colony is seriously in need of additional sugar. When the distance is gradually increased, beginning at a very few meters from the hive, it is possible to see the gradual transition from an almost instantaneous lateral vibration at the moment when the round dance is reversed in direction to an increasingly long straight wagging run. At distances of a kilometer or more the wagging run is ten or eleven millimeters in length.

The most significant of von Frisch's discoveries was that the direction of the straight wagging run was correlated with the direction that the dancer had flown from the hive to the source of food. This insight was possible only when he was able to observe waggle dances performed by bees that had returned from known food sources lying at considerable distances in various directions. Ordinarily this is not possible, because honeybees are too small and fly too far and too fast to permit direct observation. In his early experiments, von Frisch saw no reason to move his artificial food sources more than a few meters from the hive. In the 1920s it would have required a truly superhuman level of enterprising imagination to suggest that an insect might indicate the direction to a food source by some form of communicative behavior. Even a brilliant scientist who had already challenged some of the established "nothing but" dogmas of his time did not make the leap of inference necessary to imagine that the waggle dances which he and many others had observed could conceivably serve to communicate direction toward a source of food. This leads one to wonder what other versatile ingenuities of animals might be staring us in the face but waiting to be correctly interpreted.

The specific correlation between direction of the wagging run and direction toward the food takes the following form: When the food is in the direction of the sun, outside the hive, the waggle dances are oriented straight up on the vertical surface of the honeycomb. If the food is located in the opposite direction from the sun, the dances point straight down, and under other conditions the angle between the direction to the food and the azimuth bearing of the sun corresponds to the orientation of the wagging run relative to straight up. If the food is 90 degrees to the right of the sun, the dances are 90 degrees to the right of vertical.

This relationship between a direction taken by a flying bee outside the hive relative to the sun and the direction of its communicative wagging run inside the pitch dark hive is more truly symbolic than any other known communication by nonhuman animals. But one should not take this correlation to mean that the directional communication is perfectly precise. At gradually increasing distances, as the bees change from round dances to waggle dances, the straight runs do not at first point directly in the appropriate direction. Instead they alternate between being several degrees to the right or to the left of the correct distance according to the rule stated above. As the distance increases this deviation diminishes, so that by several hundred meters each waggle run points in almost exactly the same direction. The accuracy of information transfer has been measured by Towne and Gould by setting out test feeders in a variety of directions and distances with careful control for the complicating effects of odors. The majority of the recruited bees came to feeders within about plus or minus 15–20 degrees and plus or minus about 10–15 percent of the distance indicated by the dances.

Thus the system is far from perfect, but the waggle dances communicate three types of information that are all important to the bees. These are (1) the direction toward the food, expressed relative to the position of the sun, (2) the distance to the food which is correlated with the duration and perhaps length of the wagging run, and (3) the vigor of the dances which conveys the desirability of whatever the bee is dancing about. The detailed nature of distance communication has been difficult to determine. The number of waggling movements is correlated with the distance a bee must fly. But since the rate of waggling and the rate of forward movement are quite constant, both the length of the waggling run and its duration are also closely correlated with distance. While statistical analysis suggests that the duration is a better indication of distance, it is not possible from currently available data to be certain which property of the waggling run is actually perceived by other bees and used to determine the distance they will fly.

The vigor or intensity of waggle dances is easily recognized by experienced observers, for some dances seem clearly more energetic than others, and these ordinarily result when foragers have found a rich source of sugar solution or something else that is important to the bees. The concentration of sugar in an artificial feeder can be more easily manipulated by experimenters, so that it has been studied more thoroughly. But it is clear that under relatively constant conditions when carbohydrate food is scarce and dancing is actively under way, the dances are much more intense when the foragers have visited sugar solutions with a high concentration. One difference is that dances from rich sources are continued for a long time while dances from less concentrated sugar solutions may be continued only for a few cycles and then broken off, as discussed by Seeley and Towne.

Another important point about the waggle dances is that they serve to convey information to other bees inside the totally dark beehive when the subject of the communication is something entirely different from the immediate situation, namely the direction a bee should fly out in the open air. Thus the communication has the property of displacement; the bees communicate about something displaced in both time and space from the immediate situation where the communication takes place.

Wenner and Esch discovered that faint sounds accompanied the waggle dances. These are not ordinarily loud enough to hear through the glass window of an observation hive and require that the glass be removed so that the dances can be observed directly. This is obviously somewhat hazardous, since several hundred bees are completely free to fly out. But if the glass is gently removed most of the bees stay in the honeycomb, and many detailed experiments have been carried out with observation hives that can be opened. Ordinarily this is done with a darkened room around the hive, but even this is not entirely necessary, and quite bright lights can be used for photography or video recording, provided that the bees are not disturbed excessively. The sounds accompanying waggle dances are brief pulses with a fundamental frequency of about 250–280 Hz; each lasts only for a few waves but the pulses are repeated as a sort of interrupted buzzing. The fundamental frequency is nearly the same as the wingbeat frequency when the bees are flying, but the wings move only a fraction of a millimeter. The sounds and waggling movements are not always synchronous; in long-duration dances reporting highly desirable food, the sound may be delayed by a substantial fraction of a second.

Esch concluded the desirability of food or other commodities seems to be conveyed, at least in part, by the intensity or temporal pattern of these dance sounds, which seem to vary with the desirability of the food being gathered. But Wenner, Wells, and Rohlf did not find such variations, and it remains unclear whether the dance sounds convey information about food quality. One possibility is that the temporal relations between waggling movements and sound emission might convey this or other types of information. Further evidence of their importance has recently been provided by Towne, who studied two closely related species of bees in India, *Apis dorsata* and *A. florea,* which use waggle dances but nest in places where light is available. They make no dance sounds, although in other aspects of their dance communication they are similar to European *Apis mellifera.* It seems clear that only honeybees and their close relatives that dance inside dark cavities make sounds during their waggle dances.

Interpretation of these observations was initially difficult because honeybees appeared to be deaf. Despite several attempts to do so, no one had been able to demonstrate any responses to airborne sounds. This suggested that the sounds heard by human observers might be an incidental byproduct of a mechanical signal that was transmitted either by vibratory motions of the surface on which the bees were standing or in some other manner. This situation has been considerably clarified through quantitative acoustical experiments by Michelsen and his colleagues. The most important point clarified by these recent experiments is that the changes in air pressure which we detect as sounds, either by hearing them directly or via microphones, are only one physical aspect of the signals generated by dancing bees.

When any solid object oscillates against the air, whether it be a dancing bee or a loudspeaker diaphragm, air in the immediate vicinity moves back and forth at the frequency of the movement. Close to the vibrating object this motion of the air is the primary physical process, but magnitude of air movement falls off with distance very rapidly. The oscillating movements of the air also generate traveling sound waves which are areas of very slight compression and rarefaction that spread outward at the speed of sound (approximately 344 meters per second in air). It is of course qualitatively true that in order to produce a region of higher pressure some air molecules must move into such a region, but the amount of air that moves back and forth in a traveling sound wave is very small compared to that which moves about close to the vibrating source. This difference leads to the physical distinction between near field and far field sounds. In the near field the air motion is large and in the far field it is very small, because its magnitude falls off rapidly with distance from the source.

The bees that are stimulated by a dancer are ordinarily within less than one body length, and this is definitely in the near field. The sounds with which we are most familiar are far field pressure waves. Bees and most insects lack specialized auditory receptors for sound pressures, although a few specialized insects such as some of the moths that respond to the orientation sounds of bats do have sense organs adapted for responding to sound pressure. What bees and many other insects have instead are very sensitive hair-like sense organs that respond well even to feeble air movements, whether these be unidirectional or oscillating at frequencies of a few hundred Hz. Thus it is not surprising that honeybees do not respond to far field sounds, but such insensitivity tells us nothing about their sensitivity to near field acoustic stimulation which consists primarily of oscillatory air movements.

Michelsen, Towne, Kirchner, and Kryger measured the near field component of the dance sounds and found these to be quite intense, provided the measuring devices were within a few millimeters of the dancer. In further experiments Towne and Kirchner and Kirchner, Dreller, and Towne found that honeybees respond to the near field, air movement component of sound similar to the dance sounds. Using sound sources that generated either primarily near or far field sounds of either 265 Hz or 14 Hz, they were able to show clear responses to the near field signals but not to the far field sound pressure stimulation. The latter had been used in previous attempts to learn whether bees could hear.

The situation is further complicated by the fact that honeybees also transmit acoustic signals as vibrations of the substrate. With appropriate vibration detecting sensors Michelsen, Kirchner, Andersen, and Lindauer showed that the dance sounds are transmitted only very feebly into the honeycomb. But another sound at about 320 Hz is emitted by bees following a dancer; this is believed to serve as a request for the dancer to stop and regurgitate food samples. These sounds are transmitted through the substrate. When the honeycomb was set into vibration by artificial devices at this frequency dancing bees stopped dancing as they do when this so-called begging signal is generated by other bees clustered around a dancer.

Another quite different type of acoustical signaling is carried out by honeybee queens. When larvae have been fed appropriately by workers they develop into queens, and often several queens are present in separate cells. At this time, it has long been known that the queens emit two kinds of sounds called tooting and quacking. These are transmitted through the honeycomb substrate and are sensed by other queens. The tooting is produced by a queen who has emerged from her cell and the quacking signals come from queens still confined within cells. Usually the original queen together with a very large number of worker bees departs from the hive and forms a new colony. One of the new queens then cuts a hole in the side of the other queen cells and kills the occupants. Thus this exchange of acoustical signals is an important part of the social behavior of a honeybee colony around the time of swarming.

Michelsen, Kirchner, and Lindauer have succeeded in constructing a model honeybee that can transmit signals which direct recruits to search for food in particular directions. This exciting development is still at a very early stage, and all that can be said is that in a general sense the model does work. The reason it has worked better than previous attempts is probably that the near field acoustical signals have been reproduced more accurately. It will be of great interest to follow future developments in this new technology for studying the symbolic dance communication of honeybees.

It has been so astonishing to find insects communicating in such a versatile and symbolic fashion that some skeptics have remained unconvinced that the system really functions as von Frisch described it. Both Rosin and Wenner consider von Frisch's discoveries suspect because they imply, for Wenner, that bees are "capable of human-like communication (language)," or because, according to Rosin "a hypothesis which claims a human-level 'language' for an insect upsets the very foundation of behavior, and biology in general." These critics, and doubtless others, are so certain that symbolic communication is a unique human capability that they go to great lengths to deny the significance of the many experiments demonstrating that the dances convey information about distance, direction, and desirability. They do not deny that the pattern of the dances is correlated with the location of a food source, but claim that recruited bees simply search for the odor they have learned from the dancer is associated with the food. In some ingenious experiments by Gould, bees were induced to point their dances in a direction different from the actual direction from which they had returned. Test feeders were set up in the form of traps that allowed counting of the number of recruits arriving at different places, but prevented them from leaving and possibly introducing complications. The results were that most of the recruited workers flew to test feeders in the direction indicated by the dance rather than the direction from which the dancer had returned. The more recent experiments with an effective model bee, which certainly did not convey location specific odors, abundantly confirm that information about distance and direction can be conveyed from the dancer to her sisters.

The accuracy of this distance and direction information has been studied by von Frisch and more recently by Towne and Gould. The basic approach was to set out test feeders with the same odor as that associated with a rich source of concentrated sucrose and, after removing the original source of food about which bees had been dancing, to measure how many recruits arrive at the test feeders. In both von Frisch's original experiments and these more carefully controlled tests, the majority of the recruits went to feeders in approximately the same direction and at approximately the same distance as the location indicated by the dances. There is, however, considerable variation and the situation is complicated by the possibility that the odors marking the location of the original feeder or the test feeders may diffuse widely enough that bees are attracted to them even though they may not have flown very accurately to the distance and direction indicated by the dances they have followed. In general terms, it seems that the directional indication is accurate within approximately plus or minus 20–30 degrees and the distance indication perhaps plus or minus 10 or 15 percent. Evidently the finding of the exact location requires response to the odors conveyed at the time of the dancing. But the symbolism of communicating direction and distance is very significant, even though its accuracy is not all that one might ideally desire.

One misunderstanding of the dance communication of honeybees that is very widespread is the belief that it is rigidly linked to food. As mentioned above, the dances are also used to communicate about water needed to cool the colony by evaporation, but water can be viewed simply as a very dilute sugar solution. Waxy materials are sometimes collected and their location indicated by waggle dances. But the most enterprising use of the dance communication system occurs when bees are swarming. Although these dances were discovered many years ago by Lindauer, ethologists have devoted very little attention to them despite their implications concerning cognition.

Swarming occurs when a colony of honeybees increases to the point that the hive is crowded. Workers then feed some larvae a different sort of food, which causes them to develop into queens. Under ordinary conditions the bees also prepare to swarm, and part of the behavioral changes that accompany this sort of preparation is a change in the searching images of the older workers that have previously been gathering food. They now begin to investigate cavities. As new queens develop, the older queen stops laying eggs and usually moves out of the hive along with a large portion of the workers. Initially these aggregate in a ball of bees clinging to the surface of the hive or to vegetation. In normal bee keeping practice the beekeeper either enlarges the hive at the first sign of swarming so that the colony can grow further or else he provides a new hive immediately below the swarm. A beehive is an ideal cavity and the bees usually move directly into it.

Many colonies of bees flourish away from carefully tended apiaries, and when they swarm no beekeeper provides an ideal cavity in the immediate vicinity. Under these conditions many of the older workers, rather than searching for flowers or other sources of food begin to search widely for cavities. Often they must search over a very large area, crawling into innumerable crevices in trees, rocks, or buildings. Their central nervous systems must recognize a searching image for an appropriate sort of cavity. A cavity is of course something totally different from food, and these workers that now search for cavities have never in their lives done anything of the kind. Swarming ordinarily occurs at intervals of many months, and workers live only a few weeks during the warmer months when they are active. While the queen may have participated in swarming many months before, the workers have never experienced anything remotely like the movement out of the old hive and the aggregation of thousands of bees in the open.

It is not easy to find an appropriate cavity. It must be of roughly the right size and have only a small entrance near the bottom. It must be dry and free from ants or other insects. In one of the few investigations that has followed from Lindauer's discovery Seeley has studied how the scout bees investigate cavities. He established colonies on small islands where no suitable cavities were present and induced swarming by the simple procedure of shaking the queen and numerous workers out of the old hive and leaving them to their own devices in the open air. He then provided experimental cavities of different types at some distance. The workers found these and eventually induced the colony to occupy one of them. In their preliminary visits these workers crawled back and forth through most of the interior of the cavities and spent considerable time investigating them.

When Lindauer studied swarming bees he observed that workers carried out waggle dances on the surface of the swarm. These are similar in some ways to dances that are occasionally carried out on a horizontal surface in front of the hive entrance. When bees are dancing on a horizontal surface in the open they point directly toward the food or whatever the dances are about. The symbolic transfer to gravity with upward pointing dances meaning toward the sun apparently does not occur when the dancer and any of her sisters who follow the dances are on a horizontal surface and can see the sun.

The waggle dances executed on the swarm indicate the distance, direction and desirability of the cavity which the dancer has visited. This means that the dance communication system, with all its symbolism, is employed in this totally unprecedented situation. The same code indicates the location and quality of something as different from food or water as one can imagine. Worker honeybees that have been gathering nectar from flowers during the past few days, and which may even continue to do so to provide food for the swarm, utilize the totally different searching image of a dry, dark cavity of appropriate size to guide both their searches for such cavities and their communication about one that they have visited. If we accept specialized communicative behavior as suggestive evidence of thinking on the part of the communicating animal, we may infer that these worker bees think about either food sources or cavities, according to the needs they have perceived at the time.

In his classic experiments during the 1950s, Lindauer discovered that the waggle dances executed on a swarm lead to a group decision on the part of the colony about where they should move to establish a new colony. Ordinarily dozens of scout bees that locate cavities dance on the swarm about different locations. Furthermore, the intensity of the dances is correlated with the quality of the cavity. Small, damp, or ant-infested cavities produce only a few feeble dances whereas others that are dark, dry, and of suitable size lead to prolonged and vigorous waggle dances. Since different scouts have visited different cavities, a wide variety of locations are described by the numerous dances that go on over the surface of a swarm.

Lindauer spent many hours in laborious observations of the dances on the surface of numerous swarms, climbing ladders, or doing whatever was necessary to reach a suitable observation point. He found that although a wide variety of locations were signaled in the first few hours after a swarm emerged from the original hive, the dances gradually came to represent a progressively smaller number of locations. As time passed, the few cavities described by the dances that were carried out by increasing numbers of bees were those that had originally elicited the most vigorous dances. In other words, the cumulative effect of extensive dance communication was a progressive reduction in the number of cavities described. And those that continued to be danced about were the best ones available.

In further experiments Lindauer varied the suitability of particular cavities. If their quality was lowered, the dances became less enthusiastic; and in some situations at least, dances about other cavities became more numerous. This process of repeated dancing about the more desirable cavities went on for a few days, and finally almost all dances were about a single cavity. After this had been going on for several hours, a different sort of behavior occurred, which Lindauer describes as a "buzzing run." The bees making these buzzing runs moved for fairly long distances over the swarm while emitting a buzzing sound. When this had been going on for some time the swarm took wing and flew fairly directly to the cavity that had been described by the concentration of enthusiastic dances over the past day or so.

These dances on the swarm lead to a sort of consensus whereby the colony selects out of many possible cavities the one that has been judged by the scout bees to be the best. There seem to be adaptive advantages to prolonging this process of evaluation, because cavities may change their desirability as conditions change. For example, one that has been dry in good weather may be damp on a rainy day. Thus it seems that the bees do not reach this crucial decision until dancers have been, so to speak, singing the praises of a particular cavity for a considerable period of time. Many factors probably play a role in the evaluation of cavities by individual scout bees. In addition to size, dryness, and a dark interior with a small opening, distance from the original colony is important. It seems that other factors being equal the bees prefer a cavity a few hundred meters from the old colony. This presumably has the advantage of avoiding competition for food sources with the thousands of bees that remain in the original cavity and continue to search vigorously for food.

Although we can only speculate about what, if anything, these dancing bees and their sisters who follow the dances on swarms are thinking, their vigorous communication suggests that they are thinking about a suitable cavity, perhaps similar to the one from which they have recently emerged. Lindauer also observed another feature of the communication on swarms that has significant implication for a cognitive interpretation of the communicative behavior. In the first day or two, after the swarm had emerged and a number of different cavities were being described by various returning scouts, one might suppose that the less desirable cavities dropped out of the communication process because scouts that had visited them simply stopped dancing, while those returning from the better cavities continued. Yet the individual bees return repeatedly to the cavities after a bout of dancing, so that the same individuals continue to describe the cavities they have located over many hours or even a few days.

By marking individual dancers, Lindauer discovered that something even more interesting was going on. Bees that had visited a cavity of mediocre quality sometimes became followers of more enthusiastic dances than their own. Then some of them visited the better cavity they had learned about as followers of vigorous dances, returned, and danced appropriately with respect to the superior cavity they had now visited. Lindauer was only able to observe this in a handful of cases, and it is not clear how large a role this change of reference of the dances of individual bees plays in the whole process of reaching a group decision. Nevertheless the fact that any bees change from dancing about one cavity to another, after switching roles from dancer to follower, means that the whole process of communication by means of waggle dances is not a rigid one linked tightly to the stimulation received during visits to the first cavity a particular individual has located. It seems reasonable to infer that under these conditions bees are thinking about cavities, and are able to change their "allegiance" from one they have discovered themselves to a better one they have learned about as recipients of symbolic information from the dances of one of their sisters. Unlike the weaver ants studied by Hölldobler and Wilson, however, Lindauer did not see any signs of chain communication. When dancers changed the cavity about which they danced they did so only after visiting the second cavity.

Honeybees gathering pollen and nectar. Illustration by Mary Wellman (*American Insects* by Vernon L. Kellogg, Henry Holt and Co., New York, 1908).

All this communicative versatility certainly suggests that the bees are expressing simple thoughts. . . . One significant reaction to von Frisch's discovery was that of Carl Jung. Late in his life he wrote that, although he had believed insects were merely reflex automata, "this view has recently been challenged by the researches of Karl von Frisch; . . . bees not only tell their comrades, by means of a peculiar sort of dance, that they have found a feeding place, but they also indicate its direction and distance, thus enabling beginners to fly to it directly. This kind of message is no different in principle from information conveyed by a human being. In the latter case we would certainly regard such behavior as a conscious and intentional act and can hardly imagine how anyone could prove in a court of law that it had taken place unconsciously. . . . We are . . . faced with the fact that the ganglionic system apparently achieves exactly the same result as our cerebral cortex. Nor is there any proof that bees are unconscious."

Griffin, Donald R. 1992. *Animal Minds*. Chicago and London: The University of Chicago Press.

Acknowledgments

We would first like to thank our patient editor, Kate Bradford, for her enthusiasm for the project and her skill in helping us edit and determine the final selections. Her assistant, Diana Madrigal, shared in the enthusiasm and provided much needed support and extra help in preparing the final manuscript. The following made good suggestions for selections that we considered, some of which we used: Jon Coddington, Stuart McKamey, Tom Eisner, Ed Wilson, Janice Harayda, and Jane Holtz Kay. Also thanks to Steve Koselke and Marj Holme for introducing TRS to *Silver Wings and Golden Scales.*

At the Smithsonian, Beth Norden provided considerable research support, for which we are grateful. We thank John Steiner, Don Hurlbert, and Shanna Moore of the Smithsonian's Center for Scientific Imaging and Photography of the National Museum of Natural History. We would also like to acknowledge Marc E. Epstein, Carl C. Hansen, Pamela M. Henson, Eric Grisell, and Ted Suman.

A special thanks is due to Rebecca Wilson for researching the illustrations and for her fine eye in giving us so many to choose from. A big thank-you to Sarah Wedden for her support and, specifically, help to research and obtain the large number of permissions.

For permission to reprint copyright material, we gratefully acknowledge the following:

Berenbaum, May: to Perseus Books for "Insects in carrion (of maggots and murderers)" from *Bugs in the System* by May Berenbaum. Copyright © 1995 by May R. Berenbaum. Reprinted by permission of Perseus Books Publishers, a member of Perseus Books, L.L.C.

Berenbaum, May: to the Entomological Society of America and May Berenbaum for "Fatal attractions" from *American Entomologist* (1996). Reprinted with permission.

Buchmann, Stephen L. and **Gary Paul Nabhan:** to Alexander Hoyt and Island Press for "Need Nectar, Will Travel" and "Bees in the Bestiary, Bats in the Belfry" from *The Forgotten Pollinators.* Copyright © 1996 Stephen L. Buchmann and Gary Paul Nabhan.

Buck, John and **Elisabeth Buck:** for "Biology of synchronous flashing of fireflies." Reprinted by permission from *Nature,* vol. 211, pp. 562–564. Copyright © 1966 Macmillan Magazines Ltd.

Byatt, A. S.: for "Morpho Eugenia" selection in *Angels & Insects.* Reprinted by permission of The Peters Fraser and Dunlop Group Limited and Chatto and Windus on behalf of A. S. Byatt, © A. S. Byatt 1992.

Dethier, Vincent G.: for excerpt from *To Know a Fly* © 1962 Holden-Day, Inc. Reprinted with the permission of The McGraw-Hill Companies.

Edwards, F. W.: to The Linnean Society of London for "The New Zealand Glow-worm" from *Proceedings of the Linnean Society of London* (1934).

Eisner, Thomas: to *BioScience* and Thomas Eisner for "For love of nature: exploration and discovery at biological field stations" from *BioScience.* © 1982 American Institute of Biological Sciences.

Evans, Howard Ensign: to Smithsonian Institution Press for "Enjoying Insects in the Home Garden" in *The Pleasures of Entomology,* © 1985 by Smithsonian Institution, used by permission of the publisher.

Forsyth, Adrian: to Adrian Forsyth, *Equinox,* and Malcolm Publishing for "Jerry's Maggot" by Adrian Forsyth from *Equinox* (1984).

Gordon, David George: to Ten Speed Press for "Fancy Footwork," "A Day at the Races," and "Harborage Horrors." Reprinted with permission from *The Compleat Cockroach* by David George Gordon, Ten Speed Press, Berkeley, California (1996).

Griffin, Donald R.: to the University of Chicago Press and Donald R. Griffin for "Symbolic Communication" from *Animal Minds.* © 1992 by The University of Chicago.

Heinrich, Bernd: to Bernd Heinrich and the Xerces Society for "Insect Architects: Caddisfly Larvae" from *Wings* (1996).

Hunkin, Tim: for "Insecticides" and "Parthenogenesis" © Tim Hunkin from *The Observer* Colour Magazine and in *Almost Everything There Is to Know* (Pyramid Books, Octopus Publishing Group Ltd., 1988).

Kelly, Kevin: for "Hive Mind" from *Out of Control* by Kevin Kelly. Copyright © 1994 by Kevin Kelly. Reprinted by permission of Perseus Books Publishers, a member of Perseus Books, L.L.C.

Larson, Gary: to FarWorks Inc. and Creators Syndicate International for two Gary Larson cartoons. The Far Side® by Gary Larson. © 1984 and 1985 FarWorks Inc. Used with permission. All rights reserved.

Lloyd, J. E.: to Jim Lloyd for "Mating Behavior and Natural Selection" (text and photographs) from *The Florida Entomologist* (1979).

Mertins, Jim: to the *Food Insects Newsletter* 3(3):2 for "Insect Extravaganza" (1990).

Milne, Lorus J. and **Margery Milne:** for "The Social Behavior of Burying Beetles." Reprinted with permission. Copyright © 1976 by Scientific American, Inc. All rights reserved.

Newell, Irwin M. and **Lloyd Tevis, Jr.:** to the Entomological Society of America for "*Angelothrombium pandorae* N.G., N. Sp. (Acari, Trombidiidae), and notes on the biology of the giant red velvet mites." *Annals of the Entomological Society of America* (1960). Reprinted with permission.

Oldroyd, Harold: for "Swarms of Flies" from *The Natural History of Flies* by Harold Oldroyd. Copyright © 1964. Reprinted by permission of W. W. Norton & Company, Inc., and Joan Oldroyd.

Quammen, David: to The Lyons Press for "Sympathy for the Devil" in *Natural Acts* (1985). Reprinted by permission of The Lyons Press.

Schell, Jonathan: to Alfred A. Knopf, Inc., and Jonathan Cape for an excerpt from *The Fate of the Earth* by Jonathan Schell. Copyright © 1982 by Jonathan Schell. Reprinted by permission of Alfred A. Knopf, Inc., and Jonathan Cape.

Shafer, George D.: Reprinted from *The Ways of a Mud Dauber* by George D. Shafer with the permission of the publishers, Stanford University Press. © 1949 by the Board of Trustees of the Leland Stanford Junior University.

Snodgrass, R. E.: to Cornell University Press for evolution illustration from *Principles of Insect Morphology.* Copyright © 1993 by Ellen Burden and Ruth Roach. Used by permission of Cornell University Press. Originally published 1935 by the McGraw-Hill Book Company.

Starr, Christopher K.: to the *Journal of Entomological Science,* Georgia Entomological Society, for "A simple pain scale for field comparison of Hymenopteran stings" (1985).

Stewart, Marilyn Hoff: for male wood-boring beetle illustration from *Sonoran Desert Summer* by John Alcock. Copyright © 1990 Marilyn Hoff Stewart. Reprinted by permission.

Swain, Roger B.: for "Bee Bites" from *Field Days,* published by Scribner's and Lyons & Burford. Reprinted by permission of Don Congdon Associates, Inc. Copyright © 1983 by Roger Swain.

Topps Company: for "Mars Attacks" cards. © The Topps Company, Inc. 1999.

Treat, Asher: for *Mites of Moths and Butterflies* excerpt (text and illustration) by Asher Treat. Copyright © 1975 by Cornell University. Used by permission of Cornell University Press.

Von Frisch, Karl: for "Construction of the comb." Excerpt from *Animal Architecture* by Karl von Frisch, translation by Lisbeth Gombrich, copyright © 1974 by Karl von Frisch and Otto von Frisch, reprinted by permission of Harcourt Brace & Company.

Wheeler, William Morton: for "Studies of the Mediterranean wormlion" from *Demons of the Dust* by William Morton Wheeler. Copyright © 1930. Reprinted by permission of W. W. Norton & Company, Inc.

Wiggins, Glenn B.: to Doug Currie, Glenn B. Wiggins, and Royal Ontario Museum, Centre for Biodiversity and Conservation Biology, for caddisfly larvae illustration by Anker Odum in *Larvae of the North American Caddisfly Genera (Trichoptera)*, 2nd edition, 1996.

Wilcox, Jayem: to Victor Dricks, for *The Solitary Hunters* illustration by Jayem Wilcox (Weird Tales, 1934).

Wilson, Edward O.: for "The Importance of Social Insects." Reprinted by permission of the publisher from *The Insect Societies* by Edward O. Wilson, Cambridge, Mass.: Harvard University Press, Copyright © 1971 by the President and Fellows of Harvard College.

Yates, George Worthing and **Ted Sherdeman:** to Warner Bros. for *THEM!* Story by George Worthing Yates, screenplay by Ted Sherdeman, directed by Gordon Douglas, and produced by David Weisbart (1954).

Every effort has been made to track down the copyright holders of each selection and illustration, and the editors will be pleased to honor any inadvertent omissions.

Illustrations not credited in the text: Frontispiece: A rich harvest (*Silver Wings and Golden Scales* by Anon., Cassell Petter & Galpin, London, 1889); Chapter 1: Mayflies *(Silver Wings and Golden Scales);* Chapter 2: Cockchafers *(Silver Wings and Golden Scales);* Chapter 3: Diligence attacked by cockchafers *(Silver Wings and Golden Scales);* Chapter 4: Processional caterpillars *(Silver Wings and Golden Scales)* and caterpillar border *(Silver Wings and Golden Scales);* Chapter 5: Hanging wasp's nest *(Silver Wings and Golden Scales);* Chapter 6: Nest of the Armadillo Wasp *(Silver Wings and Golden Scales);* Chapter 7: A pair of katydids (*Tenants of an Old Farm* by Henry C. McCook, George W. Jacobs & Co., Philadelphia, 1884); Chapter 8: Water-beetle and larvae *(Silver Wings and Golden Scales);* Chapter 9: An insect free-booter and an insect beggar by Lancelot Speed (*The Romance of Insect Life* by Edmund Selous, Seeley and Co., London, 1907); Chapter 10: Burying beetles by Theo. Carreras (*Insect Artisans and Their Work* by Edward Step, Dodd, Mead and Co., New York, 1919).

Index of Authors

Subject Index